Foundations of
Complex-system Theories

Complex behavior can occur in any system made up of large numbers of interacting constituents, be they atoms in a solid, cells in a living organism, or consumers in a national economy. Scientific theories about this behavior typically involve many assumptions and approximations. *Foundations of Complex-system Theories* analyzes and compares, for the first time, the key ideas and general methods used in studying complexity in the physical, biological, and social sciences. It highlights the features common to the three areas, clarifies conceptual confusions, and provides a nontechnical introduction to the way we understand and deal with complexity. The book begins with a description of the nature of complexity. The author then examines a range of important concepts: situated individuals, composite systems, collective phenomena, emergent properties, chaotic dynamics, stochastic processes. Each topic is illustrated by extended examples from statistical physics, evolutionary biology, and economics, thereby highlighting recurrent themes in the study of complex systems. This detailed yet nontechnical book will appeal to anyone who wants to know more about complex systems. It will also be of great interest to philosophers engaged in scientific methodology and specialists studying complexity in the physical, biological, and social sciences.

Foundations of Complex-system Theories

in Economics, Evolutionary Biology, and Statistical Physics

SUNNY Y. AUYANG

CAMBRIDGE
UNIVERSITY PRESS

PUBLISHED BY THE PRESS SYNDICATE OF THE UNIVERSITY OF CAMBRIDGE
The Pitt Building, Trumpington Street, Cambridge, United Kingdom

CAMBRIDGE UNIVERSITY PRESS
The Edinburgh Building, Cambridge CB2 2RU, UK   http://www.cup.cam.ac.uk
40 West 20th Street, New York, NY 10011-4211, USA   http://www.cup.org
10 Stamford Road, Oakleigh, Melbourne 3166, Australia
Ruiz de Alarcón 13, 28014 Madrid, Spain

First published 1998
First paperback edition 1999

Typeset in Meridien

A catalog record for this book is available from the British Library.

Library of Congress Cataloging-in-Publication data
Auyang, Sunny Y.
Foundations of complex-system theories : in economics,
evolutionary biology, and statistical physics / Sunny Y. Auyang.
        p.   cm.
Includes bibliographical references and indexes.
1. System theory.   2. Statistical physics.   3. Biological systems.
4. Economics.   I. Title.
Q295.A95   1998
003'.7 – dc21   97-27006

ISBN 0 521 62167 4 hardback
ISBN 0 521 77826 3 paperback

Transferred to digital printing 2003

*To the memory of
my brother Haychi*

Contents

Preface

Einstein once said that "thinking without the positing of categories and of concepts in general would be as impossible as is breathing in a vacuum." His remark echoes a long tradition of Western philosophy arguing that our experience and knowledge are structured by a framework of categories or general concepts. The categorical framework contains our most basic and general presuppositions about the intelligible world and our status in it. It is not imposed externally but is already embodied in our objective thoughts as oxygen is integrated in the blood of breathing organisms. Since the categories subtly influence our thinking, it is as important to examine them as to test whether the air we breathe is polluted. Philosophers from Aristotle to Kant have made major efforts to abstract them from our actual thoughts, articulate, and criticize them.

This book continues my effort to uncover the categorical framework of objective thought as it is embedded in scientific theories and common sense. Scientific theories contain some of our most refined thoughts. They do not merely represent the objective world: They represent it in ways intelligible to us. Thus while their objective contents illuminate the world, their conceptual frameworks also illustrate the general structure of theoretical reason, an important aspect of our mind.

As a physicist who turns to philosophize, I naturally started by examining relativity and quantum mechanics. Many general concepts, including the familiar notions of object and experience, space–time and causality, seem problematic when physics pushes beyond the form of human observation and analyzes matter to its simplest constitutive level. Quantum and relativistic theories have been used by many philosophers to argue for the impossibility of invariant categories. I contested their arguments in *How Is Quantum Field Theory Possible?* There I compared the conceptual frameworks of quantum field theory underlying elementary particle physics, general relativity for the large-scale structure of the universe, and our everyday thinking about the world. Their topics vary widely, but they share a common categorical structure. Modern physical theories reject many specific everyday opinions about particular objects and properties but retain the general commonsense notions of object and property. They do not abrogate the seemingly problematic categories. Instead, they make the categories explicit, precise, and incorporate them within themselves, effectively clarifying and consolidating the general presuppositions that we tacitly understand and unreflectingly use in our daily discourse.

After quantum and relativistic physics, the next obvious stop is statistical mechanics. Most advancement in statistical mechanics occurs in condensed-matter physics, which investigates the microscopic structures of solids and liquids. A solid or a fluid is a many-body system, a complex system made up of a great many interacting constituents of similar status. The many-body theories in condensed-matter physics provide a conceptual framework that unites the microscopic and macroscopic descriptions of large composite systems without disparaging either.

Guided by the philosophical aim of seeking general patterns of thought, I promptly realized what eluded me when I worked in condensed-matter physics. Many-body theories address a major problem that has concerned philosophers since the debate between the Greek Eleatics and Atomists. *How do we explicitly represent the composition of a large system while preserving the integrity of the system and the individuality of its constituents?* The importance of the problem is evident from the diversity of circumstances in which it arises. Hobbes's *Leviathan* and Leibniz's *Monadology* offer disparate solutions in different contexts, and their answers have many competitors. The problem persists today, as does the polarity in solutions. Large-scale composition is highly complex. Crude models unable to handle the complexity conceptually sacrifice either the individual or the system. Even worse, the expedient sacrifice is sometimes interpreted as "scientific" justification for various ideologies. Such simplistic models and interpretations have practical consequences besides undermining the credibility of science. The modern individualistic and egalitarian society is a many-body system, and the tension between the individual and the community lurks beneath the surface. What concepts we use to think about the situation affect our perception of ourselves and the society in which we participate, and our perception influences our action. We can clear up some conceptual confusion by a careful analysis of scientific theories to see how they represent composition and what assumptions they have made in their representations.

Many-body systems are the topics of many sciences, for they are ubiquitous in the physical, ecological, political, and socioeconomic spheres. If there is indeed a categorical framework in which we think about them, then it should not be confined to physics. To maintain the generality and broad perspective befitting philosophy, I decided to diversify. I looked for sciences of many-body systems that have developed reasonably comprehensive theories. After a brief survey I chose to learn economics and evolutionary biology to sufficient depth that I could analyze their theories, compare their conceptual structures to that of statistical physics, and extract the common categorical framework. I always have a strong interest in these fields and seize the opportunity to find out what their researchers are doing.

The parallel analysis of theories from economics, evolutionary biology, and statistical physics does not imply that the social and biological sciences ape physics or are reducible to it. Science is not like a skyscraper in which the upper floors rest on the lower; it is closer to an airport in which the concourses cooperate as equals in an integral system. I draw analogies among the sciences. Analogies are instructive not because they pattern the strange in terms of the familiar but because they prompt us to discern in the familiar case general ideas that are also applicable to unfamiliar situations. For example, the comparison of the perfectly competitive market theory in microeconomics and the self-consistent field theory in physics does not explain consumers in terms of electrons, or vice versa. It brings out a general theoretical strategy that approximately represents a complex system of interacting constituents as a more tractable system of noninteracting constituents with modified properties responding independently to a common situation jointly created by all. The theoretical strategy is widely used, but the rationale behind it is seldom explained in the social sciences. Consequently social theories using it often spawn ideological controversies over the nature of the independent individuals and their common situation. The controversies can be cleared up by drawing an analogy with physical theories in which the theoretical transformation between the original and approximate representations is explicitly carried out, so that we can see plainly the assumptions involved and the meaning of the resultant individuals.

A significant portion of this book is devoted to the presentation of scientific theories and models that serve as the data for philosophical analysis. Because of the complexity of many body systems, the sciences rely heavily on idealization and approximation, and each splinters into a host of models addressing various aspects of the systems. I try to lay bare the assumptions and presuppositions behind the models so that readers can assess their claims, which are often culturally influential. Besides clarifying general concepts, I hope the book will stimulate more dialogue among scientists in various fields, not only about what they are studying but about how they are proceeding with it. Therefore, I try hard to make the material intelligible to a general reader, presenting the conceptual structures of the sciences as plainly as I can, using as little jargon as possible, and explaining every technical term as it first appears. Since the book covers a wide range of material, I try to be concise, so that the major ideas stand out without the cluttering of details.

The root of the project reaches back to a conversation between my brother Haychi and me in his home on a windswept foothill overlooking the glow of Denver. One night in June 1985, as we talked about Nietzsche, the first Western philosopher we read, we came across the question: What would you do, if you could do without the security of a steady salary and constant peer reassurance? We toyed with the idea of an interdisciplinary endeavor difficult to undertake in departmentalized academia. The conversation would have been forgotten, if Haychi had not died accidentally three months later. Suddenly questions about the meaning of life became real. Critical reflection and the practical preparation for reorientation took a long time, and so did the search for a significant problem on which I really cared to work long and hard. In its course I come to appreciate how much I am indebted to society's infrastructure, including freely accessible libraries and universities. For this book, I thank the many economics professors at M.I.T., whose first lectures in each term I attended and whose reading lists led my research.

Cambridge, Massachusetts

Introduction

1. Synthetic Microanalysis of Complex Composite Systems

According to our best experimentally confirmed physical theory, all known stable matter in the universe is made up of three kinds of elementary particle coupled via four kinds of fundamental interaction.[1] The homogeneity and simplicity at the elementary level imply that the infinite diversity and complexity of things we see around us can only be the result of that makeup. Composition is not merely congregation; the constituents of a compound interact and the interaction generates complicated structures. Nor is it mere interaction; it conveys the additional idea of compounds as wholes with their own properties. Composition is as important to our understanding of the universe as the laws of elementary particles, and far more important to our understanding of ourselves, for each of us is a complex composite system and we participate in complex ecological, political, and socioeconomic systems. How does science represent and explain the complexity of composition?

Large-scale composition is especially interesting because it produces high complexity and limitless possibility. Zillions of atoms coalesce into a material that, under certain conditions, transforms from solid to liquid. Millions of people cooperate in a national economy that, under certain conditions, plunges from prosperity into depression. More generally, myriad individuals organize themselves into a dynamic, volatile, and adaptive system that, although responsive to the external environment, evolves mainly according to its intricate internal structure generated by the relations among its constituents. In the sea of possibilities produced by large-scale composition, the scope of even our most general theories is like a vessel. Theories of large composite systems are complicated and specialized and lack the sweeping generality characteristic of theories in fundamental physics. To explore their peculiar approach, structures, and results is the purpose of this book.

Large composite systems are variegated and full of surprises. Perhaps the most wonderful is that despite their complexity on the small scale, sometimes they crystallize into large-scale patterns that can be conceptualized rather simply, just as crazy swirls of colors crystallize into a meaningful picture when we step back from the wall and take a broader view of a mural. These salient patterns are the emergent properties of compounds. Emergent properties manifest not so much the material bases of compounds as how the material is organized. Belonging to the structural aspect of the compounds, they are totally disparate from the properties of the constituents, and the concepts about them are paradoxical when applied to the constituents. Life emerges in inanimate matter; consciousness emerges in some animals; social organization emerges from individual actions. Less conspicuous but no less astonishing, the rigidity of solids and turbulence of fluids emerge from the intangible quantum phases of elementary particles; rigidity and turbulence are as foreign to elementary particles as beliefs and desires are to neurons. Without emergent properties, the world would be dull indeed, but then we would not be there to be bored.

One cannot see the patterns of a mural with his nose on the wall; he must step back. The nature of complex compounds and our ability to adopt different intellectual focuses and perspectives jointly enable various sciences to investigate the world's many levels of organization. Things in different organizational levels are so different each science has developed its own concepts and modes of description. What are the general conditions of our mind that enable us to develop various sciences that operate fairly autonomously but share an objective worldview? What are the general conditions of the world that make it possible for us to use disparate concepts to describe structures of the same stuff on various scales and organizational levels? Given that various organizational levels are causally related by composition, what are the theoretical relations among the corresponding descriptive levels? More specifically, what are the relations between theories for large composite systems and theories for their constituents, theoretical relations that give substance to the notion of composition?

This book tries to answer these questions by extracting, articulating, and comparing the conceptual structures of complex-system theories from several sciences specialized in the connections between different organizational levels. We examine economics; evolutionary biology, especially its theoretical core known as population genetics; statistical mechanics, especially its application to condensed-matter physics that studies the microscopic mechanisms underlying the macroscopic behaviors of solids and liquids. In addition, we investigate three mathematical theories that find extensive application in the three sciences and beyond: deterministic dynamics, the calculus of probability and stochastic processes, and the ergodic theory connecting the two. The sciences and mathematics have separately received much philosophical attention, but I know of no systematic comparison and few that focus on composition.[2]

Theories in economics, evolutionary biology, and statistical physics cover a wide range of topics, which is further extended by the applications of the mathematical theories. Our analysis focuses not so much on *what* the theories address but on *how* they address it. Despite the diversity in topics, their theoretical treatments share an abstract commonality, which makes possible interdisciplinary workshops in which biologists, economists, and physicists work together and pick each other's brain.[3] This book tries to show that the recent interdisciplinary exchange is only the tip of an iceberg. Beneath, on a more abstract level, the commonality is foundational, for the subject matters of the sciences have a general similarity, and the scientists all share the general theoretical reason of human beings.

The subject matters of economics, evolutionary biology, and statistical physics are all complex systems made up of many interacting constituents: national economies made up of millions of consumers and producers bargaining and trading; evolving species comprising billions of organisms competing for resources; solids constituted by septillions of electrons and ions attracting and repelling each other. The sciences aim to study the properties of the systems as wholes and connect them to the behaviors of and relations among their constituents: the causal relations between the performance of an economy and the decisions of consumers; between the changing composition of a species and the adaptedness of organisms; between the ductility of a metal and atomic bonds. Economics, evolutionary biology, and statistical physics are not the only sciences of complex systems generated by large-scale composition. They are outstanding for being sufficiently theoretical to illustrate the structure of theoretical reason in accounting for the wholeness of large systems, the individuality of their constituents, and the maze of causal relations binding the system and its constituents. Hence they are the benchmark for thinking about more complicated phenomena of composition that have so far eluded theorization. A fuller introduction to the sciences is given in the remaining sections of this chapter.

Properly generalized, the idea of composition applies to the mathematical theories we investigate. The probability calculus essentially studies the structures of certain types of large composite systems, for instance, long sequences made up of the repeated independent tossing of a coin, given the characteristics of a single toss. It provides a conceptual framework for us to grasp a sequence of coin tosses as a whole, to enumerate all its possible states, and to compute the frequencies of possible states with certain gross configurations. The synthetic view of the whole gives the calculus much of its analytic power. It is widely employed in the sciences of composite systems and contributes much to their success.

A dynamic process consisting of many stages is a kind of one-dimensional composite system. In a deterministic process, the successive stages follow one another according to the iteration of a rule, akin to the constituents in a compound interacting with each other according to certain laws. Unlike classical dynamic theory, which is satisfied to find the behavior of a particular process given a specific initial condition, the modern formulation of deterministic dynamics grasps entire processes and studies the general features of classes of processes. It again proves the power of the synthetic view of wholes. Chaos, bifurcation, attractor, and strange attractor are properties of processes as wholes. Chaos is an emergent character of deterministic dynamics. The contrast between a chaotic process and the determinate succession of its stages illustrates the disparity between the emergent concepts for wholes and the concepts for their parts. Such disparity is common in the theories for large composite systems.

Contrary to popular opinion, deterministic and stochastic dynamics are not the antagonistic dominions of law and chance. Ergodic theory, which has its root in the foundational research on statistical mechanics, shows that the two can be fine-grained and coarse-grained descriptions of the same process. By filtering out insignificant details, coarse graining often brings emergent properties into relief. The distinction and relation between fine-grained and coarse-grained descriptions pervade all the sciences we examine: mechanics and thermodynamics, microeconomics and macroeconomics, population genetics and macroevolution. They illustrate the importance of multiple focuses and perspectives in scientific investigation. Furthermore, when the theoretical relations among various perspectives are obscure, illusions such as a mysterious force of chance can occur.

The subject matter of the sciences and mathematics we examine covers physical, biological, social, and abstract systems. When we cut through the diversity of their topics, however, we discover a common synthetic microanalytic approach. Synthetic microanalysis institutes a broad theoretical framework in which concepts describing constituents and composite systems join forces to explain the complexity of large-scale composition. It stands in contrast to microreductionism, whose narrow theoretical framework has no room for system concepts.

Microreductionism is part of an influential philosophy that conceives ideal science as a seamless web of logic based on a few universal laws, from which all knowledge can be deduced. Focusing on composition, microreductionism assumes that once we know the laws and concepts for the constituents, mathematical prowess and large computers are all we need in principle for the complete knowledge of everything the constituents make up, no matter how complex they are. Concepts and theories about systems as wholes are reducible, which means in principle dispensable, because they are nothing but the logical consequences or definitions in terms of constituent concepts. Emergent properties, whose descriptions require system concepts, should therefore be outlawed from science.

The bottom-up reductive method is very successful for small and simple systems. In universalizing it, microreductionism tacitly assumes that large systems are simply more of the same and can be treated by the same theoretical framework and method. This assumption, encapsulated in the slogan "The whole is nothing but the sum of its parts," is correct if the parts do not interact, but unrelated constituents make trivial systems. Interaction and relation among the constituents make the whole more than the sum of its parts so that a larger whole is not merely a larger sum. They form structures, generate varieties, produce complexity, and make composition important. Microreductionism thinks that interactive effects can be accounted for by the addition of "and relations" in its slogan. Without pausing to consider how relations are summed, the breezy addition is a self-deception that blinds it to the effort of many sciences, including the largest branch of physics. The theoretical treatment of structure formation in large composite systems with interacting constituents is tremendously difficult. It introduces a whole new ball game in science.

Systems with a few million interacting constituents are not magnified versions of systems with a few constituents. Their structures differ not only quantitatively but qualitatively. Consequently they engender different kinds of questions and demand different theoretical approaches. We can adequately describe the solar system in terms of individual planetary motions, but we cannot comprehend a galaxy with billions of stars solely in terms of individual stellar motions. To understand galaxies we need new theoretical apparatus, including galactic notions such as spiral arms.

Small compounds share the same organizational level as their constituents; large systems constitute a higher organizational level. That makes a big difference. Entities on a single level are characterized by a single set of concepts; entities on different levels are often characterized by different concepts. Thus theories connecting two levels differ greatly from theories for a single level. This point is missed by microreductionism, which admits only single-level theories and assumes their adequacy in interlevel explanations. The assumption is unwarranted. Interlevel explanations require a theoretical framework that simultaneously accounts for the behaviors of systems on one organizational level, the behaviors of their constituents on a lower level, and the relations between them. This is the framework of synthetic microanalysis, which connects the descriptive levels for the system and the constituents, not by discarding system concepts but by enlisting them to join constituent concepts in explaining composition, thus allowing multiple perspectives.

The laws governing the constituents are important, but knowledge of them is only a small part of knowledge about large compounds. The immediate effects of the laws pertain to the tiny forces among individual constituents, which form an intractably complex network of mutual interaction. Constituent concepts focus on the minute details of the network, which quickly become overwhelming and obscuring. We overlook or average over the details and shift our attention to the network's salient structures.

What are the structures worthy of attention? Because the number of possible configurations generated in large-scale composition is overwhelming, it is not humanly possible to predict many system structures by the pure bottom-up method stipulated by microreductionism. Unpredictability, however, is not inexplicability, for explication has the help of hindsight. We can first observe the system structures and then microanalyze them in terms of the laws of the constituents plus suitable idealization and approximation. For this we need a top-down view of the whole. The holistic perspective is provided by the conceptual framework of synthetic microanalysis, in which scientists intellectually leap to the system level, perform experiments to find out nature's emergent properties, delineate macroscopic phenomena with system concepts, then reach down to find their underlying micromechanisms.

In studying large-scale composition, scientists microanalyze complex wholes instead of putting together myriad parts. They seek the constituent behaviors responsible for specific system properties, not the system patterns resulting from specific constituent behaviors. Synthetic microanalysis still uses bottom-up deduction but guides it by the top-down view of composite systems as wholes. The system and constituents views, each confirmed by its own experiments, inform scientists about the approximations, idealizations, and possible contingent factors required for connecting them. Without this intelligence, blind deduction from constituent laws can never bulldoze its way through the jungle of complexity generated by large-scale composition.

Perhaps the best example of synthetic microanalysis is statistical mechanics, which is not simply a flexing of muscle by particle mechanics but a new theory with an elaborate conceptual structure developed specially to connect the mechanical and thermodynamic levels of description. Thermodynamics and particle mechanics are single-level theories; the former describes a composite system on the macroscopic level, the latter on the microscopic level. Statistical mechanics is an interlevel theory. It employs the probability calculus, not to reduce thermodynamics to a colony in the empire of particle mechanics, but to integrate the two in the federation of physics. The probabilistic framework and the postulates justifying its employment in physics are not derivable from the laws of classical or quantum mechanics. They are the irreducible theoretical contribution of statistical mechanics. Without the probabilistic framework, physicists would be unable to invoke the laws of mechanics in the substantive explanations of thermodynamic phenomena. Since the idea of composite system as a whole is fundamental to the probabilistic framework, definitions made within the framework do not make system concepts dispensable.

Synthetic microanalysis and microreductionism differ in their views on the roles of experiment and mathematics, the unity of science, and the nature of human reason. The laws governing the constituents of solids are well known. If all solid-state phenomena followed mathematically, then experiments would be mere rubber stamps for verifying the derived results. Nothing can be further from the truth. Many interesting solid-state phenomena are experimentally discovered before they are theoretically explained. Important factors, for instance, the lattice structures of crystals, are experimentally measured and then put into theoretical models "by hand" for the prediction of other phenomena. Experiments often lead theories in research. Scientists, realizing that nature is subtler than they are, design experiments to elicit hints on what complex systems can achieve. These experiments and their results contribute to the top-down view in synthetic microanalysis.

When microreductionists say that all phenomena are mathematically derivable from certain laws, they usually regard mathematics as a mere calculational tool. Mathematics is a mighty calculational tool, but the calculational capacity hardly exhausts the meaning of Galileo Galilei's remark that the grand book of nature is written in the language of mathematics, nor of Willard Gibbs's remark that "mathematics is a language," which Paul Samuelson chose as the epigraph of his classic economics treatise. More important than calculation is the power of mathematics to abstract and articulate precisely significant ideas with wide implications, to construct complicated conceptual structures, and to analyze exactly the interconnection of their elements. This conceptual capacity of mathematics is fully utilized in the sciences, for instance, in microeconomics to spell out clearly the ideal conditions underlying the feasibility of a perfectly decentralized economy. The move of mathematics to higher abstract construction, which started in the nineteenth century, yielded many large conceptual structures that find applications in twentieth-century scientific theories. Differential geometry, which underlies general relativity, and group theory, which is fundamental to elementary particle theories, are examples beyond the scope of this book. Like deterministic dynamics and the probability calculus, they provide expansive synthetic theoretical frameworks for the analysis of complex systems.

The history of science witnesses not the successive reduction of theoretical structures by the expulsion of system concepts but the successive introduction of more encompassing synthetic conceptual frameworks that facilitate microanalysis. Another example of synthetic frameworks is the modern formulation of dynamics, which underlies the study of chaos by including the notions of both dynamic processes as wholes and their successive stages. Our conceptual structures expand, but their expansion rates are much slower than the explosive rate increases in the number of phenomena that they explain. The disparity in rates of expansion, not the purging of ideas, manifests the unifying power of theoretical reason.

Unity has many styles. Synthetic microanalysis unites without reducing, achieving a federal unity of science in whose broad constitutional framework both system and constituent theories enjoy a certain degree of autonomy. It contrasts with the imperial unity advocated by microreductionism, which, by dispensing with systems concepts in the description of composite systems, subordinates everything to the authority of constituent theories.

Galileo distinguished "scientists" from "calculators." Calculators care only about grinding out the mathematical consequences of given laws. Scientists want to understand the world. People often say that scientists solve problems. They do, but the universe is not a university and scientists are not students assigned homework problems. Scientists have to frame the questions themselves. To discern interesting phenomena and to introduce appropriate concepts that represent them as definite problems are the most important steps in the frontier of research. A phenomenon totally opaque in one theoretical representation may become obvious in another; the Copernican revolution exemplifies the difference made by a model from a better perspective. Synthetic microanalysis, in which scientists jump ahead to discover and delineate important system behaviors whose micromechanisms need explanation, provides the freedom in representation that microreductionism stifles.

Microreductionism extols instrumental reason, which is engrossed in deduction and calculation. Theoretical reason, which drives science, goes way beyond proficiency in technical procedures. We will see in the following chapters how scientists maneuver their positions, shift their viewpoints, idealize judiciously, postulate creatively, discard irrelevant details about individual constituents, introduce novel concepts to represent organizations of wholes, and repeatedly reformulate their problems to make them more manageable. In the process they use logical reasoning and mathematical deduction, but they also think realistically and intuitively, trading off between detail and generality, authenticity and tractability. They rely on robust common sense, familiarity with the subject matter, active observation, and experimentation on the objective world. This informal and creative thinking marks theoretical reason from mere instrumental reason.

2.   Topics, Theories, Categories

Apples and oranges are comparable only on the general level in which they are both fruits. Theories from the physical, biological, and social sciences are comparable only on the level of philosophical generality. Obviously the theories are very different, if only because their subject matters are so different. In examining them in parallel and comparing them philosophically, we try to abstract from the specifics that differentiate them and discern a general conceptual framework.

Consider a political system comprising citizens who debate with each other, organize into groups to vie for power, and occasionally erupt in spontaneous protest or revolution that may topple the power structure. The polity also levies tax and declares war as a unit. How do the actions of and the relations among citizens constitute the structure of the political system? How does the system influence the expectations and behaviors of the citizens? Can the conflicting claims of various interest groups be balanced? What are the forms of compromise? Is the political system stable? How does it evolve? How can it be modified?

Consider a solid, say a bar of gold. It comprises ions and electrons that electrically repel and attract each other, engage in collective motions, and under certain conditions rearrange themselves so radically that the solid melts or evaporates. The solid, with its rigidity and elasticity, is a unit subjected to its own laws. How do the behaviors of and relations among the ions and electrons constitute the structure of the solid? How does the crystalline structure influence the behaviors of electrons? Can a balance be struck between the conflicting forces of the bond among lattice ions and the thermal agitation of individual ions? How is the compromise expressed? Under what conditions is the solid stable? How can its characters be modified?

Comparing the sets of questions about polities and solids, we find that the substances expressed by the nouns, verbs, and adjectives are totally different. However, the ways in which the substantive words are logically used and connected in the specific questions are similar. A citizen is more complex than an electron, but logically both are individuals. A polity is more baffling than a solid, but both are composite systems made up of interrelated individuals, both are themselves individuals, and both influence the behaviors of their constituents, through taxation or constraint of motion. In either case, we have a composite system characterized in its own concepts and a set of constituents characterized in their own concepts, and questions are raised about the relation between the two. This is the relation of composition, a complicated concept including not only material aggregation but also causal interconnection, self-organization, and structure formation. The recognizable structure spanning a composite system clinches its integrity.

The comparison of polities and solids reveals two kinds of concepts, *substantive concepts* such as citizen and debate and *general concepts* such as individual and relation. Substantive concepts and their interconnections are represented explicitly by words in speech and definite terms in scientific theories. They constitute the content of theories and discourses and are the proper concern of the sciences. General concepts are also called categories, and their inter-connection constitutes the *categorical framework* of our thinking. Some general concepts are thing, event, process, property, causality, quantity, quality, relation, possibility, observability, part and whole, space and time. They usually do not appear as definite terms in scientific theories, which address consumers and electrons, wealthy and charged, but not individual in general or property in general. However, whenever we talk about consumers and wealth we always tacitly understand that they are individuals and properties of individuals. Without the general concepts of thing and relation, "The book is on the table" becomes meaningless. Since the world is intelligible to us only through the lens of general concepts, these concepts also account for the most general structure of the world. We can quarrel over what kinds of things the world contains, but we tacitly agree in our dispute that the world generally contains things or individual entities.

The categorical framework is general and not biased toward any theory. It abstracts from scientific theories, which in turn abstract from the specifics of wide ranges of phenomena. The shared categorical framework of different theories reveals both the general way of our thinking and a general commonality among the topics of the theories. Thus our consideration must include the phenomena the sciences try to understand. In this introductory section, the complex systems that are the subject matter of the scientific theories to be reflected on are described briefly. Then the general characteristics of their theories are sketched, with emphasis on the peculiar conceptual structure that marks them from the more familiar theories of simple systems. Then I explain the aim, nature, and significance of our categorical analysis, to clear up some philosophical problems engendered by the peculiarities of the theories.

The cycle of topic, substantive theory, and categorical framework will be repeated, in various orders, throughout the book as we examine the sciences in more detail. The cycle is not a concatenation of segments but a continuum characteristic of the conceptual web of human understanding. It is impossible to discuss the topics without using some theories or to present the theories without invoking some categories. By sliding the spotlight over what we try to understand and how we conceptualize it in various degrees of abstraction, I hope to illustrate how topics, theories, and categories mesh into each other and thereby how our thoughts relate to the world.

Complex Systems

Simplicity may have a unified form, but complexity has many varieties. Pure mathematics has a vast repertoire of complex abstract systems, which we do not consider because our aim is to investigate our objective thought or thought about the world. Concrete complex systems spread across a whole spectrum of complexity. For systems on the high-complexity end of the spectrum, such as brains or persons, our current sciences offer catalogues of facts but no comprehensive theory. Of course articulation and classification of facts have already involved many concepts and theories, but these theories concern only small regions of the systems and are usually known as empirical generalizations. To investigate how we intellectually grasp complex systems as wholes, we have to be content with systems on the low-complexity end of the spectrum, for which the science has gone beyond the stage of empirical generalization to be called theoretical.

We will concentrate on the complexity arising from large-scale composition, understood in the broad sense that includes dynamic processes composed of stages. There are several broad types of large composite systems. We are mainly concerned with the simplest, many-body system, the name of which originates in physics.

A *many-body system* is made up of a large number of constituents belonging to a few kinds and coupled to each other by a few kinds of relations, for instance, an economy of consumers and producers, an evolving species of organisms, or a fluid of atoms. The cohesive strength of the interconstituent relations may vary, but it does not vanish totally. A many-body system is not a heap of sand; its constituents are so combined that it has enough integrity to stand as an individual in a larger system, as a steel beam stands as a part of a building. Many-body systems are particularly interesting because of their ubiquity and their susceptibility to theoretical treatment. Many-body theories are perhaps the most advanced theories of complex systems and large-scale structure formation. Beside economics, evolutionary biology, and statistical physics, they are found in many areas, including nuclear and elementary particle physics. Society is conceived as a many-body system in theories stressing its individualistic and egalitarian nature, for example, in rational-choice and communicative-action theories. Many-body theories will be the major topic of our analysis.

An *organic system* consists of components of many different kinds that are highly specialized and tightly integrated. Organic systems are conducive to functional descriptions, in which the components are defined and characterized in terms of their functional roles in maintaining the system in a desired state, so that they are totally subordinated to the system. The paradigm of organic systems is an organism, but one can also think of jet fighters and other advanced types of machinery. Society is conceived as an organic system in functionalism and structuralism, against which methodological individualism revolts. We will not consider organic systems, for theories about them are at best sketchy. Organisms will be mainly treated as the constituents of evolving species; as such they are highly simplified and the functions of their parts are represented only in optimization models. Optimization is common to theories of simple systems in many sciences and engineering, where it usually has no functional interpretation.

Cybernetic systems such as neural networks combine the complexity of many-body and organic systems. Persons are so complex and unitary many people balk at calling them composite systems. Behaviors suffice for the characterization of other systems but not for the characterization of persons, for persons act intentionally. These highly complex systems are way beyond the scope of this book. Since the free-market economy is the unintended consequence of human action, economic theories can abstract from intention and work with behavior instead of action. The constituents of economies they depict are so idealized and simplified as to be closer to things than persons.

The Complexity of Many-body Systems

Many-body systems are more homogeneous than organic systems, but they are far from monotonous. There are several sources of complexity in many-body systems: the variety and intricacy of the constituents, the variety and strength of their mutual interactions, and the number of constituents.

Consider for brevity a many-body system made up of constituents of the same kind coupled to each other via the same kind of relation. This does not mean that the constituents are all the same with the same relation. We are all human beings, but each of us is unique and relates uniquely to others. The constituents have their individual characters and relations, which can vary greatly. A law of physics forbids any two electrons to be in the same state,[4] evolution depends on the variation of organisms within a species, and the market economy thrives on the diversity of consumer tastes. The variation of constituents contributes greatly to the diversity of systemic configurations. Even more diversity stems from the variation in individual relations. On the other hand, the concept of a kind implies certain rulelike generalities that apply to all members of the kind. Thus the variation among the constituents of a many-body system is more susceptible to theoretical representation, for it falls under the general rules specific to their kind.

Realistically, the constituents of a many-body system are themselves complicated, and the basic relations between two constituents can be tangled. In most theories, the constituents and their basic relations are greatly idealized to make the problem of composition tractable. The internal structures of atoms and molecules, which are complicated enough to engage atomic and molecular physics, are neglected in most theories of solids and liquids. Similar simplifications on the properties of organisms and humans are made in biological and economic theories. The idealized consumers featured in economic theories are simpletons in many respects, and their genius in solving calculational problems is actually a simplifying assumption.

Suppose the constituents and their basic relations are so simplified that the behaviors of a system with a few constituents are well understood. Say we know how two electrons repel, two organisms mate, or two consumers trade. In a many-body system, each constituent is coupled to many if not all others. The multilateral interconnections constitute a relational network that makes the system highly complex even if its constituents and their basic relations are quite simple. The relational network is chiefly responsible for the great variety of phenomena in the world. It is also responsible for the difficulty in studying large composite systems.

Consider a simple composite system made up of constituents with only two possible states, say an array of pixels either black or white, related by their relative positions. If we neglect the relations among the pixels, then a system with n pixels is trivially the sum of m black pixels and $n - m$ white ones. What makes the system interesting is the relative arrangement of its constituents. If the pixels are arranged in a one-dimensional sequence, then a system with two constituents has four possible configurations: black–black, black–white, white–black, white–white. There are 1,024 possible configurations for a system with 10 constituents, 1,048,576 for a system with 20, 2^n for a system with *n*. As the number of a system's constituents increases linearly, the number of its possible configurations increases exponentially and quickly exceeds even the astronomical. The numbers increase further if the configurations are two-dimensional as in pictures, for the same sequence can be arranged in different spatial arrays. The *combinatorial explosion* explains why there are practically infinite varieties in pictures of any size.

The number of possible states for a large system with billions or trillions of constituents is staggering. It is still tremendous even if it is severely limited by the system's history or environment. Frequently only a small portion of the possibilities for a kind of many-body system are realized or will ever be realized. For instance, the geometry of shells can be characterized by three parameters according to the way the shells grow. Taking proper consideration of size limitations, biologists have figured out the range of possible shapes of shells. They found that only a tiny portion of the possible shapes are realized by all known shells, living and fossilized.[5] Shells are among the simplest of complex things, yet their possibility already far outstrips their actuality. Faced with abundant possibility, the important questions become what particular possible states are realized under certain conditions, what gross structures the actual states have, how they evolve, and how their past limits their future possibilities.

Viewed microscopically, causal factors act on individual constituents: An electric field accelerates individual electrons, an epidemic kills individual organisms, an inflation hurts individual savers. Since the number of constituents is large, the causal effect of each is minute. The behavior of a many-body system is the gross manifestation of numerous tiny forces tugging and pulling in all directions. Sometimes the constituents are so organized that the change in the behaviors of a few may tip the balance, as a few drifting flakes trigger an avalanche or a few swing votes decide a close election. These are the critical situations that present a junction in the system's history, with each fork leading to a different set of possibilities. Which particular fork the system takes may be accidental in the sense that the causes influencing the behaviors of the critical few are beyond the purview of existing and foreseeable theories and conceptualization. The contingency gives the histories of complex systems a peculiar temporality absent in the predictable trajectories of simple systems.

The exponentially increasing complexity of large systems is not as hopeless as it appears, for large numbers can also spawn simplicity if viewed from proper perspectives. The laws of large numbers in the probability calculus show rigorously how gross regularities precipitate in large systems. More intuitively, consider the simple system made up of black and white pixels. As the number of pixels in the array increases, sometimes new regularities appear that make the idea of "picture" significant for the first time. We recognize patterns describable concisely in their own terms. Of course this need not be so. Often all we can do to describe the array is to cite the colors of individual pixels. Occasionally, however, we can describe it as gray, or salt-and-pepper, or even as a landscape or a portrait. The landscapes, portraits, and other recognizable patterns are the emergent properties most interesting to many-body theories.

Two Notions of Complexity

There is no precise definition of complexity and degree of complexity in the natural sciences. I use *complex* and *complexity* intuitively to describe self-organized systems that have many components and many characteristic aspects, exhibit many structures in various scales, undergo many processes in various rates, and have the capabilities to change abruptly and adapt to external environments. To get a more precise notion of complexity, definitions in the information and computation sciences would be helpful.

The idea of complexity can be quantified in terms of information, understood as the specification of one case among a set of possibilities. The basic unit of information is the *bit*. One bit of information specifies the choice between two equally probable alternatives, for instance, whether a pixel is black or white. Now consider binary sequences in which each digit has only two possibilities, 0 or 1. A sequence with n digits carries n bits of information. The *information-content complexity* of a specific sequence is measured in terms of the length in bits of the smallest program capable of specifying it completely to a computer. If the program can say of an n-digit sequence, "1, n times" or "0011, $n/4$ times," then the bits it requires are much fewer than n if n is large. Such sequences with regular patterns have low complexity, for their information content can be compressed into the short programs that specify them. Maximum complexity occurs in sequences that are random or without patterns. To specify a random sequence, the computer program must repeat the sequence, so that it requires the same amount of information as the sequence itself carries. The impossibility of squeezing the information content of a sequence into a more compact form manifests the sequence's high complexity.[6]

Science is usually not so much interested in the definite description of a specific system as in the classes of systems that satisfy certain general criteria. For instance, an important physical problem is to find the configurations of an n-body system that have the lowest total energy, for these are the configurations in which the system settles. The difficulty in solving such problems is also an indication of the system's complexity. It can be expressed in terms of the idea of computation-time complexity.

Suppose we have formulated a problem in a way that can be solved by algorithms or step-by-step procedures executable by computers and now want to find the most efficient algorithm to solve it. We classify problems according to their "size"; if a problem has n parameters, then the size of the problem is proportional to n. We classify algorithms according to their computation time, which, given a computer, translates into the number of steps an algorithm requires to find the worse-case solution to a problem with a particular size. The *computation-time complexity* of a problem is expressed by how the computation time of its most efficient algorithm varies with its size. Two rough degrees of complexity are distinguished: tractable and intractable. A problem is tractable if it has polynomial-time algorithms, whose computation times vary as the problem size raised to some power, for instance, n^2 for a size-n problem. It is intractable if it has only exponential-time algorithms, whose computation times vary exponentially with the problem size, for instance, 2^n.

Exponential-time problems are deemed intractable because for sizable n, the amount of computation time they require exceeds any practical limit.[7]

The example of the array of pixels shows how the number of possible configurations for an n-body system increases exponentially with n in combinatorial explosion. If the best we can do to solve an n-body problem is to search exhaustively through its possibilities, then the problem has only exponential-time algorithms and is intractable. Tractable problems are usually those for whose structures we have gained deeper insight, so that we need not rely on brute-force searches. Mathematicians have proved that many practical problems in operations research, game theory, and number theory are intractable. Far from being defeatist, the intractability proofs are invaluable because they save algorithm engineers from banging their heads against stone walls. Once engineers know that a problem is intractable, they can direct their effort to reformulating the problem, relax some conditions, or find alternative tractable problems.

Problems and solutions in natural science are less rigorously definable than in computation science, but natural scientists have adopted similar strategies. If the lowest-energy configurations of an n-atom system involve specifying the state of each atom, then the problem is intractable. With n a trillion or more, physicists do not even try the brute-force approach. They wisely spend their effort in formulating more tractable and interesting problems, for which they develop many-body theories. With insight and versatile strategies, problems regarding systems with the highest information-content complexity need not be the most difficult to solve. Statistical methods deal with random systems more adequately than with systems with some regularities.

Many-body Theories

We conceptualize the intelligible world in many organizational levels and individuate entities in each level. Entities on one level are more likely to interact with each other than with entities on a different level, and causal regularities are most apparent and explanations most informative within the level. Thus many scientific theories concentrate on entities and phenomena in a single level with little reference to other levels. These theories study the interaction among a few entities, but the resultant systems are small enough to belong to the same scale and level as the constituents. For instance, the laws of classical mechanics adequately explain the motion of the planets. We have microscopic quantum laws, but their connection to planetary motion is often obscure and the information they offer is irrelevant.

Many-body theories are peculiar in explicitly taking account of and relating entities on two different scales, for the sheer number of constituents pushes a many-body system over to a different level. They encompass within a single model both solids with their rigidity and atoms with their quantum phases; both economies with their price levels and consumers with their individual preferences; both evolving species and characteristics of organisms. By explicitly treating the constituents and their interactions, many-body theories differ from theories that treat composite systems as units without referring to their constituents. By characterizing composite systems with predicates and perhaps causal laws of their own, they differ from theories that depict the relative behaviors of a few interactants, in which the system lacks individuality because it has no description independent of its constituents.

Theories encompassing entities on two scales have a special difficulty. The concepts used to characterize entities in different scales are often inconsistent. For example, water molecules are discrete particles in classical mechanics, and water is a flowing continuum in hydrodynamics. Continuity and discreteness constitute one contrasting pair; others are temporal irreversibility and reversibility, alive and inanimate, conscious and incognizant. Many other concepts, although not contradictory, are incongruous. Viscosity, heat capacity, and expansion coefficient, which are concepts for fluid water, are not applicable to water molecules.

A many-body theory conceptualizes and characterizes the same large composite system on several levels. For the moment let us ignore the intermediate structures and consider only the two extreme levels of description, generically called the *micro-* and *macrolevels*. They can employ totally disparate concepts. The fine-grained *microdescription* extrapolates from the method of small-system theories and characterizes the system in terms of the individual behaviors of its constituents. Because of the complexity of sizable systems, the amount of information it contains is overwhelming. The coarse-grained *macrodescription* characterizes concisely the gross features of composite systems in terms of system concepts. It contains less information than the microdescription, but less information is not necessarily less informative. A single piece of key datum is more informative than a ton of irrelevancy.

Many-body theories try to connect the micro- and macrodescriptions, not to eliminate one in favor of the other. The connection is not achieved by finding a predicate in the microdescription to match with a predicate in the macrodescription; simplistic type–type identification is usually futile. The task is much more difficult; it must make sense of the intricate compositional structure that exhibits variegated features with scales ranging from that of the constituent to that of the system. Realizing that many microscopic details are inconsequential to macrophenomena, scientists introduce extra theoretical means to distill the significant information that highlights the particular systemwide patterns they are studying. The better theories systematically withdraw to successively coarser-grained views, screening out irrelevant details step by step. Besides objective considerations, the theories also introduce additional postulates that represent, among other things, a shift in perspective and emphasis. The new perspective and postulates account for the conceptual incongruity of the micro- and macrodescriptions. They inject extra information absent in the microdescription. In terms of the number of bits, the additional information is negligible. Yet it crystallizes the deluge of microscopic configurations into a few informative macrovariables and transforms the crushing mass of microinformation into a tractable theory. Its introduction is a testimony to the ingenuity of theoretical reason.

The connection between micro- and macrodescriptions introduces a new kind of explanation about a new domain of phenomena, the compositional structure of the system. Its approach is *synthetic analytic*, where one does not logically construct from given parts but intellectually analyzes wholes into informative parts. Scientists do not try blindly to deduce their way up from the constituents, a hopeless exercise mired in the swamp of microscopic possibilities. Instead, they peek at the answers to the deduction by doing experiments to find out what macroscopic features nature has realized. The intelligence is like a map of a swamp, enabling them to land on the islands of certain macrobehaviors and reach down to dredge for the responsible micromechanisms. In the process they rely on system concepts to pick out the macrophenomena, articulate them, and guide the analysis. Synthetic analysis accommodates the freedom to generalize on both the micro- and macrolevels; hence its conceptual framework is broader than that of brute-force deduction from the laws of the constituents.

The broad framework of synthetic analysis also forges the conceptual unity of the theories, which are otherwise fragmented. Because of the combinatorial explosion, a typical many-body system has many facets and harbors many simultaneous processes on various temporal and spatial scales. It is impossible to treat all of them in a monolithic theory. To investigate many-body systems, scientists construct simplified models that capture various important aspects of a larger picture from various perspectives. Consequently economics, evolutionary biology, and statistical physics all branch into a multitude of models, each of which addresses a particular process or a specific aspect of composition. Various models employ various approaches and approximations as befit their specific topics. The conceptual framework of synthetic analysis holds these models together, so that we do not lose sight of the large picture.

Descriptive Philosophy and Categorical Analysis

The preceding description of many-body theories – their accommodation of incompatible concepts for phenomena on two levels, their attempt at reconciliation, the logical gaps resulting from approximations and independent postulates, their dependence on probabilistic and statistical concepts, and their fragmentation into many models – is quite abstract, for it covers theories in all three sciences. The generality indicates a common structure that can be articulated in terms of categories.

The categorical framework expresses the general structure of our theoretical reason. It includes the most general presuppositions we have made about the world and our status in it, for instance, the presupposition that the world is made up of individual things with distinct properties and interrelations. Its analysis traditionally belongs to metaphysics, for the general presuppositions of objective thought are also the general nature of reality as it is intelligible to us. Peter Strawson has distinguished two kinds of metaphysics: "Descriptive metaphysics is content to describe the actual structure of our thought about the world, revisionary metaphysics is concerned to produce a better structure." He said that most philosophers engaged in both, but on the whole, Aristotle and Immanuel Kant were descriptive; René Descartes, Gottfried Wilhelm Leibniz, and George Berkeley, revisionary; David Hume was hard to place.[8]

The analysis of the categorical framework of objective thought is still a major goal in contemporary philosophy. Equally persistent is the dichotomy between the descriptive and prescriptive approaches.[9] For instance, the philosophy of science contains general doctrines and the analyses of specific sciences, the former mostly prescriptive, the latter descriptive. Prescriptive philosophies of science stipulate revisionary structures of scientific theories, canonical forms of explanations, and criteria for theory appraisal and hypothesis testing. Reductionism and microreductionism are examples of their prescriptions. Their arguments are logical and abstract, often making little contact with actual scientific theories and practices. Descriptive philosophies such as the philosophies of biology, economics, and physics delve into the substance of the sciences, sometimes becoming so engrossed in technical matters as to invite the question, Where is the philosophy?

The present work is descriptive; it treats the categorical framework not as an external standard to which the theories are matched, but as something presupposed by actual theories, embedded in them, and playing active roles in them. Its task is to abstract from the substantive contents of various scientific theories, expose their categorical framework, and articulate the general structure of our thought about composition and complexity. It pays as much attention to the content of the sciences as the philosophies of specific sciences but differs from them in the breadth of its perspective and the generality of concepts it considers. Along the way it compares the actual structures of scientific theories to various revisionary prescriptions. I argue that the prescriptions are too impoverished to accommodate the features of reality and too simplistic to convey the thoughts expressed in scientific theories. Their claim of discharging the functions of the commonsense notions they expel is an unredeemable check.

Traditionally, philosophers who aimed to describe the general structure of objective thought investigated our everyday thinking, even if they were fascinated by science, for scientific theories were restricted in scope, incomplete in logical structure, and dependent on common sense to function. I focus on scientific theories because everyday thinking is fluid and vague, whereas scientific reasoning, although narrower, is more focused and refined. The categories are like air, essential for living but difficult to see; they are ethereal as a result of their generality and taken for granted because of their familiarity. To inquire about the importance of air, it is better to ask astronauts who carry oxygen on their space walks than to ask flatlanders who have never gasped for air. The sciences are like astronauts. Striving to gain knowledge of difficult subjects that befuddle common sense, the sciences would have explicitly incorporated a category in the logical structures of their theories if the category were essential for thinking about their topics. Modern scientific theories have distilled and absorbed many presuppositions of everyday thinking; consequently their logical structures are much strengthened.

Another advantage of analyzing scientific theories is their mathematical nature. Often the same mathematics is applicable to widely disparate topics. The topics differ drastically in substantive features, but on an abstract level they share some common structures that make them susceptible to representation by the same mathematical language. General structures are philosophically interesting. They are reflected in the structure of the mathematics, which has already abstracted from substantive specifics. Thus philosophers can save much effort of abstraction by using off-the-shelf items. We need not start from scratch with simplistic logical models; the mathematical structures of empirical theories present high-level and sophisticated models whose axioms are readily available in mathematics texts. We need not worry about the applicability of the abstract axioms; they have been thoroughly tested in scientific theories. Of course we must be careful in our analysis and interpretation of the mathematics; sometimes it is merely a calculational tool and sometimes there are several mathematical formulations of the same topic. We have to examine the mathematics in applications and carefully separate its objective, ideal, and instrumental functions, and all these from rhetoric.

For instance, the probability calculus finds applications in physics, biology, economics, and many other areas. We should not be fooled by the name of the calculus into thinking that chance rules the world; chance is not even defined in the calculus. We examine the axiomatic structure of the calculus, the definitions and relations of its terms, and their meaning when applied in various scientific theories. In this way the mathematics helps us to extract from the scientific theories a general pattern of thought, the objectivity of which is proved by its successful functioning in empirical knowledge. We take advantage of the systematic abstraction, clear thinking, and precise formulation that are already accomplished in the mathematical sciences to make a short cut in categorical analysis.

Outline of the Book

This book is organized around the interrelated categories of *individual* and *time*, which are, respectively, extracted from the equilibrium and dynamic models of complex-system theories. The two are united by the category of *possibility*. Whether an individual is a system or a constituent, it is formulated in scientific theories in terms of its possibilities, represented by the *state space* that encompasses all its possible states. The state space is often the most important postulate of a scientific theory, for it defines the subject matter under investigation. It unifies equilibrium and dynamic models; the evolution of a system traces a path in its state space. The number of possible states explodes exponentially as a many-body system increases in size. The asymmetry between the enormity of possibilities and the scarcity of actualities underlies the concepts of probability, contingency, temporal irreversibility, and uncertainty of the future.

Part I examines *equilibrium* models. Chapter 2 lays out the conceptual structure of many-body theories, concentrating on how it harnesses incongruous micro- and macroconcepts in the account of composition. It argues that their approach is *synthetic analytic*, emphasizing the analysis of wholes instead of the combination of parts. Even where some sense of "the whole is the sum of its parts" is valid, the parts are not the constituents familiar in small systems. They are customized entities obtained by analyzing the system for the understanding of prerecognized macrobehaviors, and they have internalized most interconstituent relations. *Optimization*, a general idea extensively applied in all three sciences, is also introduced.

Both the system and its constituents are individuals, the former explicitly composite and the latter situated. Thus the *general concept of individual* must incorporate the idea of integrity in view of an individual's parts and the idea of distinctiveness in view of the myriad relations binding an individual to the multitude. These general ideas are examined in Chapter 3, which introduces state space and its utility in representing various individuals. Like most chapters in the book, this one contains four sections, one of which discusses the general philosophical ideas; the other three examine how the general ideas are separately instantiated in biology, economics, and physics.

Scientists seldom try to find exhaustive substantive theories covering all aspects of a kind of many-body system. Instead, they raise specific questions targeting some salient features and make simplifications to disentangle the targeted phenomena from other factors. The result is a host of models and regional theories explaining various aspects. Chapters 4 to 6 examine three broad classes of model and approximation with an ascending degree system integrity. Going through the three classes of model, our attention shifts from individual constituents to the system as a whole. The shift in focus reflects the varying nature of the system properties under investigation; individual constituents are more prominent for the explanation of resultant properties than emergent properties. We will find that not only do the characters and interpretations of the constituents change, but new individuals such as collective excitations and social institutions appear. The varying image of the constituents and the shifting balance between the constituents and the system readily fuel ideological controversy when the nature and assumptions of various models are not made explicit.

The *independent-individual approximation* discussed in Chapter 4 is the simplest and most widely used approximation. In it, the characters of the constituents are transformed to absorb most of their mutual relations, and the residual relations are fused into a common situation determined by all but impervious to each. Consequently the system becomes a "lonely crowd" of solipsistic individuals responding only to the common situation.

The idea of *emergence* is explored in the two subsequent chapters. In the models examined in Chapter 5, the many-body system is more integrated because its constituents are no longer uncorrelated. Some constituents behave in unison and others cohere preferentially, leading to the appearance of novel entities that constitute an intermediate layer of structure. In the models examined in Chapter 6, the constituents are so tightly correlated, the system defies modularization and must be treated as a whole and represented by concepts of its own. Some models show explicitly how the details about the constituents drop out as we systematically retreat to coarse views for the conceptualization of macroscopic regularities. The insignificance of microscopic peculiarities in many macrophenomena is encapsulated in the *universality* of many macroconcepts.

Part II considers the notions of time in dynamic, stochastic, and historical *processes*. Chapter 7 analyzes the relation of the *general concept of time* to those of thing, event, and causality. It sets the sciences we are investigating in a broader context by relating their temporal concepts to the temporal concepts in relativistic physics that encompass all changes and the temporal concepts in our everyday thinking that make use of the tenses.

Deterministic dynamics, familiar since Newtonian mechanics, has recaptured headlines with notions such as nonlinearity, chaos, bifurcation, and strange attractors. Chapter 8 explores the modern formulation of *deterministic dynamics* in terms of state space, its connection to stochastic processes via the ergodic theory, and its role in the foundation of statistical mechanics. It shows how simple dynamic rules, when iterated many times, can generate highly complex and chaotic results. Yet among all the unpredictable complications, sometimes simple new structures that are susceptible to sweeping generalization, as apparent in the universality of some features, emerge at a coarser level.

In Chapter 9, I examine the structure of the calculus for *probability* and *stochastic processes* and argue that their wide application in the sciences does not suggest the dominion of chance as some inexplicable cause or propensity. The meaning of accident and randomness is explored by comparing deterministic and stochastic formulations of the same process. This analysis again illustrates that important conceptual differences arise from changes in scale, focus, and perspective, not from the intervention of mysterious forces.

Chapter 10 investigates the differences in the temporal notions of the sciences: Evolution is natural history; economics looks forward; physics has no idea of past and future. Yet the time in thermodynamics is already more complicated than the time in mechanics, which lacks even the notion of irreversibility. The emergence of temporal irreversibility is a major problem in the foundation of statistical mechanics. Time is no less a problem for biology and economics. Finding no law that covers the history of organic evolution, biologists join historians in using narratives to explain specific incidences. Economists are still wrestling with the representation of uncertainty regarding the future, which is crucial in decision making and planning.

This book considers only those categories that are embodied in the theories of economics, evolutionary biology, and statistical physics. The topics of these sciences belong to the midrange phenomena of the world: neither elementary nor too complicated. Their theories need not worry about problems that arise in the study of more extreme topics. Constrained by the data of philosophizing, I will take for granted many general concepts, notably those of object, experience, and space–time. These concepts, explicitly embodied in quantum field theory, I have analyzed in a previous work.[10] Also neglected are consciousness, intentionality, ethics, aesthetics, and other personal categories. The *Homo sapiens* in biology is an animal. *Homo economicus*, whose whole being is the maximization of the utility of marketable commodities, lacks vital dimensions of a full person. Uncovering the general nature that *Homo economicus* shares with things accentuates our awareness of what is peculiarly human.

3. Economics, Evolutionary Biology, Statistical Physics

No work of this length can possibly do justice to the rich substance of the sciences, especially when its major aim is the analysis of general concepts common to three sciences. The parallel treatment of the sciences obscures their individual internal coherence. I attempt to outline the conceptual structures of microeconomics (§ 16), macroeconomics (§ 25), equilibrium statistical mechanics (§ 11), nonequilibrium statistical mechanics (§ 38), and population genetics for microevolution (§ 12). Macroevolution lacks an adequate theoretical framework and is better addressed by historical narrative (§ 41). These structures leave out many topics because of the fragmentary nature of the sciences. To provide a context in which the reader can locate the ideas and models discussed later, this section sketches the histories and current trends of the sciences. It also introduces some concepts that are self-evident today but that took a long time to develop. The rocky historical paths illuminate the subtlety obscured by their present obviousness.

A science develops mainly according to its inner dynamics, but it is not immune to its social and historical milieu, which often channels the effort of research by stressing the importance of certain problems and allocating more research resources to them. Economics, evolutionary biology, and statistical mechanics are sciences of mass phenomena. Not coincidentally they emerged during the period in which the industrial society and the age of the masses took shape. As the French Revolution and its aftermath mobilized the resources of entire nations, the Industrial Revolution spawned organizations of increasing scale, and societal relations became ever more complex, the scope and depth of human understanding expanded tremendously. With their newly gathered statistical data, empirical social scientists showed that society, which gradually came to be recognized as a large system comprising autonomous individuals, exhibits an order whose stability seems to be unaffected by the caprice of individual actions. People began to see that large systems, hitherto too nebulous and complicated to comprehend, can be grasped as wholes with gross regularities describable in general terms.

Reflecting on the nature of history toward the end of the Enlightenment, Kant referred to social statistics and remarked: "Since marriages, births consequent to them, and deaths are obviously influenced by the free will of man, they should seem to be subject to no rule according to which the numbers of them can be reckoned in advance. Yet the annual registers of these events in major countries show that they occur according to laws as stable as those of the unstable weather, whose particular changes are unpredictable, but which in the large maintains a uniform and uninterrupted pattern in the growth of plants, the flow of rivers, and other natural events."[11] Kant may not have been aware that several years prior to his essay, Adam Smith had inaugurated scientific investigation into the causes underlying the economic aspect of social regularities. Seventy-five years later, James Clark Maxwell, impressed by the statistical method of social scientists, introduced it into physics to study large systems containing many particles whose behaviors are deemed irregular. About the same time Charles Darwin published a theory of natural history in which the minute variation in the characters of individual organisms is translated into the evolution of entire species.

Economics, Micro and Macro[12]

Economics is the oldest of the three sciences we consider. As the practical wisdom of household management and bureaucratic administration, economic thoughts have flowered in cultures all around the globe since antiquity. As a science or systematic investigation, economic analysis is more recent and regional. It concentrates on the behavior of the self-regulating market system, which itself only emerged during the seventeenth and eighteenth centuries in the Western Hemisphere.

Economic analysis focuses on several problems faced by societies in producing and distributing material goods under the sanctity of private property and the moderate scarcity of resources: the allocation of resources for efficient production, the distribution of income among the constituents of the society, the growth and stability of the economy, the effects of money and credit. Allocation and distribution belong to *microeconomics*, where the consideration of demand and supply applies to individual firms and households. Economists generally agree that allocation is most efficiently achieved by the price mechanism of the free market. Growth and money belong to *macroeconomics*, which considers the levels of income and unemployment, inflation and price level, interest rate and the quantity of money, stability and business cycles, international trade and the balance of payments, and how these factors are influenced by government policies. Microeconomics characterizes the economy by specifying the behaviors of every constituent, macroeconomics by using a handful of aggregate variables. The clear dichotomy between the two was made in the 1930s, and it remains in force today, despite efforts to connect them in microfoundations research. There is more consensus among economists regarding the postulates and results of microeconomics than those of macroeconomics.

Let me adopt the conventional characterization of Adam Smith as the father of economic science, skipping the contributions of the English mercantilists, who argued that national wealth could be increased by regulating foreign trade, and the French physiocrats, who argued that wealth was created by domestic production. Smith's *An Inquiry into the Nature and Causes of the Wealth of Nations* was published in 1776, the year in which the steam engine made its commercial debut. It predated the economic manifestation of the Industrial Revolution, which, judged by the sharp upswing in British production and trade figures, occurred around the time of the French Revolution of 1789. The flying shuttle, spinning jenny, and other machines, which enjoy extensive coverage in many histories of the Industrial Revolution, are hardly mentioned in *Wealth of Nations*, whose paradigm technology is the manufacture of pins and nails. Yet Smith divined much of the essence of the industrial society. He realized that the organization of society's productive effort is as important to economic growth as mechanization. The organizational structure he analyzed in terms of several aspects: the increase in specialization and the division of labor underlying a rising productivity; the market mechanism holding together the host of specialized producers; the accumulation of capital resulting from the reinvestment of profits and leading to further specialization. He emphasized that the organization of effort, which engenders economic growth and optimal resource allocation, is not anyone's design but the unintended outcome of myriad individual actions, each based on self-interest. Developed by subsequent economists, the idea of the self-centered *Homo economicus* and the thesis that the market's invisible hand coordinates myriad conflicting self-interests to achieve harmonious and beneficial results for the economy have both become the basic postulates of economics. They are given a rigorous mathematical formulation in modern microeconomics (§§ 13 and 16).

A key concept around which economic thinking progressed is *value*. A satisfactory theory of value must answer several questions: What determines the value of a commodity? Is there an invariant value ascribable to a commodity as its intrinsic attribute? How does the intrinsic value of a commodity relate to its fluctuating price or exchange value? How are values measured? How are they related to the distribution of income? The answers to these questions distinguish various schools of economic thought. In medieval times the value of a thing was its fair price as determined by the social norm. The classical economists – Adam Smith, David Ricardo, John Stuart Mill – generally argued that commodities have intrinsic values, which are determined by the costs of production. With subtle variations, Smith and Ricardo both subscribed to the *labor theory of value*, which Mill amended but did not discard. According to the theory, the value of a commodity is the total amount of labor required to produce it, including the indirect labor embodied in the fixed capital used up in its production.

Classical economic theories divide the constituents of the economy into three large groups: laborer, capitalist, and landlord. Each group has its peculiar form of income: wage, profit, or rent. Classical economists strived to explain the distribution of income among the groups and to relate the effect of distribution to economic performance. The problem was tackled by Smith unsystematically and Ricardo more rigorously. Besides a keen perception of economic affairs, Ricardo had the ability to distill from the maze of familiar economic notions a few precise and strategic variables and to construct tractable models that, although greatly simplified, nevertheless capture some salient elements of complex economic phenomena. The analytic method he pioneered has been emulated by economists ever since.

As is common to complex phenomena, there are too many economic variables, rendering the problem tremendously difficult. Ricardo cut through the mess by making postulates to fix several variables, thus reducing the problem to a manageable form. On the basis of Thomas Malthus's theory of population, which asserts that population grows at a faster rate than food and will increase until checked by starvation, Ricardo posited the iron law of wages that fixed wages at the subsistence level. Considering the successive utilization of lower-grade land, he posited the principle of diminishing returns, that increasing input into production yields increasing output, but at a decreasing rate. His theory depicted the growth and final stagnation of the capitalist economy, a scenario that won for economics the epithet "the dismal science."

Many classical economic concepts, including the labor theory of value and the division of society into laborers, capitalists, and landlords, were taken up by Karl Marx. Marx's economic theory is more comprehensive and realistic than that of his contemporaries and predecessors. He was the only one to treat human behavior and technological advancement not as externally fixed but as variables intrinsic to the process of economic development. However, on the whole he kept to the classical framework. It is not difficult to get the ideas of exploitation and class conflict from classical economic concepts without looking into the plight of the laboring mass during the Industrial Revolution. "[R]ent and profits eat up wages, and the two superior orders of people oppress the inferior one."[13] The characterization is from Smith, not Marx.

Smith, Malthus, Ricardo, Mill, and Marx were all concerned with real economic problems of their times and debated on policies. Their attention was mainly focused on macrophenomena. Smith studied the forces governing the long-term growth of national wealth; Ricardo was interested in the relative shares of the annual product obtained by various *groups* of people. This is a macroanalysis, distinct from the income distribution among individual persons, which is a microanalysis. When the classicists examined in detail the forces underlying the phenomena and investigated relative prices, they naturally brought in microfactors such as individual trading behaviors. Such is the case of Ricardo's theory of rent. The distinction between macro- and microanalyses was not clearly made by the classicists, who shifted from one to the other without much ado.

The second wave of economic theories is distinctly microanalytic. It focuses on the allocation of resources among individuals instead of the distribution of income among groups of individuals. Also, it replaces production with exchange at the center of economics.

As the nineteenth century progressed, the flaws of classical economics became apparent. The iron law of wages and the principle of diminishing returns melted in the forge of empirical evidence, which showed that real income for the masses rose as the population expanded. In the 1870s, Stanley Jevons, Carl Menger, and Léon Walras independently but almost simultaneously developed *marginal analysis* in English, German, and French. Marginal analysis uses differential calculus extensively; a marginal quantity is usually a derivative of some sort.

The marginalists introduced a theory of value based on consumption and utility, which had been largely neglected by the classicists. They argued that a commodity has no value if no one wants it, no matter how much resource has been spent in its production. The cost of production is past and gone, but value lies in the future, in the commodity's use or *utility*, or more vaguely in its desirability to consumers. Marginal analysis replaces production in the spotlight with consumption. It abandons the idea of an intrinsic and invariant value, arguing that different consumers assign different utilities to the same commodity. The concept of utility had many difficulties that took decades to iron out: How is value as individual utility measured? Can utility be summed to obtain the total utility for the society? Can the evaluations of different consumers be compared? If not, how do we account for the distribution of income among individuals? Some of these questions are examined in § 16 and compared to conceptually similar questions in physics.

The marginalists further argued that the price of a commodity is determined not by its total or average utility but by its *marginal* utility, or the utility of the last amount of the commodity one desires. Smith had wondered why water is so cheap while diamonds are so expensive, as water is vital to life and diamonds totally useless (he was a bachelor scholar). The marginalists answered that the paradox arose because the classicists thought only in terms of total utility. No doubt one is willing to pay a bundle for the first glass of water in the desert, but the price of water is generally determined by the last glass, the marginal or differential amount for which one is willing to pay when he is bloated. "At the margin" becomes a key concept in economics and with it the *principle of marginal substitution*. If a consumer finds more utility in an extra unit of commodity A than in an extra unit of commodity B, he will substitute a unit of A for his last unit of B. Thus a consumer seeking to maximize his satisfaction will allocate his resources in such a way that the marginal utilities of all commodities are equal.

Utility is private and not observable. The observable economic variables are the *prices* and the *quantities* of various commodities. Economists are interested in the relation between the two, especially in the prices and quantities that remain stationary. The classicists argued that long-run equilibrium prices are determined by the costs of production. The marginalists assumed a fixed quantity of available commodities and argued that prices are determined by the desires of consumers. Alfred Marshall synthesized their ideas using marginal analysis. He argued that production and consumption are thoroughly interrelated like the two blades of a pair of scissors, so that prices are determined jointly by *supply* and *demand*. *Equilibrium* obtains when demand equals supply or when the market clears. Finally, almost a century after *Wealth of Nations*, demand equals supply, which kindergarten kids nowadays can parrot, was first enunciated. Marshall's synthesis came to be called *neoclassical*, and it forms the basic framework of present-day *microeconomics*.

Microeconomics reckons with the price of each commodity and the quantities of each commodity demanded by each consumer and supplied by each firm. It seeks the equilibrium state of the economy in which the prices and quantities of all commodities are simultaneously determined to the maximal satisfaction of all. A central microeconomic theory is that equilibrium can be achieved by a *perfectly competitive market* in which the desires of individuals can be reconciled in costless exchange. This "invisible hand theory" is presented in § 16 to illustrate an approximate method of solving many-body problems. In the approximation, individuals do not interact with one another but respond independently to a common situation, which is characterized by the commodity prices in the case of economics.

The theory of the perfectly competitive market is based on a host of idealistic assumptions. It also takes economic institutions such as business firms as basic units without worrying about their internal organization. In recent decades, economists are making much effort to relax some of the assumptions. They are developing *information economics* and *industrial organization theories*, which use game theory to study so-called market imperfections. These theories take account of the direct interaction among individuals, the asymmetry of their bargaining positions, and the internal structure of economic institutions. Some of these theories are discussed in § 20 to illustrate the formation of intermediate structures in many-body systems.

The neoclassicists gave sharp formulations to concepts such as demand, supply, price, and equilibrium. At the same time, they narrowed the scope of economics to the efficient allocation of scarce resources. Equilibrium theories provide a static picture. Many important problems in classical economics, such as the dynamics of economic growth and the distribution of income, were brushed aside. The classicists called their discipline "political economy," highlighting the relation between economics and policy. Policies concern aggregate phenomena and find little place in the microscopic view. Marshall dropped the old name in favor of *economics*, which caught on in the English-speaking world. For sixty years economics essentially meant microeconomics. Then the Great Depression, a macrophenomenon, forced itself on the attention of economists.

The price for neglecting macrophenomena under microeconomic hegemony is heavy. When prolonged and severe unemployment plagued the capitalist world in the early 1930s, economists were hamstrung for the want of suitable conceptual tools to analyze the situation. The policies they recommended were ad hoc and often wrong.[14] John Maynard Keynes challenged the neoclassical orthodoxy. Among his criticisms, two stand out: The first is "its apparent conviction that there is no necessity to work out a theory of the demand and supply of output *as a whole*."[15] One of the first tasks in Keynes's *General Theory of Employment, Interest, and Money* is to define the appropriate aggregate variables and their interrelations. The causal network among variables such as national income, national expenditure, aggregate supply, aggregate demand, quantity of money, and general price level enables economists to grapple with the behaviors of the economy as a whole. Since Keynes's work, *macroeconomics* has become a branch of economics in contradistinction to *microeconomics*. The basic conceptual framework of macroeconomics is discussed in § 25 and compared to that of thermodynamics.

Keynes's second criticism of the neoclassical doctrine is far more controversial. Keynes challenged the tenet of market perfection, in which prices and wages always adjust rapidly to clear all markets and keep the economy near the full employment level. He criticized it for ignoring the *effect of money* and people's *expectation, uncertainty,* and *level of confidence*. These factors are taken into account in the theory he developed. The notions of risk, uncertainty, and expectation in economic models are examined in § 42.

For twenty years the construction of macroeconomic models dominated the work of economists, spiced by the heated debate between the Keynesians and neoclassicists. Finally, economists found that many basic assumptions of Keynesian and neoclassical theories are compatible; both use the concepts of demand, supply, and equilibrium, although one applies them on the aggregate level and the other on the individual level. In the mid-1950s,

Paul Samuelson, who had earlier translated many economic ideas into the language of mathematics, gave the name *neoclassical synthesis* to the emerging consensus in macroeconomics.[16] The neoclassical synthesis puts the micro- and macro- branches of economics under the same neoclassical label but leaves open the gap between them. Each branch has its own style, concepts, and techniques.

The capitalist economy suffers recurrent slumps on the heel of booms. The phenomenon of *business cycles* is a major topic in macroeconomics, the deterministic and stochastic modeling of which is examined in § 35. The Keynesian idea that recession and overheating can be prevented by government stabilization policies slowly gained acceptance. In the early 1960s, the Kennedy administration started to take an active role in managing the economy. When the economy landed in a stagflation in the 1970s, Keynesian economics came under heavy fire. Olivier Blanchard and Stanley Fischer said: "The Keynesian framework embodied in the 'neoclassical synthesis,' which dominated the field [macroeconomics] until the mid-1970s, is in theoretical crisis, searching for microfoundations; no new theory has emerged to dominate the field, and the time is one of exploration in several directions with the unity of the field apparent mainly in the set of the questions being studied."[17]

There are now several contending schools of macroeconomics: the neoclassicists, Keynesians, monetarists, new classicists, new Keynesians, and others. One economist has counted seven.[18] Economists generally believe in the efficacy of the free market and its price mechanism in coordinating productive efforts. However, they disagree sharply about whether market coordination is perfect, whether slight imperfection can generate significant economywide phenomena, whether the free market by itself always generates the best possible economic results, and whether public policies can improve the economy's performance. Various brands of classicists deny or underplay market imperfection, arguing that prices and wages are always flexible enough to clear all markets instantaneously and achieve the most desirable results. Hence they are skeptical of government stabilization policies. Various brands of Keynesians maintain that in reality prices and wages are sticky and such market imperfections can be magnified by the economy's complexity to generate undesirable results; therefore there is some role for public policies. Robert Hall has divided economists into "freshwater" and "saltwater" breeds; the classically oriented, including the monetarists, are mostly located around the Great Lakes and in the Midwest; the Keynesian oriented along the East and West coasts. We will examine how the two breeds address the relation between micro- and macroeconomics in the so-called microfoundations research (§ 25).

Probability, Statistics, and Statistical Inference[19]

The theories of probability and statistics are fundamentally involved in the sciences we consider. Their relevance is apparent in the very name of statistical mechanics. The statistical theories of biometrics and econometrics are the empirical limbs of evolutionary biology and economics. Darwin did not think quantitatively, but the modern synthetic theory of evolution is statistical.

Probabilistic and statistical ideas are not pervasive in theoretical economics, yet they are important in many regional problems, including stochastic business cycles and decision making under risk. Joseph Schumpeter put statistics with history and theory as the elements that distinguish the thinking of economists from casual talks about economic matters.[20]

The probability calculus originated in the correspondence between Blaise Pascal and Pierre Fermat in 1654. Its classical form received a definitive expression in Pierre Simon Laplace's treatise of 1812. The classical theory was motivated by various practical problems, including games of chance, finance of risky enterprise, and weighing of evidence in civic life. As the mathematics continued to abstract from its original motivations, many classical assumptions were criticized and the meaning of probability itself debated. In the modern axiomatization of the probability calculus introduced by Andrei Kolmogorov in 1933 and accepted by most mathematicians, probability is by definition a *proportion* or *relative magnitude*.

Statistics has a different root. Originally "statistics" meant numerical information about the political state; it was once called "political arithmetic" in England. As an avalanche of numbers and tables poured out of government bureaus after the Napoleonic War, social scientists began to construct theoretical models to understand them better. In 1835, Adolphe Quetelet imported techniques from probability calculus. He took the *normal distribution* or the bell-shaped curve, originally developed by Laplace and Karl Friedrich Gauss to describe errors in astronomical measurements, and showed that it fits social data equally well. His exposition demonstrated that diverse human and social characteristics are susceptible to quantitative treatment and conceptualization. Quetelet was fascinated by statistical means or averages and invented the idea of the "average man." Other people were quick to explore the additional information carried in the distribution he popularized.

There are many distributions besides the bell-shaped curve. A *distribution* systematically gives the relative quantity of items with a specific value for a variable. For instance, if the variable is annual income, then the income distribution gives the number or percentage of households in each income bracket. From the distribution we can compute the average income as a measure of the level of prosperity, or the spread in income as an indicator of equality. If the income distribution "broadens," that is, if successive distributions show that the percentages of households with high and low incomes increase but those with middle incomes decrease, we say economic inequality increases. Darwin and Maxwell were both familiar with Quetelet's work. Variation became a key concept in the theory of evolution. Maxwell modified the normal distribution to describe the distribution of molecular velocities in gases.[21]

Applying the calculus of probability, scientists came to grips with mass phenomena, first in the social then in the natural realm. Sections 35 and 39 examine the structure of the calculus and its many interpretations, which show how vague the meaning of probability is. Much confusion arises from the refusal to distinguish probability from statistics and the automatic association of probability with chance. The technical definition of probability in the probability calculus is a relative magnitude. Thus I will use *proportion* or its synonyms in technical contexts and reserve *probability* for contexts where its commonsense meaning applies. A distribution is a systematic summary of proportions.

Evolutionary Biology[22]

There are roughly 1.4 million known and described species of existent organisms, and biologists estimate that these are only 5 to 25 percent of the total. The adaptedness and specialization of organisms to their ways of life put clumsy machines to shame. A major question in biology is, in Darwin's words, to show "how the innumerable species inhabiting this world have been modified, so as to acquire that perfection of structure and coadaptation which justly excites our admiration."[23] Before the nineteenth century, the diversity and adaptedness were interpreted in terms of divine design. Today they are generally explained in terms of evolution.

The idea of evolution was familiar by Darwin's time. Cosmological and geological evolution has been extensively discussed. The first theory of organic evolution was put forward by Jean-Baptiste de Lamarck in 1809. Lamarck regarded species as variable populations and argued that complex species evolved from simpler ones. He postulated that evolution resulted from the bequeathal of acquired characters. Organisms respond to the challenges of the environment, change their characters, and pass on the characters acquired through their adaptive effort to their descendants. The postulate has many difficulties. for instance, the protective colors of many insects are adaptive, but they cannot be acquired by the effort of the insects.

In 1831, Darwin set sail on H.M.S. *Beagle* as a naturalist for a five-year voyage in the waters of South America, during which he observed the geographic distribution of organismic characters and the significance of living conditions on the differentiation of species. He worked out the theory of evolution by natural selection in the first few years after his return but did not publish it until he learned that Alfred Wallace, working independently, had come up with a similar idea. *The Origin of Species* appeared in 1859, replete with detailed examples and factual accounts accumulated in the interim years.

Darwin summarized his major ideas in the introduction of the *Origin* and expounded them in the first five chapters. His theory of evolution by natural selection asserts that the variation in the survivorship and fertility of organisms in their environments gradually leads to the evolution of species and the diversity of life forms. The theory has five main premises:

1. *Variation*: there are variations in the morphological, physiological, and behavioral characters of organisms that make up a species.

2. *Heredity*: the characters are partially heritable, so that on the average offspring resemble their parents more than they resemble unrelated organisms.

3. *Proliferation*: organisms reproduce and multiply; their population will explode until it is checked by the limitation of resources.

4. *Natural selection*: some characters are more favorable to the conditions of living than others, and organisms with favorable characters tend to leave more offspring.

5. *Divergence*: gradual changes in the structure of a species caused by natural selection lead to distinctive varieties, which eventually diverge to form new species.

Because of natural selection, more organisms in the succeeding generations have characters that are favorable to life and the organisms become better *adapted* to their environments. The formation of new species from the gradual divergence of varieties in an existing species underlies the ideas of *common descent* and the *branching phylogeny* of the "tree of life."

The first four postulates form a group, which can be summarized in the general ideas of *variation* and *differential proliferation*. A composite system must evolve if there is variation in its constituents and the relative abundances of different variants change with different rates. For instance, if there is variation in the wealth of citizens in a society, and if the percentages of poorer citizens grow faster than the percentages of richer citizens, then the society evolves, a phenomenon not lost on the eyewitnesses of the Industrial Revolution. The major questions are the causes of the variation and the differential proliferation in the biological context. One of Darwin's insights is that competition is more fierce among kindred than between predators and preys. Thus the basic source of evolution is the change in the relative abundance of varieties *within* a species. Darwin took for granted intraspecific variation, which had been popularized by social statistics. He explained differential proliferation by his postulates 2, 3, and 4. Heredity compensates for the mortality of organisms so that long-term processes are meaningful. Proliferation, the idea that Darwin got from Malthus, creates a scarcity of resources and generates the pressure that eliminates the less adapted variants, the process known as natural selection. The change in the distribution of a single species or population is now called *microevolution* and addressed by *population genetics* (§§ 12, 17).

The fifth postulate, divergence or *speciation*, the origin of new species, becomes part of *macroevolution*, which studies major evolutionary trends, the emergence of novelties, the rate of evolution, and extinction.[24] The process of speciation is less well understood than population genetics, but Darwin noted the importance of geographic isolation. Suppose organisms in a species segregate into isolated populations in areas with different conditions. Each population evolves as organisms in it adapt to their environment, until organisms in different populations are reproductively incompatible. Thus the original species splits into several distinct species; the variation in one form of life is amplified into the diversity of life-forms.

Sewall Wright remarked that the major difference between the Lamarckian and Darwinian evolutionary processes is that the former is explained in direct physiological terms whereas the latter is presented as a *statistical* process. Francis Galton, a cousin of Darwin, founded *biometrics*, which uses statistical methods to study organismic variation and the spread of variation in the population.

Darwin erected a broad conceptual framework involving many complex mechanisms and processes, the substance of which must be fleshed out for it to succeed as a scientific theory. He had no good explanations for the source of organismic variation and heredity, and despite the effort of biometricians, the mechanisms underlying natural selection remained obscure. After twenty years of popularity, the influence of Darwinism, the doctrine that evolution is mainly driven by natural selection, declined. Evolution was still upheld, but biologists found nonselectionist mechanisms more attractive. Lamarckism and use adaptation were revived. Even the rediscovery of Mendelian genetics, which answered many questions about the mechanism of heredity, at first created a competitor rather than an ally for Darwinism.

Gregor Mendel, using controlled experiments on the hybridization of garden peas and mathematical reasoning, established several rules of heredity in 1865. Mendel focused on discrete characters that an organism either possesses or does not, such as the smoothness of seeds. From his experimental results he inferred that the heredity of characters must be carried in units, each of which has two forms, and each form is expressed in a distinct character. Hereditary units we now call *genes* and their forms *alleles*. For the characters studied by Mendel, each organism carries two genes, which it inherits separately from its father and mother. It in turn passes one of the two to its offspring. When the two genes of an organism are of different forms, only one of the forms is expressed. Thus there is no blending of discrete characters; a pea is either green or yellow, never yellowish green.

Mendel's results laid buried for thirty-five years. With their revival came the *Mendelian school* or genetics of evolution. Since Mendel's results are for discrete characters, Mendelians thought that only discrete characters were hereditary and denied they were influenced by external environments. They argued that natural selection was not effective and evolution proceeded by *mutation*, or the random appearance of new alleles and consequently new discrete characters. Mendelians quarrelled bitterly with biometricians, who focused on continuously varying characters such as height or body weight. The hostility quelled attempts at reconciliation for decades.

Like the synthesis of supply-side classicism and demand-side marginalism leading to modern microeconomics, the rapprochement of Mendelism and biometrics inaugurated the *modern synthetic theory* of evolutionary biology. The foundation of the synthesis was laid in the early 1930s by Ronald Fisher, J. B. S. Haldane, and Sewall Wright, who developed *population genetics*. Acknowledging that genetic mutation is the source of variation but arguing that a mutation can only have a small physiological effect, Fisher showed that genetic variability can generate the kind of continuous variation of characters described in biometrics. Natural selection on the variation leads to a gradual evolution of a species's distribution. Wright went further and considered the evolution of small populations. In sexual reproduction, each parent contributes to an offspring only one of its two genes, randomly picked. The random factor is insignificant in large populations because of the law of large numbers. It is important in small populations and becomes an additional source of evolution called *genetic drift*. Thus inbreeding in a small group can lead to accelerated evolution.

Population genetics is a statistical theory, whose mathematics was unfamiliar to most biologists. Theodosius Dobzhansky explained the central concepts in prose and integrated them with the vast amount of data from laboratory experiments and field observations. He applied Wright's results on small populations to isolated or semiisolated colonies of organisms and discussed various isolation mechanisms that could lead to speciation. His work stimulated much research activity. Ernst Mayr, George Simpson, Ledyard Stebbins, and others extended the ideas to systematics, paleontology, botany, and other branches of biology. The resultant modern synthetic theory received a great boost from the burgeoning field of molecular genetics after the discovery of the double-helix structure of deoxyribonucleic acid (DNA) in 1953. The conceptual structure of population genetics is presented in § 12.

To cope with the complexity of organisms and their genomes, the modern synthetic theory makes drastic approximations to obtain concrete results. A major approximation analyzes the genome into noninteracting genes and the organism into uncorrelated characters. Another severs genes from organismic characters. Yet another forsakes the dynamics of evolution and concentrates on equilibrium, which uses various *optimization* models to argue for the adaptedness of organisms. These approximations and the conceptual confusions they engender are discussed in § 17. The ongoing controversy surrounding group selection and the effect of population structure is examined in § 21.

In the early days of the modern synthesis, biologists generally agreed that many mechanisms contribute to evolution. During the 1940s, the pluralist view gradually gave way to the orthodoxy that natural selection predominates, sometimes accompanied by the corollary that all organismic characters are optimally adaptive because suboptimal characters are eliminated by selection. Stephen Gould called the change the "hardening of the modern synthesis" and said he could not explain why it occurred in view of the available empirical evidence.[25]

Since the 1960s, the orthodoxy is being increasingly challenged as many biologists explore alternative evolutionary mechanisms. Experiments on the molecular structures of genes and proteins reveal a surprisingly large degree of variation. To explain the data, Motoo Kimura picked up Wright's idea and argued that most changes on the molecular level are not adaptive but are the result of random genetic drift. He introduced the slogan "the survival of the luckiest," which is staunchly resisted by the adherents of natural selection chanting "the survival of the fittest." His arguments became part of the *neutral theory of evolution*, which is presented in § 37 with other stochastic factors in evolution.

The development of the dynamic theory of complexity in the past two decades (Chapter 8) is another source of new ideas in evolutionary biology. It provides the theoretical tool to investigate the possibility of evolution driven by the inner dynamics of a population instead of by the external environment. One of the inner dynamics is *self-organization*, or the spontaneous appearance of order, which is common in complex systems. In the orthodox view, evolution is a tinkering process, in which natural selection preferentially and successively preserves minute modifications produced by random mutation. As Stuart Kauffman remarked, organisms are viewed as "ad hoc contraptions cobbled together by evolution." Arguing that the tinkering process obscures the holistic nature of organisms and fails to explain the emergence of novelties such as the bird's wing, Kauffman put forward models showing how the self-organization of organisms can be a major force in evolution. These models expand the conceptual framework of the evolutionary theory and aspire to seek universal statistical laws of organic evolution. They are still heretical to most evolutionary biologists (§ 24).

The existent theoretical models in evolutionary biology are too crude to answer many questions in the history of evolution confidently. That is why historical narrative is the dominant mode in accounting for specific evolutionary events of life on earth. Much controversy, including the dispute over human sociobiology, results from the reckless extrapolation of crude theoretical models to explain complicated historical events and to justify the status quo (§ 41).

Statistical Physics, Equilibrium, and Nonequilibrium[26]

The idea of equilibrium is crucial to all three sciences we investigate. Crudely, an isolated system is in *equilibrium* if its state does not change with time. Isolated systems out of equilibrium generally change; most, although not all, evolve toward their equilibrium states. The study of *nonequilibrium* phenomena, especially their evolution, is more difficult than the study of equilibrium phenomena, and it is less well developed in all three sciences.

The macroscopic systems investigated in statistical physics are often called *thermodynamic systems*. Their study began in the seventeenth century, when an international cast of scientists were drawn to the phenomenon of air pressure. Let me skip the work in the following century, in which the nature of heat and the flow of energy were intensively studied. In the 1840s, physicists realized that energy can be transformed from one form to another with its quantity unchanged. The *conservation of energy* refines and expands the concept of energy. Rudolf Clausius and William Thomson (Lord Kelvin), investigating the conversion of energy between heat and mechanical work, began to develop the theory of thermodynamics in 1850.

The *first law of thermodynamics* is a special case of energy conservation. It states that the change in the internal energy of a system, defined as the difference between the heat it absorbs and the work it performs, is the same for all transformations between a given initial state and a given final state. The *second law of thermodynamics* has several forms. Clausius's statement is that there exists no thermodynamic process whose sole effect is to extract heat from a colder system and deliver it to a hotter one; extra energy must be provided in order for refrigerators to work. To account for heat conversion between systems of different temperatures, Clausius introduced a thermodynamic variable called *entropy*. Another way of stating the second law is that the entropy of a thermally isolated system never decreases when the system changes from one state to another. The one-way change in entropy ushers in *irreversibility*, a temporal concept absent in mechanics.

Clausius originally called his theory the "mechanical theory of heat." He postulated that a thermodynamic system is made up of molecules and heat is the motion of the constituent molecules. The kinetic theory had a venerable pedigree. Atomism started with the ancient Greeks. Isaac Newton argued in the *Principia* that many observed behaviors of air pressure can be explained if gases are assumed to consist of particles that repel each other with an inverse-square force. Fifty years later Daniel Bernoulli constructed a model in which gases consist of tiny hard spheres all traveling with the same velocity, the square of the velocity proportional to the temperature of the gas. From the model he derived the relation among the pressure, volume, and temperature of gases. The atomistic models were not without competitors: the wave theory of heat and the caloric theory in which heat is a fluidlike substance.

In 1860, 173 years after the *Principia* and 5 years before he wrote down the equations of electromagnetism now bearing his name, Maxwell introduced the *kinetic theory of gases* featuring gases made up of molecules with various velocities. The variation in velocity is important because some phenomena depend more on the behaviors of the "elites" than those of the "average molecules." For instance, the rate of loss of atmospheric gases to outer space depends on the number of exceptionally fast molecules. Maxwell used a distribution similar to the normal distribution to account for the number of molecules with each velocity. From the distribution he derived the relation between the temperature of a gas and the average kinetic energy of the gas molecules (the kinetic energy of a molecule is proportional to the square of its velocity). Maxwell's distribution, extended by Ludwig Boltzmann to include not only the kinetic energy but also the potential energy of the molecules, became the centerpiece of the kinetic theory of gases.

The kinetic theory postulates that the behaviors of gases are the causal results of the behaviors of their constituent molecules. It regards the molecules as classical point particles and takes for granted that momentum and kinetic energy are conserved in molecular collisions, as classical mechanics dictates. These hypotheses about molecular behaviors have to be justified, as does the postulate of the velocity distribution. Direct experimental verification was decades away. Maxwell and Boltzmann both tried to justify the postulate of the velocity distribution by theoretical argument and a miscellany of indirect empirical evidence, not all of which was favorable. Their two approaches led separately to equilibrium and nonequilibrium statistical mechanics.

One of Boltzmann's arguments is that the Maxwell–Boltzmann distribution, as we now call it, is the most probable state for a gas at equilibrium. He distinguished what we call microstate and macrostate, both describing the gas as a whole. A *microstate* of a system specifies the position and velocity of each of its constituent molecules; a *macrostate* is characterized by a few macroscopic variables such as energy, temperature, and pressure, which can be readily measured without referring to molecules. A velocity distribution, which gives the number of molecules with each velocity, is a refined characterization of a macrostate of a gas. There are many possible microstates compatible with the same macrostate, although only one microstate is realized at each moment. Using concepts from the probability calculus, Boltzmann calculated the numbers of possible microstates compatible with various macrostates. He proved that by far the largest number of possible microstates correspond to the macrostate characterized by the Maxwell–Boltzmann distribution. Thus it is overwhelmingly probable that the gas will exhibit the distribution *if* each microstate has an equal probability of being realized. The *if* is a fundamental and independent postulate of statistical mechanics. Attempts to justify it led to *ergodic theory*, which has grown into a sophisticated branch of mathematics. Nevertheless, the justification of the postulate falls short of being totally satisfactory (§§ 11, 33).

Boltzmann further defined a macrostate's *entropy* as proportional to the logarithm of the number of possible microstates compatible with it. If we think that a system is more ordered when it has fewer possible microconfigurations, then a larger entropy means greater *disorder*. At equilibrium, a system attains maximum entropy or maximum disorder as it stays in the macrostate with the maximum number of compatible microstates. The definition of entropy establishes a bridge between the micro- and macrodescriptions of the thermodynamic system. Boltzmann's method of enumerating possible microstates and Maxwell's idea of an ensemble, which is a collection of microstates compatible with a specific set of macrovariables, were generalized, developed, and given mathematical formulation by J. Willard Gibbs in 1902. It becomes the framework of our *equilibrium statistical mechanics*, which applies not only to gases but to solids and liquids. It is presented in § 11 and compared to the framework for the kinetic theory of gases.

Boltzmann's second approach to justify the Maxwell–Boltzmann distribution is to derive a dynamic equation for velocity distributions based on the laws of molecular motion and show that its time-independent solution is the Maxwell–Boltzmann distribution. By making certain randomization approximations, he obtained a dynamic equation with the desired equilibrium solution. The Boltzmann equation is one of the most successful *kinetic equations* in the study of *nonequilibrium and transport phenomena*, for example, thermal and electrical conduction. With further approximations, the *hydrodynamic equations* for the macroscopic motion of many-body systems are derived from the kinetic equations. The laws of mechanics, the kinetic equations, and the hydrodynamic equations separately describe a many-body system in three spatial and temporal scales and degrees of detail. The derivation of the kinetic and hydrodynamic equations and the approximations involved are discussed in § 38, together with the meaning of randomness.

Unlike the equations of motion in mechanics, the Boltzmann equation is *temporally irreversible*. Boltzmann also tried to show that it explains the second law of thermodynamics, although the effort is less successful than he had hoped. The irreversibility problem is considered in § 40.

Condensed-matter Physics[27]

The statistical framework developed by Maxwell, Boltzmann, and Gibbs is sufficiently general to accommodate both classical and quantum mechanics. The application of statistical mechanics mushroomed after the infusion of quantum mechanics in 1925, which gives the correct treatment of microscopic particles.[28] Statistical mechanics underlies the study of stellar atmospheres and other astrophysical and plasma phenomena. The major impetus of its continuing development, however, comes from *condensed-matter physics*, which relates the macroscopic characteristics of solids and liquids to the quantum mechanical behaviors of their constituents. Condensed-matter physics is by far the largest branch of physics, producing twice as many doctorates as the runner-up, elementary particle physics. It is also the branch most closely associated with technological innovation. With contributions such as the transistor, semiconductor, computer chip, liquid crystal, solid-state laser, high sensitivity detector, and an array of other electronic devices, it forms the bedrock of the electronic industry.

In gases, the molecules are far apart; hence their interactions are relatively weak and can be treated with simple approximations. The situation is different in condensed matter, the high density of which implies that each constituent interacts strongly with many neighbors. Physicists have to tackle the difficult situations of many-body systems. The laws of quantum mechanics and the conceptual structure of statistical mechanics apply, but the solutions of concrete problems require much ingenuity and approximation.

As in other many-body systems, the major source of difficulty in condensed-matter physics is the interaction among the constituents. A most common approximation is to transform the problem of interacting constituents into a problem of *weakly coupled or noninteracting entities*, the aggregate of which can be regarded as a kind of "gas" and treated by the statistical techniques developed for gases. For example, the system of electrons in a metal is often modeled as an "electron gas," but the noninteracting electrons in such a "gas" are new entities whose characters differ drastically from those of electrons in free space. A considerable amount of effort goes into the definition and characterization of these weakly coupled or noninteracting entities. Sometimes, as with metallic electrons, they are modifications of the original constituents, and they interact not with each other but only with a common environment, formally analogous to the passive consumers in a perfectly competitive market. This method, called the *independent-particle approximation*, is presented in § 15. In other cases, discussed in § 19, physicists discern collective excitations. A *collective excitation* involves the motion of many original constituents. It is treated as an individual with a corpuscular name such as *phonon* or *plasmon*, for it is excited independently and interacts only weakly with other collective modes. These models show the multifarious meanings of "individuals" in many-body theories.

One of the major areas of research in condensed-matter physics concerns *phase transition*, for example, the transition of water to ice. Phase transitions change the integral structures of thermodynamic systems and cannot be described solely in terms of their parts. In one kind of phase transition, the *critical phenomenon*, unlike in the water–ice case, no discontinuous jump of density occurs. Critical phenomena were first reported in 1822 for liquid–gas transitions and qualitatively explained in 1873 by the macrotheory of Johannes van der Waals. A detailed understanding of their micromechanisms emerged only in the 1960s, and lively research continues today. Phase transitions are the clearest examples of qualitative changes at the macroscopic level. They show the emergence of macroscopic features such as rigidity, which is characteristic of our familiar world of solids (§ 23).

Equilibrium

2 Theories of Composite Systems

4. Organizational and Descriptive Levels of the World

The physicist Richard Feynman closed his lecture on the relations among various sciences by contemplating a glass of wine. He said if we look at the wine closely enough we see the entire universe: the optics of crystals, the dynamics of fluids, the array of chemicals, the life of fermentation, the sunshine, the rain, the mineral nutrient, the starry sky, the growing vine, the pleasure it gives us. "How vivid is the claret, pressing its existence into the consciousness that watches it! If our small minds, for some convenience, divide this glass of wine, this universe, into parts – physics, biology, geology, astronomy, psychology, and so on – remember that nature does not know it!"[1]

Feynman did not mention the price tag, the symbol of the market economy. The division of intellectual labor he highlighted, however, underlies the success of science just as the division of labor generates the prosperity in which common people can enjoy fine wines from around the world. How is the division of intellectual labor possible? What are the general conditions of the world that allow so many scientific disciplines to investigate the same glass of wine? What does the organization of science tell us about the general structure of theoretical reason?

Our scientific enterprise exhibits an interesting double-faced phenomenon. On the one hand, scientists generally acknowledge that everything in the universe is made up of microscopic particles. On the other, various sciences investigate macroscopic entities with concepts of their own, which are drastically different from those of microscopic particles, with little if any reference to the latter. The phenomenon is often noticeable in a single science. For instance, hydrodynamics depicts a fluid as a continuous whole governed by its own dynamic equations with no mention of molecules, but when asked, hydrodynamic physicists readily admit that the fluid is made up of molecules.

The division of science into many disciplines is not merely a matter of academic politics or intellectual fashion. It is objectively underwritten by the structure of the world. One can hardly imagine that our many sciences are supported by a world of perfect simplicity like that which prevailed shortly after the Big Bang or a world of utter randomness like that projected for the end of the universe. The plurality of science suggests a world with great diversity, but not so great as to overwhelm our cognitive capacity and be regarded as totally chaotic; a world that permits the discernment of phenomena at various levels of scale, organization, and complexity, each of which can be conceptualized and investigated with little reference to the others.

A diverse world alone does not guarantee the plurality of science. We could have been one-track-minded and insisted on explaining everything by a single set of substantive concepts and principles, be they divine laws or the concepts of things observable by our naked senses or the laws of elementary particles. Then science would be like a centrally planned economy in which everyone labors to work out the consequences of the supreme directive. It is doubtful whether such a science can prosper. Fortunately, the human mind is more versatile. Actual science is more like a free-market economy. Various sciences find their special niches and open their domains of inquiry, each raising its own questions, positing its basic assumptions, adopting a suitable method, individuating a set of entities, and setting forth predicates to characterize a specific group of phenomena. The specialization of the sciences does not imply that they are totally disjoint and proceed regardless of each other. Just as players in the market observe legal and social rules that enforce contracts and promote cooperation, the sciences implicitly observe some general principles, which are revealed, among other things, in the acknowledgment of composition.

The Subject Matter of the Sciences

Our sciences present the world as a hierarchical system with many branches, featuring individuals of all kinds. The individuals investigated by various sciences appear under different objective conditions. Despite the apparent independence of individuals in various levels, the hierarchy of individuals is not merely a matter of classification. It is underwritten by the *composition* of the individuals, although the idea of composition is tacit in many sciences. All individuals except the elementary particles are made up of smaller individuals, and most individuals are parts of larger individuals. Composition includes structures and is not merely aggregation. To say that something is made up of others is not to say that it is nothing but the others. A brick house is more than a heap of bricks; it has its architecture. A natural compound differs from a house in that its structure results from the spontaneous interactions among the elements that become its components.

There are four known kinds of basic physical interaction: gravity, electromagnetism, and the two nuclear interactions. The weak nuclear interaction plays little role in the formation of stable matter. The strong nuclear interaction is crucial below the nuclear level but insignificant beyond, where the characteristic distances far exceed its range. Gravity, extremely weak but having purely cumulative effects, is responsible for the formation of astronomical objects and the large structures of the universe. The electromagnetic interaction produces the awesome diversity of medium-sized things we see around us. Like gravity, it has infinite range; unlike gravity, it is strong but its effects are not cumulative. When positively charged protons and negatively charged electrons combine to form an atom, they almost neutralize the electromagnetic interaction among them. The residue electromagnetic force among the atoms is weak and assumes various forms such as the hydrogen bond, the covalent bond, and the van der Waals potential, which are responsible for chemical bonds and reactions.

Relations among organisms and humans take on forms that are drastically different from the forms of physical and chemical interactions. They have some kind of physical base. However, their connection to the physical interactions is best articulated in the negative sense that they will disappear if the electromagnetic force is turned off, in which case the organisms themselves disintegrate. We know from experience that the behavior of one organism or human affects the behaviors of others within a certain range, and the relations between the behaviors and effects have enough regularity to be characterized in general terms. Thus we posit some kind of interaction among organisms and humans. Historically, the "power" one organism has on another was recognized before the physical interactions.

Let us start with elementary particle physics, which aims to study the elementary building blocks of the universe. It presents the elementary ontology as a set of quantum fields, three of which, the electron and the two quark fields, constitute all known stable matter of the universe. The quanta of excitation of the quantum fields are called *particles*, which are not immutable but can easily transform into each other when fields interact.

Up the ladder of complexity, we have nuclear and atomic physics. Nuclear physics investigates the structures of nucleons (neutrons and protons) and atomic nuclei made up of nucleons. The strong nuclear interaction, which binds quarks into nucleons and nucleons into nuclei, has a range of about 10^{-13} centimeter. Beyond this distance it is unable to overcome the electric repulsion among the protons. The short range of the nuclear interaction limits the size of the nuclei, hence their variety. Constrained by the variety of nuclei, there are only 109 kinds of atom, the heaviest of which, unnilennium, contains 266 nucleons and 109 electrons. A kind of atom can have several isotopes, which have the same number of electrons but vary in the number of neutrons. They are the topics of atomic physics.

The diversity of composite systems on larger scales is helped by the availability of more than a hundred kinds of atom instead of a few kinds of elementary particle as building blocks. More important, the electromagnetic and gravitational interactions both have infinite ranges and impose no basic limit on the size of the systems. Variety explodes, as the number of possible combinations of building blocks increases exponentially with the number of blocks. There are 1.5 million known kinds of stable chemical compound, even more kinds of organism. Responding to the diversity, the sciences diverge. Very roughly, physics branches into cosmology, general relativity, the studies of solids, fluids, and plasmas in which atoms are decomposed into electrons and positively charged ions. Chemistry studies the reactions among a great variety of molecules, organic and inorganic. It also enables us to synthesize long chains of molecules that constitute various plastics. Further up the slope of complexity, the miracle of life occurs in entities that grow and reproduce themselves. Biology divides into molecular biology, genetics, cytology for cells and cellular organelles, histology for tissues, ethology for organismic behaviors, physiology, ecology, evolution, and more.

Within the diversity of life, the organisms of one species are conscious of their existence and have the capability to think in symbolic terms. They become the topic of the human sciences including psychology and cognitive research. People engage in meaningful communication, exchange, and conflict; they form groups and institutions that are meaningful to them. To study these phenomena the social sciences spring up.

Individuals in a certain level of scale and complexity interact most strongly among themselves. They are often posited independently and described without mentioning individuals in other levels. Concepts pertaining to one level bear little or no relation to concepts in remote levels, and most satisfactory explanations of an individual's behavior use concepts of its own level. Elementary-particle concepts are irrelevant to theories of solids, nuclear concepts are not pertinent in biology, and biological notions are beside the point in explaining many human actions. When a camper asks why the food takes so long to cook, physicists answer in terms of pressure at high altitudes. Physicists use macroscopic thermodynamic concepts although they have microscopic theories in reserve, for the microscopic information is irrelevant and would obscure the explanation. The prominence of explanations on the same level secures a significant degree of autonomy for the sciences, for they investigate disparate entities and use logically independent concepts.

Coherence and Disruption

Adhesive interaction among individuals is the basis of composition and structure formation, but not all interactions are adhesive. The electromagnetic force among electrons is repulsive; hence a group of negatively charged electrons will fly apart and never form a unit by themselves. They can participate in a system only if the system contains other ingredients, notably positively charged protons, that neutralize their antagonism. Besides antagonistic relations, excessive idiosyncrasy in individual behaviors also frustrates unity. No unity can be maintained if each individual goes its own way. A system can crystallize only if the individuals somehow harmonize their behaviors through agreeable relations, as a democracy exists only when a significant degree of consensus prevails among its citizens. Furthermore, a system is endangered if it is subjected to external forces that attract the constituents, as villages dwindle when outside opportunities entice young people to leave.

To succeed in forming a system, the cohesive interaction among the individuals in a collection must be strong enough to overcome internal disruption and external impediment. If the cohesion exceeds the forces that pull the individuals apart or attach them to outside elements, the collection acquires an

Table 2.1. *An entity is stable only at temperatures below its binding energy. The thermal energy at room temperature (25°C) is 0.02 eV (electron volt). All bonds cited except the nuclear interaction are electromagnetic.*

Entity and bond	Binding energy (eV)
Proton and neutron (strong nuclear interaction)	10^9–10^{10}
Nucleus (residual strong nuclear interaction)	10^6–10^7
Atom (first ionization)	3–10
Small molecule (covalent bond)	1–5
Biological molecule (hydrogen bond, van der Waals bond)	0.1–1

integral structure and becomes a composite system on a larger scale. Thus the criterion of cohesion is not absolute but relative; it is a tug-of-war between the ordering and disordering forces. The war can often be represented in terms of some variables suitable to the particular systems under study.

In physics, the crucial variable is *energy*. The cohesiveness of a physical system is determined by the relative importance of two kinds of energy: the *kinetic energy* characterizing the agitation of each constituent and the *binding energy* characterizing the bond among the constituents. The kinetic energy is mainly determined by the ambient temperature. The higher the temperature, the higher the kinetic energy, and the more agitated the constituents become to go their own ways. When the kinetic energy matches or exceeds the binding energy of an entity, the constituents break free and the entity disintegrates. Thus an entity is stable only in temperatures below its binding energy.

Table 2.1 lists a few composite entities together with their characteristic binding energies, which must be supplied to break an entity into its constituents. At the high temperatures found within split seconds of the Big Bang, quarks were free entities. Quarks interacted, but the interacting energy was small relative to the kinetic energies of individual quarks, so that each quark went its own way. As the universe cooled, the kinetic energy of the quarks decreased. When the temperature dropped below the characteristic energy of the strong nuclear interaction, the quarks were bound into protons and neutrons. At still lower temperatures, but no less than tens of billions of degrees Celsius, which are the temperatures found in the interior of stars, protons and neutrons fuse into nuclei. When a collection of elements coalesces into a stable composite entity such as a nucleus, it gives off an amount of energy equal to its binding energy. The energy released in nuclear fusion fuels the sun and the stars. Physicists are trying to harness it in nuclear fusion reactors.

As the temperature falls further, nothing can happen to the nuclei. They are so stable and tightly bound that the motions of their components are negligible; thus they are often approximately treated as simple units without internal structures. New entities, the atoms, overcome the destructive force of thermal agitation and emerge to take over the center stage. In the low energy environment typical of condensed-matter physics and chemistry, a large part of an atom is tightly bound and only electrons on the outermost atomic shell can break free, leaving a positively charged ion. The atoms, ions, and electrons become building blocks of molecules, chemical compounds, and solids. In the even lower temperatures favorable to life, large chunks of molecules can be regarded as units connected by weak bonds. Since these weak binding energies are not much higher than the thermal energies at physiological temperatures, they are easily formed and broken, facilitating the multitudinous biological processes of growth and decay.

The idea that relative cohesiveness determines the formation of entities is not limited to the natural sciences. In economics the major variable is *cost*. The formation of composite units such as large corporations depends on the relative importance of two kinds of cost: *transaction costs* that impede market coordination of productive effort and favor managed coordination within corporations; *administrative costs* that impede managed coordination and favor decentralization. When transaction costs were low and technology permitted only simple production processes in the early stages of industrialization, firms were small, each deciding for itself what to produce, buying its material and parts from others, and selling its product as parts to still others. With the rise of large-scale industries, especially the railroads and telegraphs, and the high-volume production and distribution they facilitated, it became more cost effective to pay for professional managers to organize the productive efforts of many individuals. Large corporations with hierarchical managerial organizations appeared. These enterprises routinize the transactions within their parts, lessen the friction between production and distribution, accelerate the flow of goods and cash, improve the schedule of delivery, increase the utility of capital and facilities, and dodge anticollusion laws. All these actions reduce costs. As long as the saving in transaction costs exceeds administrative expenditures, large corporations will thrive. Conversely, when transaction costs are reduced by information technology and administrative costs increased by union wage and employee benefits, outsourcing becomes more popular as corporations downsize and purchase parts from outsiders instead of producing them in house.[2]

The conflict between cohesion and disruption is not clearly represented as a theoretical variable in biology, but it is no less important. The history of organic evolution consists of several major transitions in which entities previously reproduced independently became parts of wholes reproduced as units. For instance, ancient unicellular protists survived and reproduced as individuals, but their descendants, cells of plants and animals, are integrated into larger wholes. Why and how the transitions to more complex organisms occurred are still not understood. However, it is clear that for the wholes to survive the winnowing of natural selection, the internal rivalry of their parts must be subdued. The development of the myriad cells of a multicellular organism from the division of a single cell can be seen as one way to ensure the genetic homogeneity of the cells and hence the minimal reproductive competition among them.[3]

Atomism and Physicalism

The hierarchical worldview described has several tacit assumptions, notably physicalism and a weak form of atomism. These are *ontological* assumptions concerning the way the world is. They should not be confused with *episte-mological* issues concerning the nature of our knowledge about the world or *methodological* doctrines such as reductionism and microreductionism, which argue for the specific forms of scientific theory and approaches of research.

The philosophical doctrine of atomism, first enunciated by the Greek Atomists, asserts that the world is made up of indivisible and immutable particles. The atomism we have assumed is weaker. It asserts only that most entities are analyzable into parts with characters of their own, and a macroscopic entity examined on the microscopic level exhibits microscopic entities and processes. The atomism is weak on two counts. First, it denies that an entity is no more than the sum of its parts, for the structure of the entity, in which the parts find their places as parts, is equally important. The structure is destroyed in decomposition and overlooked in microscopic examination. If we look at a human brain on the atomic level, we find phenomena characterized by atomic physics. However, on this microscopic level we are studying atoms in certain situations, not neurons and not brains. It is as if we are looking at a mural through a bathroom tissue-paper tube: We cannot claim that what we see is the picture.

Second, weak atomism does not claim that the parts of an entity can always be detached and investigated in isolation, and it admits that even if the parts are detachable, the isolated elements usually lose some of their characters as parts. Thus weak atomism is not incompatible with a weak holism asserting that the characters and interactions of the parts can only be studied in situ, as the behaviors of persons can only be studied within societies. The compatibility of weak atomistic and holistic ideas are apparent in physics, especially in the lowest level of organization. Quantum field theory for elementary particle physics depicts a holistic world in which the parts are defined only as parts of the whole, the field.[4]

Physicalism asserts that all entities and processes in the world are physical in the widest sense. People and organisms are all physical; they are made up of atoms not different from those that constitute inanimate things. The chemical processes occurring in organisms are the same as those occurring in test tubes. There is no nonphysical vital force and no spiritual substance. A man dies when his heart stops beating or his brain stops functioning, not when his soul leaves the body; there is no soul or mental substance that can leave the body.

Note again that physicalism is an ontological and not a conceptual doctrine. It does *not* assert that all entities and processes must be conceptualized exclusively in the vocabulary of the physics department. Physicalism denies entelechy and soul substance, but it does *not* exclude vital and mental concepts, the appropriateness of which is determined by the biological and human sciences and not by philosophical ideology. It is compatible with theories that attribute "being alive" and "being conscious" to organisms and persons, if the attribution does not depend on the postulation of nonphysical forces or substances. In these theories, the vital or mental predicates describe emergent characters of complex physical systems. It is characteristic of emergent properties of complex systems that they differ from the properties of their constituents. For instance, the characters of classical objects that we routinely handle are emergent and are paradoxical in quantum mechanics that governs the behaviors of their atomic constituents; illustrations of the paradox can be found in most interpretations of quantum mechanics. The meaning of emergence will be substantiated in the following chapters.

Weak atomism and physicalism are the general presuppositions we make about the world. They underlie the general idea of composition and unify the sciences as studies of various aspects and levels of the same glass of wine, the same universe.

5. Deductive Construction of Small Systems

Although entities in each scale and organizational level can be studied on their own, the connection between theories on two levels, if available, consolidates our knowledge and opens a new area of study. We gain a deeper understanding of an individual if we also grasp the relation between its behavior and the behavior of its constituents or the behavior of the system in which it participates. For instance, metallurgy has been with us for a long time, but it is revolutionized by infusing the results of chemistry and molecular physics, which lead to an array of materials undreamed of before. The connections between levels belong to the theories of composition.

Under the ontological assumption that all things in the world are made up of microscopic particles obeying a few fundamental physical laws, we turn to the main concern of this book: How does theoretical reason represent and investigate the complex and diverse forms of composition? More specifically, how do scientific theories handle large-scale composition in which a system and its constituents are entities of different scales and are characterized in disparate concepts?

There are two general approaches in studying composite systems: *deductive construction* and *synthetic microanalysis*, the former more suitable for small systems, the latter for large systems. The constructive approach is more familiar. Although it is impractical for large systems, the extrapolation from small systems helps to cement the idea of composition. For large systems, scientists do not construct from elements to find the resulting macroproperties but analyze systems to find their microstructures. The synthetic microanalytic approach and the general conceptual structure of the many-body theories that employ it are presented in the following section and substantiated by many examples in the subsequent chapters. This section examines some general concepts shared by the two approaches, with an emphasis on the constructive approach.

The Myth of Bare Elements

Composition presupposes the idea of ingredients. Let us start with the conceptualization of the ingredients. A formulation favored by revisionary philosophies with a set-theoretic bend is the positing of a set of bare elements followed by the imposition of relations for the construction of compounds. The approach is intuitive at first sight but reveals deep problems on scrutiny.

Elements are not constituents. *Constituents*, as parts of composite systems, are situated microindividuals that cannot be understood apart from the systems of which they are parts. *Elements* are not parts of composite systems, at least not the systems under study. When elements enter certain situations and become constituents, their behaviors are substantially transformed by their causal relations with other constituents. There are two ways of viewing elements, as being *bare and free-floating* or as being *situated in a limiting or minimal context*. Bare elements, stripped of all relations and excluded from all situations, are posited independently of the ideas of relations and systems. Participants of minimal situations presuppose some ideas of relations in wholes, albeit tiny wholes. I argue that the notion of bare and free-floating elements has no realistic significance; a real thing is not merely a bare x in abstract logic.

The concept of individuals unencumbered by explicit relations is paramount in ordinary language and scientific theories, for they stand as the subjects of subject–predicate propositions. In actual discourse, however, the individuals are always understood within certain situations or implicit relations, and the contexts make them constituents rather than bare elements. The implicit relations introduce a holistic sense that mitigates atomism by making the atoms relatable. This is one reason why I subscribe only to a weak form of atomism in the previous section.

Many revisionary philosophies commit the common error of pushing a valid idea too far, thereby making it fallacious. They push the idea of individuals to the extreme and turn individuals into bare elements by demanding the elimination of *all* relations. When the idea of bare elements is sharply formulated, however, its difficulty proves to be fatal, for it ignores the preconditions of relation. The *relatability* of concrete individuals is not a trivial condition. One cannot arbitrarily apply glue to unprepared materials and expect them to stick; a glue works only for certain kinds of material under certain conditions, and surface preparation of the material is mandatory for proper bonding. Bare elements are incapable of supporting relations because they lack the proper precondition, just as chimpanzees raised in isolation are neurotic and incapable of entering into social relations. Therefore the bare elements that were once posited in scientific theories were promptly replaced. I cite two examples.[5]

The "free electron" or "electron in free space" in physics is *not* a bare electron. Physicists did postulate bare electrons in the late 1920s, but the postulate led to great difficulties that took twenty years of research to overcome. The solution, known as renormalization, rejects the idea of bare particles as physically meaningless. An electron's charge, which is its intrinsic property, is also the source of the electromagnetic field, the mechanism of interaction. Consequently the "free space" is actually filled with electromagnetic field, if only that generated by the electron itself. Furthermore, the electromagnetic field can generate electron–positron pairs. Thus the electron is an inalienable part of the interactive electron–electromagnetic field system. Its intrinsic relation to the electromagnetic field is built into the monadic description of its mass and charge by renormalization.[6] The free electron and other free particles are not bare elements but participants in minimal situations in which other particles are so far away that their effects are negligible.

As another example, consider social-contract theories that posit a state of nature, the individuals of which come together to form a society by agreeing on its structure and the duties and obligations of its citizens. In early renditions, the individuals who negotiate the social contract are bare elements, as Thomas Hobbes described: Those who come together to form a community are "men as if but even now sprung out of the earth, and suddenly, like mushrooms, come to full maturity, without all kinds of engagement to each other."[7] The inadequacy of bare individuals is duly acknowledged by later contract theorists. In more carefully argued theories, the state of nature is an abstraction from present societies and represents the fair conditions of negotiation that would lead to the fair terms of social cooperation. Individuals in the state of nature are situated citizens with a general sense of sociability; all they temporarily put aside are certain privilege and personal knowledge so that no one has a bargaining advantage over the others. They are free and equal, not bare and asocial.[8]

Theories of composition study the structures of large and complex systems on the basis of the knowledge of constituents of small systems, not on the assumption of bare elements. In physics, the small systems may be minimal and realizable in controlled experiments. In the social sciences, so-called small situations may be quite considerable and not susceptible to engineering, as social scientists insist that persons are meaningful only within certain social contexts. In the following, unqualified "element" always refers to a participant of a minimal situation and never to a bare element.

Type and Value

It is intuitive that a large composite system is made up of constituents that are familiar to us as the participants in smaller or minimal systems. We have a rough idea of how people deal with each other, and the interaction of particles in binary systems has been thoroughly studied. We are interested in the results of similar interactions in large systems. The familiar constituents and relations are featured in the basic formulation of many-body problems. However, they do not stay long; the constituents will be transformed beyond recognition as the problems are reformulated with the development of many-body theories. Thus it is important for us to be clear about their representation.

Constituents interact with each other, and their mutual relations cement the composite system. The relations among the constituents are not arbitrarily imposed. They are the balance to the characters attributed to the entities related in the process of individuating the constituents (§ 10). Thus they are not merely logical but objective or causal in the broad sense, meaning they are the physical, biological, and economic relations occurring in the world and not on paper only. The relations are also the major source of complexity in many-body systems and of the difficulty in understanding them. In the following chapters, we will see that in many models, the constituents are redefined so that their monadic predicates absorb most relations. Thus the boundary between individual characters and mutual relations can shift. To prevent confusion I will stick to logical definitions. *Characters* or *monadic properties* are those qualities that are represented by monadic predicates, which are ascribed to individuals in subject–predicate statements of the form "It is thus and thus." Characters include implicit relations such as "being a son." *Relations* mean explicit relations represented in relational statements in forms such as "This and that are so coupled" or "*A, B,* and *C* are so related."

An individual has various characters, the summary of which is called the individual's *state*. The general concept of a character can be analyzed into two elements, a *type* and a *value*. Blue and red are values of the character type color; big and bigger, values of size. Other characters are temperature, weight, and annual income; their values – 25°C, 250 kg, and $25,000 – are the definite predicates attributed to individuals, and their types are represented in the units: degree Celsius, kilogram, and dollar. The number systems provide refined and systematic ways of assigning predicates for many types of character, but they are not alone in these virtues. Some character types have exotic values such as vectors and other mathematical objects with complicated structures. In theoretical economics, for instance, the character type for households is the preference order, a value for which is a specific linear ordering of various commodity bundles that represents the idiosyncratic taste of a household.

The same type–value duality holds for relations. Gravity is a type of binary relation, the strength of which is proportional to the product of the masses of two bodies and the inverse square of the distance between them. Given definite values of the masses of and the distance between two particular bodies, we obtain the definite value of the gravitational force between them.

Theoretically, a type is a *rule* that systematically assigns values to individuals. Being a rule, a type has a certain *form* and a certain *range* of possible values. The color spectrum delimits the range of the type color, and a colored object realizes one of the possible values in the range. Thinking in terms of types, we open ourselves to the *possibilities* of characters, thus in a sense anticipating the possibility of change. A value attributed to an individual is recognized to be the realization of a specific possibility. The general concept of possibility will play a crucial role in the analysis of individuals.

Character types are often called *variables* in mathematical science. A character type often takes the form of a function whose argument picks out various individuals and whose value ascribes character values. "Character," "variable," or more specifically "energy" or "height" can refer to the type or a value. In daily conversation, it often refers to a value, for we are ordinarily more interested in particular cases. In science, it more often refers to a type, for science is more interested in general rules that cover classes of instance.

Often several character and relation types are related in a rulelike fashion, which can also be expressed in a functional form, as the energy of a classical particle is proportional to the square of its momentum. The rulelike relations among various types are the crux of a scientific theory. If such relations are well established and form a coherent network, they are called laws. They inform us how the values of various types are systematically related, so that given some values for a specific system we can derive the others.

Theories of Small Systems and Their Extrapolation

At the beginning of a problem of composition, the character and relation *types* of the constituents are assumed to be known, *not* their values. If the types are coherent and general enough, we say we know the laws governing the constituents. The specification of constituent types is not trivial. Often drastic idealization is required to distill typical behaviors from the maze of reality and articulate them clearly and concisely.

What do we want to find out in theories of composition? For small systems with a few parts, we usually solve for the character and relational values of the constituents, for instance, the trajectories (values of position at various times) of the planets in the solar system. The solutions, derived, perhaps approximately, from the laws for the constituents, describe the relative behaviors of the constituents in purely constituent terms. Except aggregative quantities such as the total energy or momentum, no concept for the system as a whole is required. The bottom-up constructive and deductive method is very successful for small systems; much of Newtonian mechanics employs it.

Physically, the character and relation types do not change as more constituents are added. Thus the formulation of the problem of composition can be straightforwardly extrapolated to large systems. We simply add more terms in our equations as we add more constituents to a system. This is the formulation that initiates a many-body theory. It confers a conceptual unity to composite systems of all sizes and shows that large systems are made of the same familiar ingredients coupled by the same ordinary relations as small systems. Thus whatever strange and novel features a large system exhibits stem not from its constituents but from their large-scale organization.

The extrapolation of a formulation does not imply that we can successfully extrapolate its solution. The bottom-up deductive method suitable for small systems does not go far. As the size of the system increases, descriptions in pure constituent terms quickly become impractical and uninteresting. Even for particularly simple large systems such as gases whose constituents interact only weakly, it is humanly impossible to determine simultaneously the position and velocity values of some 10^{23} molecules. Frankly, who cares about them?

Our aim changes as the size of the systems increases. In many-body problems, we are no longer interested in the character values of individual constituents. We want to find out the characters of the system as a whole: What are its character types? Can they be correlated with the character types of the constituents? How? What character values do the system types assume under certain external conditions? To pose and answer these questions, we must expand our conceptual framework to include system concepts for a top-down view that guides the bottom-up construction. This synthetic microanalytic approach is discussed in the following section.

Reductionism and Microreductionism

Philosophy seeks generality. Reckless generalization, however, is a common mistake in philosophy, as Kant painstakingly explained. Concepts and ideas usually have definite ranges of validity, generalization beyond which lands in illusion. Such an illusion results when some revisionary philosophies, noting that the bottom-up deductive method is successful in studying small systems, generalize it to cover the study of all systems. The prescription engenders the ongoing debate on reductionism and microreductionism.

The debate on reductionism is often muddled because "reductionism" has become an amorphous term with many ambiguous meanings. Some people conflate reductionism with atomism, for ordinarily *reduce* or *reduction* can denote the decomposability of things and causes. Others conflate it with physicalism, for the rejection of nonphysical causes reduces the number of admissible causes. Still others uphold the metaphysical worldview of grand reductionism, asserting that the ultimate explanations for everything in the universe are encapsulated in a few principles. I will not use the term *reductionism* in any of these senses, which are either vague or more clearly expressible under other names, so that we can focus on the distinctive and controversial doctrines of reductionism.

The word *reductionism* seems to appear first in 1951, when the philosopher Willard Quine used it to label a dogma of empiricism. The reductionist dogma itself is much older and had received careful exposition from many philosophers.[9] Reductionism and microreductionism are doctrines about the relations among concepts, theories, and sciences. They argue there is a primary science to which secondary sciences can and should be reduced. A predicate in a secondary science is reduced if it is defined solely in terms of the vocabulary of the primary science. A law is reduced if it is deduced from the laws of the primary science. A theory is reduced if each of its sentences is translated into a sentence in the primary science that expresses the same fact without any loss of meaning. The reduced predicates, laws, and theories, together with the secondary science itself, are dispensable, for they are replaceable by the superior primary science. For example, psychology is reducible to physics if psychological concepts and theories are in principle eliminable in favor of physical concepts and theories. To emphasize the dispensability of reduced theories, the redundant qualification *eliminative* is sometimes added to reductionism. It underlies the saying that the reduced theory is *merely* or *nothing but* the application or the logical consequence of the reducing theory. Since concepts make things intelligible, the elimination of the reduced concepts suggests that the things represented by these concepts are *nothing but* the things represented by the reducing concepts. If we want to prevent confusion with other notions of reductionism in the popular literature, we can call the philosophical reductionist dogma "nothing-butism."

There are two types of reduction, homogeneous and inhomogeneous. Homogeneous reduction, in which the domain of the reduced theory is a part of the domain of the reducing theory, is valuable in explaining the history of science. Scientific theories are often superseded by more comprehensive successors, as Newtonian mechanics is superseded by relativistic mechanics, the domain of which covers the domain of Newtonian mechanics and more. In such cases, people usually agree that the narrower theory is reducible and in principle dispensable because it is a special or limiting case of the broader theory. Most reductionist arguments are valid only for homogeneous reduction. Unfortunately, they are often extrapolated carelessly to inhomogeneous reduction, where they become problematic.

In inhomogeneous reduction, the reduced and reducing theories cover separate domains of phenomena. Usually a secondary theory contains terms and concepts not found in the primary theory, as psychological notions are absent in physics. Since logical deduction cannot produce a term that is not contained in the premise, the reduction program must posit bridge laws to connect the disparate concepts. After much argument, philosophers generally agree that for reduction to succeed, the bridge laws must be either explicit definitions for the secondary concepts or biconditionals that specify the necessary and sufficient conditions for applying the secondary concepts, both expressed exclusively in terms of the primary concepts. As we will see, these stringent conditions are seldom met in actual scientific theories, which are satisfied with more liberal connections.

Microreduction is the most important form of inhomogeneous reduction. Large composite systems naturally exhibit properties that become distinctive domains of study for various sciences and theories. Microreductionism stipulates the relation between two sciences investigating entities on different scales and organizational levels, as the topic of the secondary science is made up of the topic of the primary science. It is most clearly articulated by Paul Oppenheim and Hilary Putnam, who argued that the microreduction of a science B_2 to the science B_1 "permits a 'reduction' of the total vocabulary of science by making it possible to dispense with some terms," "'reduces' the total number of scientific laws by making it possible, in principle, to dispense with the laws of B_2 and explain the relevant observations by using B_1."[10] According to the covering-law model of explanation prevalent at the time, to explain something is to deduce it from the laws of a theory plus appropriate antecedent or boundary conditions. By opting to explain everything in the vocabulary for the constituents, microreductionism restricts our conceptual perspective to the viewpoint of the constituents, because the vocabulary delimits intellectual vision. Since microreduction is transitive, and since all matter is ultimately composed of elementary particles, the logical conclusion of microreductionism is that in the ideal science, all phenomena in the universe are deducible from the laws of particle physics and describable in the particle vocabulary. The result is a monolithic theory of everything. In short, microreductionism advocates strict bottom-up deduction from the laws of the constituents, the dispensability of system concepts, and a monolithic theory with a single perspective for phenomena on different organizational levels.

Although microreductionism has lost its once dominant position, it continues to be influential in many forms. Methodological individualism banishes social concepts and demands social phenomena be explained exclusively in terms of the desires and actions of individual persons. Eliminative materialism throws psychological concepts such as belief and desire on the junk pile to keep company with alchemy and witchcraft and demands people's behaviors be explained in purely neurophysiological terms.[11] By disparaging or even outlawing system concepts, the reductive doctrines impose an imperial unity on science by reducing the sciences for composite systems to subordination if not superfluity. Not surprisingly they antagonize scientists working in the allegedly reducible areas, who argue that the reductive equation of "is made up of" with "is nothing but" obscures the formation of structures in composition.

Reductionism and microreductionism are methodological doctrines stipulating the form of philosophically correct theories, but they often gain deceptive appeal by confusion with ontological issues. Consider methodological individualism that asserts that a full characterization of each individual's behavior logically implies complete knowledge of the society without using any social concepts. As a methodological doctrine, it is distinct from ontological individualism asserting the nonexistence of society as a supraindividual entity independent of persons. However, methodological individualists often exclude social concepts on the ground that social concepts imply the existence of mystical supraindividual entities. The argument does not hold; social concepts can be about composite systems made up of persons, just as personal concepts are about composite systems made up of atoms.[12]

That a system is physically constructed of parts lawfully interacting with each other does not imply that all descriptions of the system are logically constructable from the laws governing the parts. That the behaviors of a system are the causal consequences of the motion of its constituents does not imply that the concepts representing the system behaviors are the mathematical consequence of the concept representing the constituent motion. The implications depend on strong assumptions about the structure of the world, the structure of theoretical reason, and the relation between the two. They demand that we conceptualize the world in such a way that the alleged logical or mathematical derivation is feasible for systems of any complexity. This strong assumption is dubious. Methodology does not recapitulate ontology.

Critics always complain that microreductionism is short on concrete examples. The few examples that are recycled endlessly do not meet the stringent criterion of microreduction. System concepts are reducible only if they are defined in purely constituent terms, but the definitions in the examples invariably depend on some system notions. Consider the most often cited definition "Temperature is the mean kinetic energy of particles." The definition is generally false. Particles streaming in a common direction have a definite mean kinetic energy, but this mean kinetic energy is not the temperature of the stream of particles; temperature is not well defined for such systems. Temperature is identifiable with mean kinetic energy only if the system is in thermal equilibrium or describable by a certain velocity distribution, both of which are system concepts. Similarly, the general equilibrium theory in microeconomics, the prime case offered by methodological individualism, defines individuals in terms of commodity prices, which are features of the economy as a whole (§ 16). The cases of temperature and microeconomics do not support microreductionism. Instead, I argue in the following section that they are examples of synthetic microanalysis, in which system concepts are connected to constituent concepts without being made dispensable.

There are attempts at eliminative microreduction in the sciences, but they are usually drastic approximations that distort the phenomena so badly they become the locus of dispute among scientists. For instance, the reductive microfoundation of macroeconomics depends on the assumption that all individuals in the economy are alike, which undercuts the idea of exchange, the base of economics (§ 25). Genic selectionism, which asserts that evolutionary theory should be framed solely in genetic terms, grossly distorts the adaptation models it cites as support (§ 17). These doctrines are "reductive" both in the sense of doing away with macroconcepts and in the sense of abdicating the explanation of important phenomena.

It is interesting that methodological individualism, genic selectionism, and eliminative materialism, which are so enthusiastic in censoring system concepts, all concern sciences with poorly developed theories. Statistical and condensed-matter physicists, who have the most advanced theories connecting micro- and macrophenomena, have a different attitude. In an interdisciplinary conference on problems regarding parts and wholes, Edward Purcell discussed the Ising model. In this grossly simplified model for magnetism, the atoms sit on regular sites, each atom has only two possible states, and each interacts only with its immediate neighbors. Even this simple model defies exact solution if the atoms constitute a three-dimensional lattice. Purcell concluded: "The astonishing stubbornness of the Ising problem stands as a sober warning to anyone who attempts to carve a path of rigorous deduction from parts to the whole." The stubbornness proves to be unrelenting (§ 23). More generally, Philip Anderson has said: "At each level of complexity entirely new properties appear. . . . At each stage entirely new laws, concepts, and generalizations are necessary."[13]

6. Synthetic Microanalysis of Many-body Systems

Computers can beat humans in many games, yet they play in ways different from humans. A difference between computers and humans is analogous to the difference between microreductionism and actual scientific reasoning. A game such as chess has a definite end, a rigid set of rules, and a finite number of legal positions and is made up of successive moves. The best computer programs generate all possible plays to a certain depth according to the rules of the game and evaluate most positions according to certain algorithms. Human players do compute and search, but their search is focused and restricted, and the rules they use are more heuristic than algorithmic. They recognize patterns, organize pieces into chunks, and form strategies and general principles.[14]

The constituents of a complex system and the relations among them are like the moves in a chess game, and their combinations are like plays. For large systems, the exponentially exploding combinatorial possibilities form veritable labyrinths. Brute-force deduction from the laws of the constituents is reminiscent of the computer's style of playing. It can go a long way; although computers have not yet won the world chess champion title, they are within reach. Yet its limitation is already visible in the long delay despite the phenomenal advancement in hardware technology. Chess is a closed game. The variety in the behaviors of large composite systems is open-ended and poses much more difficult problems. To tackle them scientists think more like human chess players than computers, and they fully utilize the clues nature provides. Despite the combinatorial explosion, nature exhibits comparatively few large-scale patterns, which are the hints to the "good plays." Scientists discern these hints and articulate them in system concepts, which serve as Ariadne's thread that picks out relevant information and leads them out of the labyrinth. Concepts of the system as a whole are like the strategies that focus the human chess player's search. Just as a chess strategy turns the computation of possible plays into the analysis of situations, system concepts turn mathematical deduction into the synthetic microanalysis of large composite systems.

Synthetic Analysis and Synthetic Microanalysis

Microreductionists often argue as if their only rivals are holism and isolationism. Isolationists deny the possibility of relating system and constituent theories, and the denial turns composition into some kind of mystery that cannot be clarified theoretically. Hence isolationism becomes a relativism advocating the disunity of science. Holists look only at composite systems as organic wholes unsusceptible to analysis.[15] Microreductionism, holism, and isolationism are all unsatisfactory. There are better alternatives.

Suppose we imagine the description of systems and the description of their constituents as distinct two-dimensional conceptual planes, variously called the macro- and microplanes. Microreductionism rejects the macroplane, holism rejects the microplane, isolationism rejects the connection between the planes. They are all simplistic and deficient. Actual scientific theories of composition are more sophisticated. They open a three-dimensional conceptual space that encompasses both the micro- and macroplanes and aim to consolidate them by filling the void in between. This is the approach of synthetic microanalysis (Fig. 2.1).

Analysis and synthesis were used by the ancient Greeks in empirical inquiry and mathematical proof. They became more important with the rise of modern science and were articulated by Descartes and Newton.[16] Roughly speaking, in analysis scientists resolve a complicated problem into simpler ones, which they study thoroughly. In synthesis they proceed from the results of the simple problems to the solution of the complex problem.

Analysis presupposes some idea of the whole. Before a complicated problem becomes analyzable, it must be articulated with sufficient clarity so that specific questions can be raised; otherwise one does not even know what he is supposed to analyze. If one cannot pose a question clearly, he can hardly expect a satisfactory answer. When we take account of the articulation and formulation of complex problems, then analysis and synthesis become steps

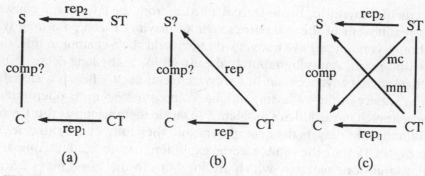

FIGURE 2.1 Ontologically, composite systems (S) are composed of (comp) constituents (C). Constituents are made intelligible by theories and concepts (CT) that represent (rep) them. We are interested in the conceptual relation between constituent theories and system theories (ST). **(a)** *Isolationism* asserts that system theories idealize and generalize in their own ways (rep$_1$ is different from rep$_2$) and are not connectable to constituent theories. Without proper theoretical articulation, the notion of composition becomes dubious. Isolationism becomes holism asserting the unanalyzability of systems as organic wholes. **(b)** *Microreductionism* asserts that the system and constituent concepts are so related that ST is dispensable and a single representation in terms of CT suffices. Without system concepts, however, the integrity of the compound and hence the notion of composition become shadowy. We are reduced to saying there is nothing but a set of interacting elements. **(c)** *Synthetic microanalysis* institutes a broad theoretical structure that encompasses two kinds of explanation for system behaviors: macroexplanations in terms of system theories (rep$_2$) and microexplanations connecting system and constituent theories (rep$_1$ rep$_2$, mm, mc, ST—CT). The connection, which includes deduction, definition, approximation, and independent postulates for structure formation, is not eliminative, although it can be quite rigorous. Unlike the purely bottom-up view of microreduction, synthetic microanalysis combines the top-down and bottom-up views; microexplanations use both CT and ST essentially. Constituent concepts in the representation of systems are usually interpreted as micromechanisms (mm); system concepts in the representation of constituents are sometimes interpreted as macroconstraints (mc).

of a single process in which synthesis both precedes and follows analysis. That process I call *synthetic analysis*.

The synthesis preceding analysis and the synthesis following it operate on two levels of abstraction and generality. The original synthesis posits a broad conceptual framework that enables scientists to set out the complex problem as a problem and prevents them from losing sight of it when they are immersed in analytic details. The synthetic framework imparts meaning to analytic notions as elements of a structure and stresses their interconnection. It may be crude and will be modified, improved, and strengthened by the results of analysis. However, it is primary even in its crudeness, because it makes possible and guides analysis. The establishment of a successful synthetic framework is always a great advancement in science. Examples of synthetic frameworks are the principles of classical and quantum mechanics, the theoretical scheme of statistical mechanics (§ 11), the overarching idea of evolution (§ 12), and the economic concepts of price and commodity (§ 16). Mathematics helps to strengthen the synthetic framework in the mathematical sciences. A mathematical structure is a complicated network of concepts the integrity of which is guaranteed by the rigor of proof. Used in a scientific theory, it keeps open an intellectual domain and makes scientists aware of the context of the concepts they are using.

A synthetic framework covers a wide range of phenomena because of its generality, but the corresponding abstractness also implies that it must be substantiated before it becomes the definite theory of particular classes of phenomena. The substantiation is the job of analysis and the subsequent synthesis, which generate concrete solutions to specific complex problems. Analysis adopts a modular approach, identifying weak conceptual links, making approximations to sever the links, and fragmenting the problem into simpler pieces for detailed investigation. Various models are developed for the fragments and tested independently. The final synthesis assembles the solutions of the fragments by restoring the links, thus tying the analytic results to the original complex problem.

In this book, we are particularly interested in problems concerning complex systems composed of parts, especially parts on a different scale from the whole. In such cases the simpler problems into which the original problem is analyzed often concern the behaviors of the parts. Such an investigation is *microanalytic*. Since the world is rich in structures of all scales, often the parts of complex systems become realms of investigation in themselves. Thus the microanalysis of matter spawns atomic, nuclear, and elementary particle physics. The establishment and pursuit of microphysics express the division of intellectual labor among scientists and do not imply the abandonment and insignificance of problems about macroscopic systems.

To study the behaviors of large composite systems in terms of the knowledge of their constituents, *synthetic microanalysis* is indispensable. A synthetic framework is posited in which microanalysis is directed so that constituents are acknowledged as parts of a whole and their relations grasped as the structure of the composite system. We delineate a composite system, raise questions in macroconcepts that capture its salient features concisely, microanalyze it into constituents, and reach down to seek the constituent behaviors and mechanisms that answer system questions. The approach neither stays on the top as holism demands nor makes a one-way journey from the bottom as microreductionism stipulates. Instead, it combines the bottom-up and top-down views, making a round trip from the whole to its parts and back. The synthesis of microcauses follows the path blazed by microanalysis leading back to the phenomenon pegged by macroconcepts, so that it is not lost in the labyrinth of complexity, as the purely bottom-up journey is wont to do. Instead of deducing from microconcepts, we cast out the net of macroconcepts to fish for the microinformation relevant to the explanation of macrophenomena. Thus system and constituent concepts join forces in explaining the behaviors of composite systems. Abundant examples are given in the following chapters.

The methods of synthetic analysis and microanalysis are used in all our intellectual pursuits. Computer programming is an example in which they are explicitly taught. We know the syntactical rules of a programming language, which are analogous to the laws governing the constituents. The finished program will contain all the microdetails instructing the computer to perform certain tasks. Yet we usually do not proceed directly to write down the details; the microreductionist approach quickly runs into trouble for programs of any complexity. We start by formulating clearly the task to be performed, then write a flow chart breaking down the required procedure into chunks, then break the chunks further into subroutines. Only when the top-down analysis yields manageable pieces do we undertake the bottom-up task of writing the details of the program, and even then we always keep the structure of the whole program in mind.

Macroexplanations and Microexplanations of Composite Systems

The theoretical framework of synthetic microanalysis encompasses two kinds of explanation for system behaviors, macroexplanation and microexplanation. They are not uncorrelated. Microexplanations answer the questions delineated in macroexplanations, and microanalysis is guided by the systems concepts clarified in macroexplanations.

Macroexplanations use macroconcepts representing systems as wholes and do not refer to their constituents. They are found in, among others, thermodynamics, hydrodynamics, classical mechanics, Mendelian genetics, and macroeconomics. Historically, macroexplanations and theories of systems as wholes appeared before their microexplanations. Atomic, nuclear, and elementary particle physics proceeded from systems to their constituents, and thermodynamics preceded statistical mechanics. Perhaps in these cases one can argue that we happen to be medium-sized beings who know medium-sized things first. But even in those cases where the constituents are on our level, theories of composition started by the formation of concepts that bring the system itself into relief. Galaxies had been studied as wholes, and stars were recognized as units before questions were raised about their internal structures. The same occurred in other sciences. Classical political economists were mainly concerned with macroeconomic phenomena; microeconomic analysis lagged by almost a century. Only after the systematic classification of biological species revealed the patterns of their origin and extinction did biologists relate their evolution to the behavior of individual organisms. People had been engaging in economic activities and the behaviors of organisms were familiar, but the knowledge of individual behaviors did not lead to economic or evolutionary theory until people had gained some conceptual grasp of the economy or the species as a whole. Theoretical science relating to large systems proceeded by analysis and not construction. Similarly, research in physiology and psychology historically proceeded in the direction of decomposition and not composition.[17]

Macroexplanations are primary for the understanding of composite systems because the recognition and description of macroscopic regularities, by which the phenomena first stand out as phenomena, necessarily involve considerable theories. Rulelike connections of system concepts often constitute laws for a specific organizational level, for instance, the laws of thermodynamics. System concepts and laws enable scientists to predict and explain many events. These are the explanations within a single organizational level familiar in many sciences.

When scientists have grasped systems as wholes and recognized their composite nature, they begin to seek underlying micromechanisms. The microanalytic effort spawns *microexplanations* of large compounds, which span two organizational levels, that of the constituents and that of the system. Microexplanations of large systems should not be confused with theories about the interaction of constituents in minimal systems. Atomic physics does not explain the behaviors of solids, for it makes no contact with solid-state phenomena. Similarly, eliminative materialists who parade the firing of neurons have not even touched the microexplanation of mind and thinking. Microexplanations must link the behaviors of the constituents and the system. For this they require specially developed theories, which are found in, among others, condensed-matter physics, molecular genetics, and microeconomics.

Microexplanations do not necessarily presuppose a clear knowledge of the constituents. Sometimes the behaviors of the constituents are clarified in microanalysis. For example, economists were convinced of the individualistic base of the market economy, but formulating a successful representation of economic individuals took decades. Physicists believed that macroscopic systems are made up of atoms before they had any empirical evidence of atomic behaviors. They envisioned atoms as tiny billiard balls knocking against each other and asked how the model helped to account for the condensation of gases and the solidification of liquids. Many details of their conjectures were proved wrong by subsequent microphysics. It is possible that the laws for the constituents, macroexplanations for systems, and microexplanations combining the two develop simultaneously.

Arguments about the relative superiority of microexplanations or macroexplanations are futile. The two kinds of explanation answer different kinds of questions. Which explanation is more appropriate depends on the phenomena we want to understand. When physicists answer why the earth is slightly flattened at the poles, they use macroexplanations about the rotation of macroscopic rigid bodies; microscopic laws would be irrelevant and obscuring. When economists in the early twentieth century were obsessed with microexplanations, they were handicapped by the lack of macroconcepts when confronted by the macrophenomenon of the Great Depression. Microreductionism is rightly criticized for jettisoning macroexplanations.

The Composite System and the World

Let us now fill in some details of the conceptual framework for synthetic microanalysis. Theories of large-scale composition start by positing a composite system to be microanalyzed, not a collection of elements to be combined, for ontologically, the constituents of a system can be in flux; epistemologically, characters of systems are what we observe and measure in most cases; historically, some understanding of systems preceded their microanalysis; conceptually, the objective meaning of bare elements is not clear; theoretically, the questions that demand explanation are framed in terms of system concepts, and the questions guide the direction of research.

It is well known that a composite system can maintain its identity even when some of its parts are replaced. Jones is the same person although he has received a liver transplantation and is constantly exchanging material with the environment by breathing, eating, and excreting. If we are committed to a set of constituents, then every time a system loses or gains an atom it becomes a new system, and theorization becomes difficult if not impossible. The persistence of a system through the flux of its constituents is especially characteristic of many-body systems and is duly taken into account by their theories. Evolutionary biology tracks the change of a species through the birth and death of organisms; economics considers the entry and exit of firms to and from a market; statistical mechanics includes the grand canonical ensemble in which the number of particles in a system fluctuates as the system exchanges matter with a reservoir. In all these theories, the system maintains its identity despite the replacement of its constituents. Generally, theories employing statistics and the probability calculus are indifferent to the identities and transience of the constituents.

How are systems delineated? Systems are embedded in the world and are often not at all obvious as individuals. Scientists have to distinguish the system they want to study from its surroundings, to isolate it partially from the rest of the world, intellectually if not practically. For instance, a thermodynamic system is intellectually separated from the heat reservoir that keeps it at a constant temperature. In the partial isolation, a large part of the world is deemed irrelevant and neglected, and the part relevant to the system but not included in it is greatly simplified and summarily represented as a set of fixed parameters that are not changed by what goes on within the system. Following economists, we call these environmental parameters *exogenous*, as distinct from *endogenous* variables that are determined by the working of the system. A system represented without any conspicuous exogenous parameter is called isolated or closed.

Exogenous parameters are also known as *control parameters*, for often they can be manipulated by scientists, in laboratories, in thought experiments, or through careful selection of observational conditions. Scientists systematically vary the exogenous parameters to study how a system responds. Sometimes patterns of regularities appear in the responses and scientists introduce broad classifying concepts for the values of the control parameters to differentiate various types of environment that generate various types of response (§ 30). Exogenous parameters can be coupled with the behaviors of individual constituents or the characters of the system as a whole. For instance, an external electric field affects each electron, and the ambient temperature constrains the distribution of the electronic system. Another example of exogenous parameters is the rate of population growth, which is considered in both biology and economics.

The ideal delineation of a system by treating the rest of the world as fixed is better than neglecting the rest of the world altogether. The approximate realization of partially isolated systems in controlled experiments contributes much to the success of the physical sciences. The boon is not universal; separating some factors and representing them by exogenously fixed parameters is not always appropriate. If the factors are intertwined with system variables and are partially determined by them, then their externalization is a bad approximation. For instance, the living environment is usually taken as an exogenous factor determining organic evolution. Organisms are vulnerable, but they are not passive; they dig holes, build dams, modify the environment, and create their own niches of survival. Thus organisms and their environment interact, and together they form an ecological system that is better treated as a whole. In regarding the environment as impervious, evolutionary biology has made an approximation that needs scrutinization. The effect of externalization is more pronounced in economics, where consumers' tastes and plans as well as all legal, political, and social institutions are regarded as fixed and unaffected by the performance of the economy. Critics complain that in forcibly disengaging economic from social and political factors, economic theories have distorted the phenomena they purport to study.

Despite the reservation, partially isolated systems are routinely being posited in the sciences. The reason is pragmatic; we must make the isolating approximation to get manageable topics of investigation. As they are, most partially isolated systems are already too complex to handle. Often further approximations must be made to isolate a subsystem for better understanding. Many such approximations are examined in the following chapters.

The endogenous variables of a theory define its scope of conceptualization, and the scope can be narrowed by focusing on parts of the system. A system can be reduced to a subsystem by externalizing some endogenous variables, that is, by considering some variables that should be the results of system dynamics as given and fixed, thus treating the features represented by those variables as part of the immutable external world. For instance, there are abstract economic theories describing the simultaneous variation of all commodity prices, but practical analyses of the market of a specific commodity usually assume fixed prices for most other commodities.

Microstates, Macrostates, and Distributions

Among the myriad large composite systems, we are only interested in *many-body systems*, which are made up of a few kinds of constituents with a handful of character and relation *types*. As discussed in the preceding section, the limitation leaves open the possibility of indefinitely many *values* and great variety. Minute variations in the values of a type can lead to radical differences in the systemic configuration. The young universe was a uniform hot soup, and the galaxies and other structures in the present universe arose from slight fluctuations in the density of the soup. Variations within a species lead to the diversity of species in organic evolution. Slight market imperfection can generate high unemployment and prolonged depression. To find the microexplanations of these large-scale organizations is the job of many-body theories.

There are three general ways to characterize a many-body system as a whole, via its microstate, macrostate, and distribution. Distributions are closer to the macrostate than the microstate. Microeconomics opts for the microstates of the economy, macroeconomics the macrostates. Statistical mechanics and condensed-matter physics interpolate between the microstates and macrostates. Population genetics and the kinetic theory of gases describe a system in terms of the distributions over the states of its constituents.

A *distribution* of a system gives the number or percentage of constituents with each value of a constituent character type. Consider the character type of the tax burden, each value of which is an amount of tax paid. The tax distribution of a nation gives the percentage of citizens paying each amount. A system is describable by many distributions. *Statistics* such as the mean, spread, and skewness are brief ways to describe distributions. Like the distribution itself, they characterize the system as a whole. The connection between a system's distribution and the characters of its constituents is rather transparent. Characterization in terms of distributions misses many intricacies of relational patterns and is less appropriate for systems with strongly interacting constituents.

Microstates are not to be confused with states of the constituents; like macrostates, they are states of composite systems. A microstate type of a system is the conjunction or product of the states of its constituents. A microstate value specifies the state value of each constituent as it is a part of the system, thus describing the system in complete microdetail. Intuitively, the microstate description extrapolates the style of small-systems theories to large systems, but it is not always identical to the characterization of small systems, for it often uses system concepts in the description of constituents. Microstates are impractical. Even if the microstate types for large systems can be written down, their values can be determined only in a few exceptional cases. In microeconomics, definite values are obtained only for a couple of firms or a tiny sector of the economy. For the economy as a whole, there is a proof that certain equilibrium solutions of microstate values exist, but this is a far cry from obtaining definite solutions under specific conditions.

Microstates, even known, can fail to explain the behaviors of the system. They contain too much information, most of which is irrelevant to our understanding. Irrelevancy obscures rather than explains; it covers up the point. Suppose we ask why water freezes below a certain temperature. A microstate gives the position and velocity of each water molecule. Interchanging the velocities of two molecules results in a new microstate of the water, but this is totally inconsequential for the phenomenon of freezing. Undoubtedly the microstate changes as the temperature drops below the freezing point. Buried in the mountain of details, however, the microscopic changes are hardly dramatic enough to signal a phase transition. More important, the question would be unintelligible if microstates were the only description available; temperature and freezing are macroconcepts alien to microstates.

Explanations demand the selection of relevant information, and the criterion of selection is stipulated by the question that the explanation answers. For macrophenomena, the questions are framed in system concepts. Thus to understand macrophenomena, microstates are insufficient. We also need concise system concepts, which are provided by macrostates or distributions.

Macrostates are peculiar to large composite systems. They describe a system in terms of several macrocharacter types. Examples of macrocharacters are thermodynamic variables such as entropy and pressure, or macroeconomic variables such as the rates of inflation and unemployment. All familiar predicates we use to describe the medium-size things we handle daily represent macrocharacters and relations. Macrodescriptions guide microexplanations in picking out the interesting information of a system and summarizing it in a few variables. It is in terms of macrostates and characters that we come to grip with macrophenomena and give macroexplanations. When a composite system becomes a constituent in a larger system in the hierarchical world, it is characterized in terms of its macrostates and macrocharacters. When engineers design a bridge, they consider the hardness and strength of steel beams, not the motion of their atomic constituents.

A set of macrocharacter types is complete if their values provide enough specification for us to reproduce the system with the same macrofeatures. For some physical systems, the macrovariables are so interconnected as to acquire their own laws, as in classical mechanics, thermodynamics, and hydrodynamics. Most biological and social systems are too complicated to have complete sets of macrovariables. They can be described by some macrovariables, but these variables are too sparse to constitute laws. Even macroeconomics, which succeeds in defining many powerful macrovariables, falls short of providing satisfactory macroeconomic laws.

Usually micro- and macrocharacter types differ drastically, micro- and macrodescriptions employ completely different vocabularies, and the correspondence between microstates and macrostates, if available, is many to many. Consider a system comprising five unordered cards. A constituent of the system is a card, the state of which is, say, 2♥. The microstate of the system is, say, (2♥, 4♥, 6♥, 2♦, 6♦). The corresponding macrostate can be two pairs or twenty points or all red or all even or all low. There are many possible combinations of five cards that yield two pairs or twenty points, and the same five cards can be described in several macrostates. The multiplicity of microstates corresponding to a macrostate is intuitive. The multiplicity of macrostates corresponding to a microstate is less so. It is made explicit, for example, in the various ensembles in statistical mechanics (§ 11). There are universal macrolaws that sweep across systems with widely disparate types of constituent and relation (§§ 23, 24, 38). In short, the micro- and macrodescriptions of composite systems generalize and idealize in their own ways so that microstates and macrostates nest in their respective causal networks. The macroscopic network is not a reduced image of the microscopic network.

Rarely are there type–type identities between microstates and macrostates, except aggregative quantities.

Often a type of microstate is delineated not in microconcepts but in macroconcepts, according to the conformity of the microstates to a macrocharacter. The idea of partitioning the microstate space and delineating groups of microstates according to macroscopic criteria is built into the probability calculus, hence it is common to theories that apply the calculus. For instance, it is the essential bridge that connects the micro- and macrodescriptions in statistical mechanics, where the canonical ensemble picks out all possible microstates with a certain temperature and the grand canonical ensemble picks out all possible microstates with a certain chemical potential (§ 11). Temperature and chemical potential are macroconcepts. The leading role of macroconcepts in classifying microstates is characteristic of synthetic microanalysis and antithetic to microreductionism.

Many-body theories generally do not attempt to solve for the microstates. Rather, they aim to find the microconditions underlying certain types of macrostate, or the micromechanisms contributing to certain types of macrophenomenon. The micromechanisms are often expressed not in microstates but in some suitable coarse graining of them. When we have a firm grasp of the micromechanisms, we can also predict new macrophenomena.

Composition and Causation

That a system is composed of its constituents does not imply that the behaviors of the system are all caused by the behaviors of its constituents. Surely the behaviors of the system and the behaviors of its constituents are causally related, but it is not easy to assign causes and effects. Causation (the talk about cause and effect) is a narrower notion than causal relation because of its asymmetry and crudeness. Its narrowness spawns controversy; people argue incessantly about what causes what, as they usually regard causes as more powerful and important.

As discussed in § 34, the chief asymmetry in causality is the temporal precedence of causes over effects. In equilibrium cases where we think about systems and constituents, the time element is absent and there is no general causal asymmetry. Fundamental physical interactions are symmetric; electrons mutually repel and bodies mutually attract gravitationally. When classical mechanics represents forces as directional, it appends a third law stating that action and reaction are equal and opposite. Thus strictly speaking an equilibrium system is a network of symmetric causal relations with no intrinsic differentiation between causes and effects. We talk about causation when we pick out gross factors and offer crude descriptions for the intricate causal network.

A solid made up of some 10^{23} atoms is a causal network whose behaviors are mutually determined with the behaviors of its atomic nodes. How do we assign causes and effects if we insist on talking about causation? An atom causes its neighbors to move in certain ways, but we cannot say it causes the solid to behave in a certain way; an atom is too insignificant to be a cause.

We do not say the whole assemblage of atoms causes the solid's behavior; the structured atomic assemblage *is* the solid. Micromechanisms for macrobehaviors are causal, but the causes are usually distributed among the atoms or reside in their concerted motions, so that no entity is specifically responsible. If we look at the causal network and want to pick one entity as the cause of something else, chances are our gaze falls on the system; it is the only entity with sufficient significance. An element changes its behavior when it becomes the constituent of a system, and the change is crudely explained in terms of the system's causal influence or macroconstraint. Some people refer to the system's influence as "downward causation."

Downward causation is often accused of being mystical. It is not if it is clearly articulated. There are two dimensions in the interpretation of downward causation, the first regards the effects of the system on the behaviors of its constituents, the second the source of the system's causal efficacy. Mysticism results if the system's causal power is credited to independent sources such as vitality or psychic energy that are totally detached from the forces among the constituents and stand above them. Such attribution is not necessary. For us, the system has no independent fund of causal power, although it is described in independent concepts. Whatever causal efficacy the system has is endogenously derived from the self-organization of the constituents. Thus, downward causation means no more than the tyranny of the masses. When the strongly coupled constituents of a system organize themselves into certain patterns, any maverick will be pulled back by the others and forced to march in step. Without using the horrible name "downward causation," downward causation such as a test particle in a fluid being carried away by the current is common in the sciences. We find it, for instance, in the screened potential of electrons (§ 15). More familiarly, the intuition that the price structure of the economy downwardly constrains the spending of individual consumers is rigorously represented by the notion of passive price takers in microeconomics (§ 16).

It may be argued that composition implies "upward causation," which cannot be compatible with downward causation. The assumption of necessary incompatibility is dogmatic. Both upward and downward causation are crude descriptions of composite systems, which are symmetric causal networks. Respectable scientific models have shown that the causal symmetry can be preserved in accounts of upward and downward causation. In the self-consistent field theory in physics, the equation of motion governing a single electron in an electron sea includes a term for the potential of the electron sea as a whole. The theory has additional steps ensuring that the downward potential exerted by the electron sea on individual electrons is consistent with the upward determination of the potential by all electrons (§ 15). Similarly, the general equilibrium model for perfectly competitive markets shows rigorously how the price structure of the economy, to which individual consumers respond passively, is not externally imposed but is the endogenous product of the behaviors of all consumers (§ 16).

Micromechanism and upward causation explain system behaviors by using constituent concepts. Macroconstraint and downward causation explain the behaviors of individual constituents within the system by using system concepts. The system concepts can be inconspicuous; sometimes they are folded into constituent predicates that are tailored to the system, but the effect of the system is already embedded in the customization of the predicates (§ 14). For example, downward causation is inherent in the definition of consumers in terms of commodity prices, which pertain to the economy as a whole. In sum, upward causation and downward causation represent two crude ways to look at a system, and the synthetic framework of scientific theories is broad enough to accommodate both harmoniously.

Conceptual Components of Many-body Theories

Let us assemble the preceding ideas to form the synthetic conceptual framework for the microexplanations of large composite systems. A composite system is logically an individual and is referred to as a unit. The constituents of the system are also individuals, and as constituents, they are defined within the system. To develop theories about the composition of large systems, we need a conceptual framework that accommodates and relates individuals on two scales: a macroindividual that is *explicitly composite* and many microindividuals that are *explicitly situated*. The framework is woven of several strands, each of which is a braid of several concepts:

1. A partially isolated system and the rest of the world; the world, which affects the system but is unaffected by it, is represented by a set of fixed exogenous parameters in the characterization of the system.

2. Various characterizations of the system in terms of its microstate, macrostate, or distribution over constituent states, each with its own generalization.

3. The constituents of the system, whose character types and relational types are obtained by extrapolating from minimal systems or by microanalyzing similar systems.

4. The ontological identification of the system with the network of causally connected constituents.

Various conceptualizations at the system and constituent levels are admitted in 2 and 3. The characterization of the system in 2 accommodates macroexplanations. The ontological identification in 4 assures that the macro- and microdescriptions are of one and the same thing. Since the microstates use mainly the vocabulary of the constituents and the macrostates use macropredicates independent of the constituents, their general relation makes explicit the idea that the system is a whole composed of parts.

The ontological identification of a system with the causal network of its constituents applies not to kinds of thing but to the instances of a kind. We identify this particular ladder with these particular pieces of wood nailed together in this specific way, which does not imply the identification of ladders with wooden structures. The ontological identification is much weaker than the identification of a certain type of macrostate with a certain type of microstate or a certain macrocharacter value with a certain microcharacter value. An example of character identification is the identification of a certain degree of hardness with a certain atomic configuration. Its difficulty is apparent when we recall that the same degree of hardness can obtain in metals, ceramics, plastics, and other kinds of material. State and character identification are required for microreduction, for they render the various descriptions of the system redundant. They are *not* required for the general conceptual framework of composition.

The ontological identity ensures the significance of all the elements in the conceptual framework of composition. The framework is at once holistic and atomistic, presenting a three-dimensional conceptual space that accommodates not only the macro- and microdescriptions but also their interrelations. It paves the way for rulelike connections between the micro- and macrodescriptions but does not prescribe their form or even their existence, thus opening the door to the many approaches and models discussed in the following chapters. We have already tacitly understood the general conceptual framework when we raise questions about the dependence of the system structure on constituent relations, the effects of the structure on constituents' behaviors, and the ways the system and its constituents respond to external influences. By filling in some substantive details between the levels, we not only get a more solid understanding of the world, we also open our eyes to many more micro- and macrophenomena. More important, even if many details turn out to be approximate or missing in practical cases, the general conceptual framework sustains the idea of composition and prevents us from degenerating into relativists who regard people using different descriptions as inhabitants of different worlds.

7. Idealizations, Approximations, Models

The conceptual framework of composition discussed in the preceding section is general and common to many natural, social, and engineering sciences. Its implementation in a science institutes a synthetic theoretical framework: price, aggregate demand, and aggregate supply in economics; variation, heredity, and natural selection in evolution; the laws of mechanics and the calculus of probability in statistical physics. A synthetic framework can be mathematically rigorous, with precise definitions and clear concepts bridging the micro- and macrolevels. These definitions are abstract rules. They must be further substantiated for the investigation of specific types of phenomenon. To proceed, scientists rely on various approaches, approximations, and idealizations to carve out different aspects or processes of a kind of many-body system. They develop various models for substantive synthetic microanalysis.

Idealization and approximation are also present in theories known for their conceptual unity, for instance, classical or quantum mechanics. In these theories, approximations tend to have rather standardized forms, for the laws of mechanics cover their topics inclusively. The laws of mechanics are still inviolable in statistical physics, but they operate mainly on the microlevel. Since the microdescription quickly becomes intractable, methods must be developed to combine the effects of the laws. Because of the diversity of many-body phenomena and the complexity of their microexplanations, the methods to tackle them are variegated in form and often specialized to specific types of phenomenon. They overshadow the unity of the laws of mechanics. Consequently statistical physics, like economics and evolutionary biology, is known more for its fragmentation than for its unity. On the other hand, the multifarious approaches better illuminate the ingenuity of scientists in coping with reality.

The Need for Idealization and Approximation

The philosophical doctrine of conceptual holism claims that total science, which includes all our beliefs, is a web that must face experiences as a whole and be modified as a whole. We cannot possibly test a hypothesis in a theory singly, for we can always save the hypothesis from challenges of adverse empirical data by changing other hypotheses in our conceptual framework.[18] If we must solve everything at once and change our entire conceptual structure every time experience throws up a surprise, if we must conceptualize complicated systems in their entirety, probably we are still in the Old Stone Age. Conceptual holism shares with microreductionism the assumption that our web of concepts is perfect and seamless, so that the disturbance of one hair affects the whole body. In practice, our web of substantive concepts is full of holes and rifts created by various approximations and idealizations made in various areas. Sometimes patches of the web are tied together only by general ideas such as composition. Approximations are not mere defects. They play the crucial intellectual role of damage control by modularizing our substantive conceptual web and localizing the necessary modifications.

A mathematical empirical theory can be exact in the sense of providing mathematical definitions for key concepts, precise models for a group of prominent causal factors, mathematically exact solutions to some models, or some solutions that accurately describe certain realistic phenomena. It is not exact in the sense that the mathematical models or their solutions flawlessly mirror reality. Rarely if ever can a theory claim to be complete in the sense of having exhausted all contributing causal factors, major and minor. An empirical theory that encompasses most important relevant factors and formulates them in ways susceptible to nearly exact or accurate solutions is good enough.

From Galileo's frictionless inclined plane onward, science has progressed by making judicious idealizations. The APS [American Physical Society] News published a "Physicists' Bill of Rights," the first article of which is "To approximate all problems to ideal cases."[19] I doubt if other scientists would disclaim the right to do the same. Closing our eyes to the ideal elements in scientific theories does not make them go away but only seduces us into errors of judgment. A pretentious notion of exactness in the positivist philosophy of science invites the denunciation that physical laws lie. The acknowledgment that the intelligible world is already conceptualized and that the conceptualization involves idealization and approximation does not vitiate but enhances the objectivity of scientific theories. For it puts the burden on us to examine the ideal elements, to justify and criticize the approximations, and to make clear the meaning of being objective. The ideal elements also cushion the impact of theory change by enabling us to attribute the error to our thinking and maintain the invariance of reality.

Idealization and approximation are involved in every step of scientific reasoning. They contribute to both the *formulation* of problems and the *solutions* that yield results for specific cases. Many approximations used in the solutions are technical in nature, as scientists apply standard mathematical techniques such as the perturbation expansion. I will emphasize the approximations in formulations, which accentuate the relation between our theories and the world they purport to illuminate. The world is a web of interrelations; thus the isolation of topics for investigation is already an idealization, which is partially reflected in the ceteris paribus, or "other things being equal," clause that accompanies most if not all scientific models. This idealization is included in the general formulation of many-body theories presented in § 6.

Models

In studying complex systems that defy nearly realistic representations, scientists turn to constructing *models*. *Model* as commonly used by scientists should not be confused with *model* used by philosophers in the semantic view of scientific theories.[20] In the scientists' usage, which I adopt, there is no logical difference between theories and models. Sometimes *model* and *theory* are used interchangeably; other times the usage is determined by historical convention. A model is usually less general than a theory. It is not necessarily so; for instance, the standard model of elementary particle physics is cherished as perhaps the penultimate theory about the basic building blocks of matter. Nevertheless such cases are exceptional. More often than not a model means a variant of a theory, a bit of a theory, a tentative theory, or a theory with such drastic simplification and approximation its hypothetical nature is unmistakable. For instance, the description of a harmonic oscillator is seldom called a model, for it is highly accurate under realizable conditions for a finite time span. In contrast, the representation of a radiation field as a set of harmonic oscillators is a model. Metaphorically, a model offers not a photograph of certain phenomena but a schematic or even a caricature that nevertheless enables one to recognize its object readily. Theories for large composite systems are more likely to be called models, for they often leave out many details and represent the systems ideally in terms of crude macrovariables and relations.

By using different approximations or approaches, different models can be constructed for the same phenomenon. Section 15 presents several models depicting conduction electrons in solids to various degrees of refinement. Models invoking different causal mechanisms, for instance, various models for business cycles, allow scientists to gauge the relative importance of the mechanisms for a type of phenomenon (§ 36).

Many models yield empirically testable results. Not all results are accurate, if only because the parameters that go into calculating the results are often themselves estimates. The competing claims of the models on genetic drift and natural selection are unresolved partly because it is difficult to determine the mutation and evolutionary rates used in the models to the required degree of accuracy (§ 37). Scientists are often happy to be within the right ballpark. More important than quantitative accuracy is the correct variation of one variable with respect to another. Models are also judged for their ability to explain qualitative differences. For instance, the pinball-machine model of electric conduction yields decent values of electric conductivity under many conditiohs, but we deem it unsatisfactory because it fails to explain why some materials are conductors and others are insulators (§ 15).

Other models yield no testable results. The Arrow–Debreu model of perfectly competitive markets, another model with the scope and generality of a theory, is too abstract to yield definite results that can be empirically tested (§ 16). Abstract models are nevertheless valuable for their ability to capture some essential structures of complex systems or for their susceptibility to exact or near exact solutions. Through these solutions scientists gain a thorough understanding of the factors in idealized situations, explore possibilities theoretically, and use the understanding to study similar structures in more realistic and less tractable situations. The models also serve as incubators of new concepts and laboratories for testing their theoretical efficacy.

Theory and Reality

What do approximation and modeling mean when we start with only a dim idea of how the world works and charge the model we are developing to provide a better description? What is the epistemological justification of models without predictive power? Here we come upon the basic philosophical problem about the relation of our concepts and theories to the world. The problem is also pertinent to the nature of science and scientific research.

Scientists make laborious efforts in solving problems and working out the mass of detail that makes science solid, but before that, they must first discern phenomena and find problems. To recognize the important problems and to articulate them precisely are the most important steps in scientific research, for the answers are related to the way the questions are posed. The best way to waste years of research effort is to get entangled in some frivolous problems.

The ability to formulate fruitful models manifests the scientists' mastery of their research topics. It demands the insight to penetrate the jumble of confusedly articulated factors and the miscellany of empirical data, the intellectual acuity to separate the key elements and represent them by precise and interrelated concepts, and the sound judgment to assess the adequacy of the models. The ability can only be acquired from a thorough familiarity with the subject matter. Much of the familiarity is not formalized; it is the intuitive understanding of the whole complex of phenomena in which partial formalization will follow. As Keynes said, the art of formulating models and deciding what models are relevant to the contemporary world requires "a vigilant observation of the actual working of our system." And he remarked that "good economists are scarce because the gift for using 'vigilant observation' to choose good models, although it does not require a highly specialized intellectual technique, appears to be a very rare one."[21] The remark is also pertinent to other sciences. The gift Keynes referred to lies mainly in differentiating realistic factors, not in solving differential equations, which, although requiring specialized techniques, is not scarce because the techniques can be readily learned.

In formulating problems and models, scientists maneuver themselves into positions favorable for launching assaults. The maneuver involves much compromise and judicious balance of means and ends. On the one hand, they need problems that are within their ability to solve, and that often means the need to make drastic simplification, to ignore many minor factors or minor variations of the important ones. On the other hand, they want to make sure that despite the simplification, the problem captures some salient and typical features of the phenomena.

A model brings a few salient factors of a complex system into relief coherently. The factors it leaves behind are not discarded but continue to serve as the intuitive understanding of the whole. The intuitive understanding includes facts and empirically tested theories accumulated in the past, which provide various check points to gauge approximations. Sometimes the accumulated facts and theories are relevant to the factors incorporated into the model; in such cases they provide indirect empirical support for the model. Since internal coherence is essential to a model, the support of one factor or the connection of one factor with established theories strengthens the whole model.

A leading-edge model is judged approximate not by comparison to more comprehensive theories, for none exists yet. Nor is it judged by comparison to the naked and unconceptualized reality, for we have no access to such a reality to make the comparison. The job of criticism again falls on experiments on various facets, intuitive understanding, familiarity with the topic, informal weighing of factors, and awareness of what is left out. The understanding in turn is improved by the model, for the clear articulation of some factors illuminates others. Scientists constantly go back and forth between common sense and the approximate models, discerning new patterns, introducing new concepts, and incorporating them into improved models. Thus they make more and more vague ideas distinct and expand their models to illuminate a larger and larger part of the complex situation. This is the hermeneutic circle or the circle of understanding expounded in Continental philosophy.[22] In the hermeneutic circle, or more appropriately the spiral, the meaning of individual concepts is understood only in the context of the complex whole, and the meaning of the whole is elucidated as its parts are sorted out. By moving back and forth between partial meaning and some sense of the whole, we gradually improve our understanding of both.

One appeal of models is their susceptibility to exact mathematical treatment. The mathematics has its own charm and can be explored for its own sake. However, empirical science is not pure mathematics; it languishes if it is confused with the latter. When empirical scientists are enthralled by calculational techniques and begin to forget that they are only tools for understanding the world, their models, however elegant, become mathematical games.

Models can be the source of illusions. Although they are drastic simplifications of worldly events, they can still be very complicated. Frequently years and much intellectual labor are required to work out a model's details, not to mention its variations. Prolonged struggle engenders the feeling that reality imitates theory. The unwary may forget the idealistic nature of the model and fancy a reality that is supposed to be mirrored by it.

Models are valued for the precision of their articulation, but there is a tradeoff among precision, scope, and authenticity. Each model represents a fragmented and partial understanding of the world. Different models may involve different idealizations or pick out different key factors. When scientists lose contact with the world and models are mistaken for truths, various models degenerate into incommensurate paradigms and academic fashions.

8. Federal Versus Imperial Unity of Science

Having stressed the multiplicity of models in each science, I will begin in the following chapter to survey economics, evolutionary biology, and statistical physics in parallel and look for the general ideas that unite them. I am not presenting a contradiction. The unity of science resides in general concepts, the multiplicity in substantive concepts.

To theorize is to seek the constant and the general; thus some idea of unification is inherent in theoretical reason. Our understanding of the world takes a big stride if we can grasp under a unified principle phenomena that previously had to be explained separately. Scientists never cease trying to frame powerful and unifying concepts. The value of unification as a guiding principle for theoretical reason and scientific research is emphasized by many scientists and philosophers, for instance, Kant: "In inference reason endeavors to reduce the varied and manifold knowledge obtained through the understanding to the smallest number of principles (universal conditions) and thereby to achieve in it the highest possible unity."[23] Kant emphasizes that unification is only a guiding principle for the general tendency of theoretical reason that stipulates neither any substantive unifying principle nor any substantive unified result. He points out how postulates of substantive unities lead to illusions.

Unification is often used to support reductionism and microreductionism, for the expulsion of some concepts tends to produce a tighter theoretical structure. Unity of science is desirable. What is objectionable is the kind of unity the reductive doctrines advocate, an imperial unity in which the reduced sciences are annexed if not abolished, as they are nothing but the logical consequences of a primary science. Microreductionism claims much more than being a guiding principle. It dictates a set of substantive concepts, decrees that a manifold of knowledge be united under them without showing the feasibility of unification, and thus makes grandiose claims for their unifying power that far outrun their demonstrated capability. The disparity between its claim and its substance gives reductionism its air of arrogance.

Reductionism's simplistic and extreme view of science provokes a backlash that is equally simplistic and extreme. Relativism and isolationism proclaim that science is an anarchy in which anything goes, various theories are incommensurate, people subscribing to them cannot communicate rationally, and going from one to another is akin to religious conversion. Extreme positions attract attention but often fail to convince. Many people opt for middle ground. Consequently reductionism and the unity or disunity of science become a current topic in intellectual discourse.[24] In the following I propose a more tolerant unity of science that accommodates the autonomy of various sciences.

The Federation of Science

An empire is not the only form of unity, and anarchy is not the only way to plurality. Unity can be achieved as a federation of autonomous states with their own laws agreeing to the general terms of a constitution. There can be various degrees of unity in federations. A lengthy constitutional debate has made the United States of America a much tighter union today than it was two hundred years ago, but no matter how powerful the federal government becomes, it can never reduce the states to colonies. An imperial unity of science is based on *substantive concepts*, whose authority is projected like imperial decrees to areas way beyond their apparent domain, as some microreductionists confer on neural concepts the authority to explain all psychological phenomena. A federal unity of science is based on *general concepts* that, like constitutional articles, are not biased toward any particular science. We have encountered the general unifying concept of composition in the synthetic framework of many-body theories, where micro- and macroconcepts retain their autonomy but are united by the ontological idea that they represent the same object with different focuses and degrees of detail. Unlike microreductionist bridge laws that subjugate and make dispensable certain concepts, the synthetic analytic framework accommodates and connects concepts without reducing some to subordinates.

The general idea of plurality in unity can be extended to science at large. The universe exhibits phenomena on many levels of scale and complexity. Compared to this great wide world, the domains illuminated by our most extensive theories are small, for a higher level of organization reveals a different objective domain. Sometimes the domains of two sciences overlap. Sometimes a theory connects two other theories, opening its own domain that partially overlaps but is not exhausted by the domains of the theories it connects. Many-body theories are examples of this type. More often the domains of various sciences are disjoint. The diversity of subject matter secures the autonomy of the sciences. Although scientific theories are individually becoming more comprehensive and unified, science itself is becoming more diversified as more disciplines spring up to study more phenomena.

Despite the diversity, the sciences are united by the general idea of the objective world, which is so complicated its representation requires the multiplicity of substantive concepts and theories in various sciences. The general idea of objectivity is the categorical framework of theoretical reason that sweeps across the sciences and common sense. It does not curtail the autonomy of the sciences because it is general and impartial to specific substantive concepts, abstract and tolerant of the disjointedness of the substantive concepts.

I have argued in the context of quantum field theory that the general concept of objectivity demands many representations of the same object; the multiplicity of representations is inherent in intersubjective agreement and the invariance of objects.[25] The multiple representations also join theories and models with overlapping domains. If several well-established models overlap in some domain, then in the common domain their results *converge* in the sense that the characters of individual objects depicted in one approximate those in the others. Such agreement is routinely checked by scientists, for instance, in ensuring that the results of Newtonian mechanics are approximately equal to the results of relativistic mechanics in the limit where the particle's speed is far less than the speed of light. Note that theoretical convergence is established by comparing the results of different theories, not by comparing the results of the theories to an unconceptualized reality. If the results of the theories agree reasonably, then we accept the agreement as descriptive of the object. We do not assume in advance some ready-made entity toward which the objects depicted in the theories converge. The notion of convergence without advance positing of the limit of convergence is well defined in mathematics.[26]

A Geometric Metaphor for Science

Philosophers sometimes liken a theory or a conceptual scheme to a coordinate system in geometry. A theory makes an aspect of the world intelligible just as a coordinate system makes a region of a manifold analyzable. In *finite geometry*, Euclidean or non-Euclidean, a single global coordinate system covers an entire manifold. Global coordinate systems are found inadequate for modern physical theories, but their rejection does not imply the fragmentation of the manifold. Differential geometry offers a more versatile notion of unity underwritten by a more elaborate framework of general concepts. In *differential geometry*, a manifold is defined by many local coordinate systems, each covering only a patch of the manifold. If two coordinate patches partially overlap, they are "sewn" together by the rules that identify the common points they separately coordinatize. Differential geometry dominates modern physical theories.

In the geometric metaphor, the imperial unity of science promoted by reductionism is like finite geometry. The substantive concepts of the primary science to which everything is reduced are like the global coordinate system that covers the whole manifold. Like finite geometry, the structure of the imperial science is rather rigid and its variety constrained.

Figure 2.2 schematically illustrates the federal unity of science using the metaphor of differential geometry. It contains several general ideas:

1. a multitude of regional theories for various aspects of the world, including phenomena on different scales and organizational levels;

2. the existence of several distinct representations for the same objective domain where several theories overlap;

3. the reconcilability of the descriptions;

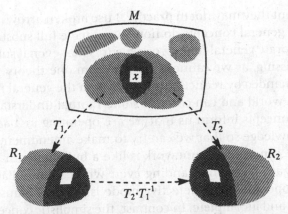

FIGURE 2.2 *The general concept of the objective world underlying the abstract unity of science.* A scientific theory T_1 is like a local coordinate map for a manifold; it represents a certain region of the objective world M in a substantive conceptual scheme R_1. The patches in M represent the aspects of the world illuminated by various theories. Many patches are disjoint; our theories do not cover all aspects of the world. When the domains of two theories T_1 and T_2 overlap, communication between their holders is made possible by the general idea $(T_2 \cdot T_1^{-1})$ that the two substantive descriptions, which may be quite different, are of the same object x, which is not changed by going from one theory to the other. The conceptual framework has enough elements to be operative in a defective mode. For instance, the substantive transformation $T_2 \cdot T_1^{-1}$ between T_1 and T_2 may be unknown or known only approximately, yet the general idea prevents subscribers of the two theories from being totally alienated.

4. objects and the objective world that are independent of theoretical representations and invariant under the transformation of representations.

The microscopic and macroscopic structures of the same stuff are different aspects, and they are disjoint unless linked by interlevel theories such as many-body theories. The categorical framework admits multiple representations (T_1 and T_2) of the same object x, makes possible the intersubjective agreement between the representations ($T_2 \cdot T_1^{-1}$), and affirms the objectivity of the representations (the analysis of $T_2 \cdot T_1^{-1}$ reveals that it refers to the object x that is unchanged under the transformation of representations). The areas not represented in scientific theories are not in total darkness but are dimly illuminated by an intuitive understanding, including the general idea that the objects represented in one theory are made up of the objects represented in another.

Ideas 1–4 belong to the general concept of the objective world that is the minimal structure of all empirical knowledge. The categorical framework comprising them is an integral whole conveying the idea that theories are *of* the objective world and that although an object can be abstracted from substantive theoretical descriptions, it cannot be comprehended absolutely without concepts.[27]

The categorical framework reserves room for possible substantive connections between theories such as $T_2 \cdot T_1^{-1}$ but does not guarantee their success. Ideally, in the overlapping region the two theories T_1 and T_2 should converge in the sense discussed, but they may not in practice. I use broken arrows in the figure to stress that the general concepts do not always have full substantive embodiments. The elaborate general framework ensures that even if substantive connections are missing, as we cannot translate from one theory to the other with the rigor demanded by reductionism, we retain the general notion that we share the same world and can come to some mutual understanding if we try. The general concepts inform us that we are operating in a *defective mode* of substantive knowledge so that we can try to make amendments.

An elaborate general conceptual framework is like a building with many supporting pillars, so that it remains standing even when some pillars are damaged. Its ability to operate in the defective mode is crucial because human knowledge is finite and incomplete. In contrast, the simplistic conceptual framework of reductionism is like a building supported by a single pillar that cannot permit any defect. It may look elegant and be fit for God's perfect knowledge but is inapplicable to us humans. Once we realize the hubris of pretending to be God, the lone pillar weakens and the whole building collapses, as reductionism falls to relativism.

I use differential geometry as a metaphor, not a model, of theoretical science. The relation between scientific theories and reality is far less rigorous than that between local coordinate systems and the manifold. Approximations and idealizations are inevitable in the theories, as discussed in the preceding section. The crux of the metaphor lies in the complexity of the conceptual framework of differential geometry, which includes Euclidean and non-Euclidean geometries (each local coordinate system is Euclidean) and much more. In its elaborate framework, we see that the antagonism between holism and microreductionism is like the dispute whether Euclidean or non-Euclidean geometry is more suitable for modern physics. Neither is, for they are both too restrictive. Differential geometry illustrates the power of higher abstract construction in mathematics. Its application in physics shows the necessity of large conceptual frameworks in articulating complex ideas. Perhaps a similar lesson is still to be learned in philosophy, in which arguments confined by ideologically specified narrow conceptual frameworks often generate more heat than light.

Science progresses by expanding the domains represented by theories and reconciling the theories where the domains overlap. However, even in its ideal limit, theoretical science does not illuminate the entire world as the local coordinate systems cover the entire manifold. Powerful as theoretical reason is, it never has a monopoly on human understanding. Besides *theoria*, Aristotle put *techne* (skill and know-how) and *phronesia* (prudent understanding of variable and contingent situations with a view toward appropriate actions). Kant's critique curbs the pretension of theoretical reason to make room for practical reason and aesthetic judgment. The arts, literature, and other forms of understanding have their roles.

Feynman envisioned a poorly linked hierarchy of the world, stretching from physical laws on the one end, through physiology and history, to beauty and hope on the other end. He used a religious metaphor: "Which end is nearer to God?" Neither, he answered: "It is not sensible for the ones who specialize at one end, and the ones who specialize at the other end, to have such disregard for each other. (They don't actually, but people say they do.) The great mass of workers in between, connecting one step to another, are improving all the time our understanding of the world."[28]

9. Equilibrium and Optimization

Part I of this book concentrates on *equilibrium* models for studying composite systems. Equilibrium models are better developed and more often used than dynamic models in economics, evolutionary biology, and statistical mechanics for a couple of reasons. First, equilibrium is simpler than dynamics. This does not imply that equilibrium models are easy; static constructions can be highly complicated, as any architectural structure testifies. Second, equilibrium models represent the macroscopic structures of many-body systems that last for a long time, which are no less important than the temporal evolution of the systems.

It happens that the equilibrium state of a system often has a character type that attains its maximum or minimum value. An equilibrium mechanical system has minimum energy; an equilibrium thermodynamic system has maximum entropy; a biological species in equilibrium consists of maximally adapted organisms; an economy in equilibrium consists of consumers enjoying maximum utilities and producers making maximum profits. To find the equilibrium states, scientists in all fields use the mathematics of *optimization*.

Since equilibrium states persist and the optimal conveys the idea of the best, the association of equilibrium and optimization in economics and evolutionary biology sometimes engenders misleading interpretations suggesting that the status quo is the best. Leading biologists and economists have criticized the Panglossian interpretations of their respective fields, which, like Voltaire's Dr. Pangloss, assert that this is the best of all possible worlds.[29]

This section presents the general ideas of equilibrium and optimization. Their application and interpretation in the sciences are frequent topics in the following chapters.

Equilibrium

The first clear articulation of the idea of equilibrium probably occurred in statics, where it meant equal weight, originally referring to the condition of balancing a lever about a pivot. The idea was generalized to the balance of many forces pulling in different directions, where force is understood broadly to be any relevant agent of change, including desire and incentive in social systems. The balance implies that the net force on a system vanishes and the system remains unchanged. Equilibrium is found in many disciplines; for instance, there are chemical equilibria in which chemical reactions going one way are balanced by reactions going the opposite way so that the relative concentrations of the reactants are constant in time.

General equilibrium is the cornerstone of economic theories. The negative qualification in *nonequilibrium* statistical mechanics is suggestive in a science with a strong tradition in dynamics. It is perhaps surprising to find equilibrium in evolutionary biology, for evolution is by definition dynamic. Although organic evolution can be abstractly represented in terms of dynamic transformations (§ 12), the substance of the formalism is seldom available because evolutionary mechanisms are so complicated (§ 17). Dynamic models are less developed in evolutionary biology than in economics and physics; evolutionary changes are addressed mostly by historical narrative, not theory (§ 41). Theoretically, biologists acknowledge that equilibrium dominates population genetics, which is the core of the modern synthetic theory of evolution. Richard Lewontin said: "A second problem is that population genetics is an *equilibrium theory*. Although we do have the dynamical laws of change in deterministic theory and are able to describe the time dependence of distribution functions in a few cases of stochastic theory, in *application* only the equilibrium states and the steady-state distributions are appealed to. Equilibrium annihilates history." Douglas Futuyma stated: "Population biologists, although sometimes paying lip service to the impact of history, tend to think in ahistorical terms. Most of population genetics, all of optimization theory, and much of community ecology are framed in terms of equilibria rapidly approached; history plays little role in the theory and is wished away whenever possible in interpreting data." John Maynard Smith wrote: "An obvious weakness of the game-theoretic approach to evolution is that it places great emphasis on equilibrium states, whereas evolution is a process of continuous, or at least periodic change. The same criticism can be levelled at the emphasis on equilibria in population genetics."[30] The preponderance of equilibrium in evolutionary theory is a source of controversy regarding many of its results and claims.

Equilibrium is a self-maintained state of an isolated or partially isolated system. It should be distinguished from the *stationary state* of an open system (a system dynamically coupled to the environment), which is maintained in an unchanging state by the continuous application of external forces. For instance, a constant temperature gradient can be maintained in a metal bar if energy is continually being fed into one end by a heat reservoir and drained from the other end in contact with a heat sink. The state of the bar is stationary but not in equilibrium, for its steadiness is imposed and not self-maintained.

An isolated or partially isolated system is in equilibrium if its state does not change with time as long as its exogenous parameters are constant. An equilibrium is *stable* if the system returns to the equilibrium state after slight disturbances. It is *metastable* if the system returns after small perturbations but not after perturbations strong enough to overcome certain barriers. If a system can be displaced from its equilibrium state by small disturbances, its equilibrium is unstable. Most equilibria we consider are stable. Metastable equilibria become important when we consider such phenomena as phase transitions (§ 23) or evolution to the edge of chaos (§ 24).

A common method in equilibrium theories is *comparative statics*, which studies various equilibrium states of a partially isolated system under various external conditions, without worrying about how the system changes from one equilibrium state to another. Physicists investigate the equilibrium volumes and temperatures of a gas under various pressures; biologists investigate equilibrium organismic characters in various environments; economists investigate equilibrium price and employment levels under various government fiscal and monetary policies.

Equilibria are not absolute but *relative to spatial and temporal scales*. Generally, several processes with characteristic speeds on different time scales go on simultaneously in a system. The system is in equilibrium in a particular time scale if the processes under study are roughly stationary in that scale. When we say a thermodynamic system is in equilibrium in the time scale of minutes, we mean its macrovariables such as the temperature and pressure do not change in minutes. We do not mean that its microstate is stationary; its microstate changes in the scale of picoseconds as the molecules constituting it jiggle and bump against each other. Nevertheless, the rapid changes can be averaged over in the time scale of minutes to yield equilibrium macrocharacters. We also allow the macrovariables to vary in a longer time scale; if the temperature changes with a characteristic rate of days, we can approximately regard it as constant in the duration of minutes. The distinction of processes with disparate characteristic time scales is crucial for nonequilibrium statistical physics, biology, economics, and history. Time scales determine the applicability of various physical equations of motion: mechanic, kinetic, hydrodynamic. Economic theories distinguish among temporary, short-run, and long-run equilibria, so that deviations from the long-run equilibrium are treated as equilibria on a shorter time scale. The equilibria in population genetics obtain only in times that are short compared to the rate of environmental change (§§ 27, 38).

Suppose the macrovariables of a thermodynamic system change at a rate of hours and we study its dynamics by measuring it at one-minute intervals. Since molecular processes can bring the system into equilibrium in split seconds, the system is in equilibrium at each measurement. Consequently our measurement yields a string of momentary equilibrium states with different characteristics, which constitute a *quasi-static process*. Quasi-static processes are often much simpler than fully dynamic processes. They are also prevalent in economics, where Rational Expectation is often assumed to be a rapid mechanism that always brings the economy into equilibrium in the time scale of interest (§ 36).

Suppose a large system is divided into many subsystems on a smaller spatial scale. If the system is in equilibrium, then each subsystem is also in equilibrium. However, the system can be out of equilibrium even if its subsystems are individually in equilibrium. The *local equilibria* of the subsystems of a system in disequilibrium are the spatial analogies of the momentary equilibrium states of quasi-static processes. Usually the global equilibrium of a system is expressed by the uniformity of some variables across all subsystems, for instance, a uniform temperature across a metal bar or a uniform price across an economy. Suppose each segment of an isolated metal bar is in equilibrium with a certain temperature, but the temperature varies from segment to segment; then the bar as a whole is out of equilibrium. A bar hot at one end and cold at the other will change until the temperature is uniform throughout, but the approach to global equilibrium is much slower than the establishment of local equilibria. Local equilibria are essential in the derivation of the hydrodynamic equations (§ 38).

Extremum Principles

A ball always rolls to the lowest ground, where the gravitational potential is the least. Generally, a physical system always seeks and stays in its ground state, the state with the lowest energy. *Lowest* is a superlative. We have engaged the idea of *extremum*, which is a maximum or a minimum of a variable.

The idea behind extremum principles is very old. Aristotle used it to argue for the circularity of planetary orbits, and Leibniz argued that some why questions must be answered by referring to what is best or optimal.[31]

In science, extremum principles first appeared in optics, when Fermat argued that light travels from one point to another in a way that minimizes the time of transit, no matter what reflections and refractions it undergoes along the way. An important extremum principle is the principle of least action, announced in 1744 by Pierre-Louis Maupertuis, developed and modified by many physicists and mathematicians, and now serving as a pillar of quantum field theory. In its modern application, the original metaphysical overtone is avoided. Physicists no longer ask why a quantity called action is minimized; they simply postulate that it is and proceed to develop their theory based on it.[32] Extremum principles are also fundamental in other sciences: The survival of the fittest is the basic premise of evolutionary theory; the maximization of profit and satisfaction is the cornerstone of theoretical economics; the maximization of entropy is a law of thermodynamics.

Optimization

Optimization is the umbrella name for many powerful methods to find the states that satisfy certain extremum principles. There are several broad types of optimization procedure. Game theory is a kind of strategic optimization (§ 20). Dynamic optimization finds the extrema for entire histories of systems; it includes the calculus of variation, optimal control, and dynamic programming (§ 42). Here we consider the general conceptual structure of basic optimization problems that, with suitable variation, can be adopted to strategic and dynamic optimization.

The method of optimization is widely used in engineering and the natural and social sciences, where it undergirds the rational-choice theories. Technically, to *optimize* is to find the state of a system that maximizes or minimizes a variable under certain constraints. An optimization problem generally consists of several factors:

1. a set of *possible states* of a system, among which the optimal states will be chosen;

2. a set of *constraints* delimiting the range of accessible possibilities;

3. the variable to be maximized or minimized, often called the *objective*, the domain of which is the set of possible states;

4. the *optimal states* for which the objective variable attains a maximum or a minimum value.

Objective and *optimal* are technical terms that generally convey no idea of intention or value judgment. All factors can in turn depend on other variables relevant to the science or engineering problem that calls for optimization. These variables, sometimes called choice variables, may be more basic to the problem, but in the optimization calculus they function only through the possible states, the constraints, and the objective. For example, in economics, a household's possible states are bundles of commodities, its objective is utility, and it is constrained by its budget. It shoots for the commodity bundle within its budget that has the maximum utility. The household's states depend on the quantities of commodities and its budget depends on commodity prices. Price and quantity are basic variables in economic theories, and they enter indirectly into the optimization problem.

Optima are not absolute but are relative to a set of possibilities. The possibility set is related to the *state space* of a system, which is the major postulate for a specific theory (§ 10). It is mandatory that all relevant options be included in the possibility set, for its range determines the extent of the optimization problem. A possibility set containing only things money can buy may be adequate in economics, where the objective is material satisfaction, but not in a problem that aims to maximize happiness.

An optimal state may be local or global. A *local optimum* is the state for which the value of the objective variable is greater than that for any other possible states within a certain neighborhood. The summit of an arbitrary mountain is a maximum in height, and the maximum is local unless it happens to be Mount Everest. A *global optimum* is a state for which the value of the objective variable is greater than or equal to any other possible states under the constraints. There may be many global optimal states; an egg carton has twelve global minima. If there is only one optimal state, the solution of the optimization problem is said to be unique.

Optimization problems can be very difficult, especially if one is concerned with global solutions for complex systems, such as the problem for "rugged fitness landscapes" (§ 24). There is no guarantee that a solution exists. Often an existence proof for a unique solution is a great step forward. A proof for the existence of optimal states for all constituents of an economy is a major result of the general equilibrium theory in microeconomics (§ 16).

The optimization procedure by which we calculate the optimal states for a system is theoretical and should not be confused with the dynamic process that leads the system to an optimal state. The solution of an optimization problem yields the optimal states but *not* how the states are realistically attained. The evolution of a system toward an optimal state is a totally different problem.

Optimal-equilibrium Propositions

Equilibrium and optimization are distinct concepts with no intrinsic relation. Optimization techniques are widely used in engineering without invoking any idea of equilibrium. However, it often happens in nature that the equilibrium and optimal states of a system coincide. These cases prompt a new conceptual structure.

The result of an optimization calculation asserts that under the conditions specified by its possibilities and constraints, a system has certain states, called its optimal states, that satisfy the general criterion of maximizing or minimizing a certain variable. When the result is combined with the assumption that the optimal states are also the equilibrium states, we get a new form of proposition. An *optimal-equilibrium proposition* has the general form, *Under certain conditions, a system settles down in a state that satisfies a general criterion.* It contains a general criterion for the equilibrium states that is absent in ordinary attributive statements. Ordinary statements assert only that a system is in a certain state under certain conditions but do not add that the state also satisfies the general criterion of maximizing or minimizing a certain variable. Usually, a general criterion opens a new dimension and alters the entire conceptual structure of our thinking.

Optimal-equilibrium propositions are tricky to interpret, because the technical concepts in their peculiar general criterion are easily misunderstood. The names *objective* and *optimum* probably originated in engineering, which is a purposive human activity. Engineers aim to design products such as the shape of a car that minimizes air resistance. It is important to note that the choice and evaluation are made by engineers, who use the optimization theories as tools and stay outside the theories. There is nothing purposive within engineering theories, where the objectives are merely variables to be maximized or minimized, just as they are in physical theories. Whether the state that maximizes a variable is the best is a separate judgment, as maximum air drag is good for parachutes but bad for airplanes. Therefore the interpretation of optimization in terms of intention, purpose, and value judgment is an extra step that must be justified on its own. Usually no justification is forthcoming in the natural sciences.

Economics features decision makers, households and business firms, each intentionally trying to find what is best for itself. They are engineers except they are designing different sorts of things. Therefore the interpretation of "optimal" and "objective" in purposive terms in economic theories seems appropriate. A literal interpretation of the optimization procedure as the actual decision-making process of households and firms is less appropriate. Economics is not concerned with the details of the deliberation processes; it uses only the resultant optimal states of the individuals to find the characters of the economy. Thus the optimization calculus can still be interpreted as theoretical.

The form of optimal-equilibrium proposition fits the form of functional explanation prevalent in biology. A functional explanation asserts that a part of an organism is in a certain state because it serves a certain function in maintaining the well-being of the organism. The function is an extra general criterion absent in ordinary descriptions, and it is often abstractly representable by the objective in an optimization model. Functional descriptions combined with the idea of the survival of the fittest lead to the widespread use of optimal-equilibrium assertions in evolutionary biology. The assertions are sometimes interpreted purposively, as genes are said to manipulate the environment to benefit their reproduction. Such interpretations, which attribute intentions to genes, damage evolutionary biology by cloaking it in mysticism. Optimal-equilibrium assertions in biology can be interpreted without invoking purpose or intention. Functional qualification can be interpreted in terms of the causal relation between the part and the whole described from the limited viewpoint of the part.

Some extremum principles in physics, for instance, the maximization of entropy, are given mechanical or statistical explanations. Others, such as the principle of least action, are left unexplained. Leibniz rightly pointed out that some why questions are special. However, questioning must stop somewhere. This is apparent in the traditional debate on the final cause; we can refer all causes to God, but what causes God to do this and that? Perhaps it is better to stop short of invoking purposive designs in the natural world. If asked why action is minimized for physical systems, perhaps the best answer is that that is the way it is.

3 Individuals: Constituents and Systems

10. An Individual and Its Possibilities: The State Space

The first question in any scientific research is its subject matter: What are we studying? The most general answer is a certain kind of system. A system, no matter how complex, is an *individual*, which is logically a particular, a unit with characters describable in general terms and a unique identity that enables us to single it out and refer to it unambiguously without invoking its characters. We must delineate the system before we delve into its properties. Thus the first general concept in the formulation of a theory is that of an individual. Individuals are represented in scientific theories by the *state spaces*, which form the keystone of the theories by defining their topics. The notion of a composite system as an integral individual with its own state space is the synthetic conceptual framework of many-body theories.

The general concept of a concrete individual in the real world is much more complicated than that of an abstract individual, which may be merely an unanalyzable *x* over which predicate variables range. The complication is most apparent in many-body theories, where both the composite system and its constituents are explicitly treated as individuals. An individual is isolated if it is not encumbered by any relation and not guaranteed to be relatable; situated, if it is an interactive part of a system; composite, if analyzable into parts; simple, if not. Most individuals are both composite and situated. A system is an individual whose composition is made explicit. A constituent of a composite system is situated and behaves differently from when it is ideally isolated. Some individuals, for instance, the collectives discussed in Chapter 5, are both systems and constituents in a single theory.

As a whole comprising parts, an individual's character is constrained by its composition; as a part of a system, its behavior is constrained by its situation. What is the meaning of individuality in view of the constraints? How does an individual maintain its integrity in view of its parts? How does it preserve its distinctiveness in view of the entangling relations in a system? Before answering these questions, we must first clarify the meaning of an individual in general. The clarification is important because, as we will see in the following chapters, the substantive characters of individuals change like chameleons in many-body theories. The general concept enables us to track the individuals through the changes. It also provides the first glimpse into microanalysis.

Carving Nature at Its Joints

It is often said that the world is the sum of individual things and organisms. What does the statement mean? Does the sum include relations? What are these individuals? How do they relate to the individuals we invoke in daily conversation and scientific theories? What are the criteria of being an individual?

In one philosophical position, the individuals are "furniture of the universe" given "out there" and absolutely independent of our concepts. The trouble is that we have no way to connect these transcendent individuals with the conceptualized individuals in our theories; we have no access to God's position to establish the correspondence between the two. We cannot say that the individuals are obvious or identify themselves by stimulating our senses. Most entities studied by the sciences are neither obvious nor visible to our senses. Revisionary philosophers can banish them as fakes; scientists can hardly do the same.

If the world is not perfectly homogeneous and uniform, then there must be clumps and clusters of matter. A lump, however, is not an individual, for it is continuously connected to its surrounding. Individuals result from our conceptualization of the world, which makes the lumps stand out by themselves. We recognize individuals in a conceptual structure with two levels of generality: its categorical framework and substantive scheme. The category of individuals presupposes the general concept of the external world that is susceptible to our representations and yet independent of them. I have analyzed the concept of the external world elsewhere and will take it for granted here.[1] By itself, it does not imply that the world consists of distinctive objects that can be singly addressed, for the world can be an unanalyzed whole, as Parmenides and other holistic thinkers maintained. Nevertheless, it provides the synthetic framework for microanalysis when we add the idea of individuals to our categorical framework. The actual microanalysis is executed when the categorical framework is embodied in various substantive schemes that distinguish between monadic characters and relations of salient lumps, ignore the relations in a first approximation, and thus carve out the clumps as individuals.

The view that we individuate entities by conceptually cutting up the world has been well known since Plato. While considering the problem of universals, Plato compared investigating nature to carving it at its joints. The trick, he said, is to observe the joints, not to mangle them as an unskilled butcher world.[2] The idea is not confined to the West. A little after Plato, the Chinese philosopher Zhuang Zhou gave the parable of the master butcher while contemplating the pursuit of potentially infinite knowledge in finite life. Novice butchers, who chop, have to change their knives monthly; the better butchers, who slice, change their knives yearly. The master's knife has been used for nineteen years to operate on thousands of oxen, but its edge is as sharp as it was when it left the grindstone, for the master discerns cavities and crevices and guides his blade through them without resistance.[3]

Several ideas scintillate in the metaphors of Plato and Zhuang. Individuals are not received passively but are results of our conceptual divisions. We conceptualize, but there are weak links in nature suitable for severance in the first approximation. Various ways of division are acknowledged, but not all are equally good. The world is a most intricate knot. We learn by trial and error, and the path of intellectual history is littered with discarded theories like so many blunted knives. Some simpler aspects of the world are loosely knit so that our intellectual blade has plenty of room to play. In these areas mature scientific theories have confidently captured the fissures of nature. Other aspects of the world are so finely meshed even our best available theories are more like axes than scalpels. Conjectures in these areas suggest the image of shifting paradigms.

Besides intellectual activity, critical tolerance, and naturalness, another idea is less obvious but no less important: Division implies prior bonds. The bonds are not totally discarded in the microanalysis of the world into individuals. The universals or monadic predicates that carve out the individuals are complemented by causal relations among the individuals that take account of the original bonds. Individuals are situated in causal networks.

The viability of an individual depends on the strength of its internal coherence relative to the strength of its external relations (§ 4). The general idea of an individual is that it is *situated* and yet *its external couplings are relatively weak* so that they can be justifiably neglected in approximations in which it is described in terms of its monadic characters. We have used the idea in ideally isolating composite systems from the rest of the world for our study (§ 6). We will see that it also dominates the microanalysis of the systems. When the constituents are bound by strong relations, scientists try to redescribe them or to repartition the system to find weakly coupled entities. The redescription is made possible by the *complementarity of monadic characters and causal relations*; it shifts the "boundary" between them to absorb some relations into the monadic characters of the new individuals (Chapter 4). Scientists adjust substantive descriptions to preserve the general idea of weakly coupled individuals. The redescribed individuals may have totally different monadic characters, but the complementarity of characters and relations enables scientists to accept them rationally without their having to be born again into another paradigm.

Spatial Location and Numerical Identity

Let us look more closely at the concepts used in individuation. Some philosophers argue that individuation is purely spatial; an individual is one that occupies a certain region of space or one that is delineated by certain spatial boundaries. Space can be replaced by spacetime without changing the doctrine. The doctrine has several difficulties. It depends on the ontological assumption of a container space, which is problematic in the light of relativistic physics. Furthermore, space is perfectly homogeneous with no boundaries between regions; therefore the individuation of spatial regions must be explained. We run into a vicious circle if we distinguish the regions by the things that occupy them. Last but not least important, spatial boundaries merely cut things up without any consideration for the characters and causal relations of what they enclose. Consequently the spatially delineated entities do not provide the ground for the systematic introduction of interaction. Naive individuations created by "obvious" boundaries are like apprentice butchers who chop: They can lead to entities that violate physical laws, for instance, spots that move faster than the speed of light.[4]

Although space or spacetime is not sufficient for individuation, spatial or spatiotemporal concepts are necessary, for they account for the *numerical identities* of individuals. Even if two individuals are qualitatively alike in all respects, they are distinguished by their spatial positions; for that reason we regard as a truism that two things cannot be in the same place at the same time. By considering the concepts of individual and spacetime together in an earlier work, I have equated the numerical identities of simple individuals to their absolute spatiotemporal positions – positions in the world, not in a container spacetime; the equation uses the concept of exclusion and not the idea of occupancy in spacetime.[5] Here I will take the spatiality and identity of individuals for granted. I assume that all individuals are either spatial or based on spatial individuals. Thus numbers are not individuals, nor are many logical and mathematical entities, because they lack numerical identity. A person is an individual, for he is inalienable from his body, which is spatial. Aggregate entities such as firms and communities are based on persons and things.

Kinds and Natural Kinds

Here is an individual, a dog. Instead of the dog, we can equally say, Here are a bunch of organs, or many cells, or even more atoms. Philosophers have long realized that spatial location is insufficient for individuation; it must be supplemented by the idea of a *kind.* It is impossible for several things to be at the same place simultaneously only if they are things of the same kind. Aristotle argued that the concept of a primary being is not simple but contains two irreducible elements; a thing is both a "this" and a "what-it-is", for example, this man or that horse. Only after we have grasped the compound idea of this-something can we proceed to describe it as "such and so"; for example, "The horse is white."[6] The general concept of an individual has two elements, its numerical identity and its kind.

Substantive kind concepts are broadly classificatory; electron and bird and bank are kinds. We recognize an individual as an instance of a kind, a tree or a ship. An "I don't know what it is" is an unfamiliar thing, not an unconceptualized bare particular, which is too abstract for comprehension. In ordinary speech, we refer to individuals with singular terms: proper names, pronouns, and common nouns preceded by definite articles such as "this tree." The common nouns represent the kinds of the individual. Common nouns show up in many proper names such as planet Earth or U.S.S. *Enterprise*, highlighting the common concept involved in the notion of individual.

We intellectually carve nature by kind and spatiotemporal concepts, but not all cuts fall on nature's joints. Philosophers and scientists distinguish among natural kinds, nominal kinds, ideal types, and capricious inventions. For some concepts, we have to learn each instance by heart and hesitate to identify further instances. These concepts we regard as arbitrary inventions. Other concepts are more successful. Once we are introduced to some of their instances, we confidently apply them in a large variety of situations to pick out unacquainted instances, and our applications enjoy wide intersubjective agreement. Even so, we may still decide that their results are *nominal kinds*, for we may be unable to find compelling causal relations among the individuals of various kinds to be sure that the classification does not stem from our conventions of thinking. Thus biologists agree that all taxonomic groups above the species level are nominal kinds. If a conceptualization partially captures some important features of the world but neglects other important ones, or if it grossly simplifies the phenomena, then the resultant individuals belong to *ideal types*. The economic units discussed in §13 are ideal types. Various kinds of atom are examples of *natural kinds*. We postulate they are the nexuses in a natural pattern that are made to stand out by our conceptualization, because they have clearly defined characteristics and nest in an extensive causal network.

We have no criterion that unambiguously distinguishes between natural kinds and ideal types, for the delineation of natural kinds also involves some theorization. There is a continuum stretching from natural kinds to ideal types depending on the level of simplification. Naturalness is a criterion with which we criticize our substantive individuating schemes. It depends on the scope, robustness, proper retention of interindividual relations, and convergence of various schemes. We judge the naturalness of a kind by how well its members fit into the broad working scheme of the intelligible world without further important qualifications.

Individuation is part of a theory with many ramifications. Successful individuating concepts retain their validity in many theories and relational contexts. The larger the scope of an individuating concept is, the less willing we are to change it; we are more prone to change the complementary relations to account for deviant phenomena. Thus the individuals represented by the concept are firmly established.

Variables and State Spaces: Possibilities

We loosely say that the members of a kind share a cluster of characters. However, the similarities are often difficult to pinpoint, for the members may exhibit considerable variation. Organisms of the same species vary so much that some biologists, finding it difficult to define species precisely, opt for rejecting species as natural kinds (§ 12). How do we grasp the idea of kinds in view of the variation? Does the variation point to some presuppositions of the idea? When we say of an individual that it belongs to a certain kind, are we not tacitly aware of the possibility of variation?

Since the idea of a kind is closely associated with the characters of its members, let us start by examining the general concept of characters. The proposition "He is red" can mean he flushes or he is an American Indian or he is a communist. Such ambiguity, usually resolved in context, points to the dual element in the general concept of characters: a *character type* and *a range of values* (§ 5). Predicates such as *red*, which describe the monadic characters of individuals, are incomplete in themselves. They are understood only as values of character types such as color or race or political affiliation. The type–value duality is made explicit in scientific theories, where character types are known as *variables* and are often mathematically represented by functions. A function is a rule that systematically assigns values to individuals. As something general, a rule opens a range of *possibilities* represented by the range of its values. When we invoke a definite predicate as a value of a rule, we have presupposed the idea of alternative values. It may happen that a character type has only one value, but that is a substantive peculiarity that does not hamper the general ideas of rule and range of possibilities.

An individual has many characters, the totality of which at a particular moment is its *state*. The state is a summary variable such that the individual retains its identity as its state changes over time. The very idea of change presupposes the idea of alternatives. Just as a definite character is an instance of a character type and one of its possible values, a definite state is an instance of a state space and one of its possible values. The *state space* of an individual contains *all possible states* that individuals of its kind can attain. (*Space* in *state space* abstractly denotes a set of entities endowed with a certain structure; it is purely theoretical and has nothing to do with the space in "space and time.")

The state spaces of the members of a kind share the same structure, signifying an equal-opportunity world. For instance, the state spaces of all classical point particles are six-dimensional Euclidean spaces. Since the state space is peculiar to a kind of individual, we can say it is part of the general concept of a kind.[7] The sciences aim to understand what is generally true about kinds of entity. Thus it is not surprising that the state space is often the primary postulate of a scientific theory, for it specifies the kind of individual with which the theory concerns itself. State spaces are prominent in most mathematical scientific theories. The term *state space* seems to be most consistently used in systems engineering. State spaces are also called *genotype and phenotype spaces* in biology (§ 12); *production and consumption possibility sets* in economics (§ 13); *phase spaces* in physics, more specifically Hilbert spaces in quantum mechanics, μ-spaces and Γ-spaces in statistical mechanics (§ 11). State spaces are no less important in mathematics. The modern formulation of deterministic dynamics presents state-space portraits that include all possible motions of the dynamic system under study (§ 29). The probability calculus introduces means to partition the state space and evaluate the magnitudes of its parts (§ 35). In all cases, the state space opens a synthetic framework in which theoretical analysis proceeds.

An individual realizes a specific state in its state space at a particular moment, and its temporal evolution traces a path in the state space. Which one of the myriad possible states is actualized at any time is often determined by various dynamic or static rules subjected to constraints peculiar to the individual, such as boundary or initial conditions. In other cases, especially in more complex systems, accidents and contingent factors affect the history and hence the actual state of an individual. The rules or accidents are posited separately from the state spaces.

Although all individuals of a kind have equal possibilities, their actual states and characters are different. Actual characters are represented by characterizing predicates in ordinary language and functions in scientific theories. Thus a categorical distinction is made between the predicates of kinds such as "horse" and predicates of characterization such as "white." Kind predicates are represented by state spaces, characteristic predicates by functions over the state spaces. Both types of predicate are involved in subject–predicate propositions such as "The horse is white," as Aristotle argued.

The Category of Possibility

Before we examine how individuals are represented by state spaces in the sciences, let us pause to reflect on the notion of *possibility*, which undergirds the state space and unifies the equilibrium and dynamic models considered in the two parts of this book. Many philosophers have argued that modal concepts such as possibility, actuality, and necessity are indispensable to our thinking. Both Aristotle and Kant included modality in their tables of categories. Kant further entitled the category of qualities "Anticipation of Perceptions." The title impressed Martin Heidegger, who elaborated on how we anticipate because we are open to possibilities. Ludwig Wittgenstein started his atomistic construction of thought with the postulate of a logical space for things that are possible constituents of states of affairs. Saul Kripke developed possible-world semantics, which is invaluable in the analysis of ordinary language. In all these philosophies, modal concepts are formal, belonging to our theories about the world and not to the world; we do not make "possibilia" into nonphysical things or "propensity" into a mysterious force.

Not all philosophers countenance modal concepts. Parmenides argued that what is necessarily exists; what is not is unthinkable and should not be admitted in any proposition; thoughts about possibility are mistakes because possibility is not. Quine argued that modal concepts are unsuitable for "scientific" discourse and promoted a regimented language that gets rid of them. If he had paid a little attention to actual scientific practices, he would have noticed that the more rigorous the theories are, the more prominent is the notion of possibilities as represented by state spaces.

Let us focus on theoretical reason and leave aside freedom, autonomy, and choice, which are impossible without the awareness of options. It is interesting that Parmenides, for whom only the actual is thinkable, rejected the individuation of the world into entities and the notion of time. The ideas of individual, possibility, and change are inseparable. In a perceptual experience, we see only a thing's unique state at that particular moment, but we often say that it was different a moment ago and will change in a moment hence. In so saying we have presupposed the possibility for it to be in a different state. An actual object is discerned only within its specific environment. If it is also bound to its environment in thought, then we have a kind of photographic mentality in which the object is an undetachable part of a picture. As such it is not outstanding and not fully individuated. The poorly individuated mentality tends to be rigid and inflexible, because an unanalyzable picture must be retained or rejected as a unit and it is difficult to replace entire images.

To individuate an object is to free it conceptually from its immediate surroundings: to see that it can be otherwise, that it has options. The idea of possibilities cuts across the plastic image and detaches the object from its actual mooring. It enables the object to exist in the original Greek sense, meaning "to stand out from" or simply to "stand out." It informs us that a thing is more than what meets the eye; it is full of threats and promises. What meets the eye is a sleeping bird, but we see a flyer and a singer; with these possible states we acknowledge the individuality of the bird as distinct from its being part of a static picture. Individuation is crucial to theorization, in which we do not merely take in what impinges on us but step back intellectually from our activities to gain a broader perspective, so that we can grasp things as they are beyond the sight of our immediate concerns. In the detachment and broadening of the vista the actual is complemented by the thought of the possible.

Dispositional attributes such as separable, flammable, or soluble refer to possible behaviors of things. These ideas are as indispensable to scientific theories as to our everyday operation. Perhaps the manipulation of things was the root of our primitive individuation schemes; although things often stick together, we realize that they are separable.

One way of distinguishing universal laws from accidental generalizations is the ability of the laws to support contrary-to-fact statements. "All metals conduct electricity" and "All coins in my pocket are quarters" have the same logical form, but the former is a law and the latter an accidental generalization. We can say of a wire that if it was metal, it would conduct electricity, but we cannot say of a nickel that if it was in my pocket, it would be a quarter. Counterfactual assertions presuppose the notion of possibilities, which enables us to inquire about the behaviors of an individual in hypothetical situations, compare it to other individuals in different situations, hence to formulate objective predicates and universal laws.

Individuals in Physics, Biology, and Economics

So far our consideration has stayed on a rather abstract level. Now we begin to examine how the general concepts are separately embodied in the substantive conceptual schemes of the three sciences. The following three sections investigate how individuals in physics, biology, and economics are represented by their respective state spaces.

Statistical mechanics has a comprehensive formulation relating a composite system's microstate space to the state space of its constituents and to its macrostates. The microstate space for a composite system is also the idea that makes the probability calculus so powerful. It is discussed in the following section. There we also examine the difference between statistical mechanics and the kinetic theory of gases, the conceptual structure of which does not include a state space for the composite system as a whole.

The conceptual structure of population genetics is similar to that of the kinetic theory of gases. Both are built on the state spaces of the constituents, and both represent a composite system by a statistical distribution over the states of its constituents. In biology, the constituents are organisms, whose state spaces include all genetic or morphological characters for organisms in the species. There is no definition for the state space of the evolving species as a whole. Consequently there is a controversy over whether the species can be regarded as an individual.

Microeconomics features two broad kinds of constituent for the economy: household and firm. Each has its state space, respectively called the consumption possibility set or production possibility set. The state spaces of all constituents, properly coordinated by a set of commodity prices, constitute the microstate space of the economy as a whole.

11. The Integrity of the Topic of Statistical Mechanics

The subject matter of statistical mechanics is a system of particles conceptualized as a unitary individual represented by a single microstate space. This section presents the synthetic conceptual framework of statistical mechanics, illustrating how it embodies the elements of many-body theories discussed in § 6.

Statistical mechanics represents a system of particles by its microstate space, which is a general conceptual structure and is independent of such specifics as the strength of interaction among the particles. Even if the system happens to be an ideal gas of noninteracting molecules, its conceptual integrity is ensured by its microstate space, which enables us to reckon with its possible microstates. The relations of the microstate space to the state spaces of the particles on the one hand and to the macrostates of the system on the other give substantive content to the idea of composition. We will see that the relations are embedded in an expanded synthetic conceptual framework with additional postulates that are not purely mechanical.

Besides the statistical mechanics formulated in terms of the system microstate space, there are kinetic theories that do not postulate any state space for the composite system but use the state space of the constituent particles to express the distributional patterns of the system. Kinetic theories, which are part of nonequilibrium statistical mechanics, are concerned with temporal evolutions and are more suitable for systems with weakly interacting constituents. They will be discussed more fully in § 38. Here we note only their way of representing composite systems, which will be compared to the formulation of population genetics in the following section.

Constituent States and System States

The state space of an individual circumscribes all possible states the individual can assume. The state space postulated by a theory encapsulates its intellectual framework, determining how its topic is conceptualized. For a composite system, we can distinguish between the *states of the system* and the *states of its constituents*, and consequently between the *system state space* and the *constituent state space*. As a system is describable in its microstate or macrostate, it has a microstate space and a macrostate space, the former more closely related to the constituent state space. Constituent state spaces are used in kinetic theories and more generally in descriptive statistics. System microstate spaces are used in statistical mechanics and more generally in the probability calculus. Let me illustrate their differences by a simple example in the probability calculus before turning to physics. The example also illustrates the meaning of probability in statistical mechanics.

Consider a flip of a coin. It has only two possible outcomes, head or tail. Thus its state space consists of two points, H for head and T for tail.

Consider *n* flips of a coin, which can be regarded as a collection of *n* constituents, each being a single flip. If we reckon the results of *n* flips using the two-point state space of a single flip, we get a histogram with two columns. The height of the column on the state H gives the number of heads, the height of the column on T the number of tails, and the two numbers sum to *n*. The histogram is a *distribution* whose variable is the possible state of a constituent, so that each flip is counted separately. Often the distribution is normalized to give the fraction or percentage of the flips that are heads or tails instead of the number of heads or tails. This sort of *descriptive statistics* is familiar in the vital, social, and opinion statistics we read in the news.

Suppose we perform many experiments, each of which consists of flipping a coin *n* times. The distributions we get vary; sometimes there are more heads, other times more tails. If the coin is fair, then the numbers of heads and tails are almost equal in most cases. If the coin is bent, then the distributions are systematically biased. Empirically, the condition of the coin can be judged from experimental data. Theoretically, it is represented by assigning a *relative statistical "weight"* or *relative magnitude* to each possible state. The statistical weights of the states H and T are equal if the coin is fair, different if the coin is bent. The relative weight assigned to a possible state is technically its *probability*, which is an undefined primitive in the probability calculus and a given quantity in specific problems.

The flips of a coin can be formulated differently. We can regard the *n* flips as a sequence or a composite system whose constituents are single flips. The system has 2^n possible *microstates*, each of which is a product of some possible states of its constituents. Thus we posit a system microstate space with 2^n points, each of which specifies a complete sequence of heads and tails. For instance, a sequence of 3 flips has $2^3 = 8$ possible microstates: HHH, HHT, HTH, HTT, THH, THT, TTH, TTT. Its microstate space has eight points. Notice that the number of possible microstates increases exponentially with the number of flips in the sequence. The formulation in terms of the *system microstate space* is the foundation of the *probability calculus* and facilitates the proofs of many important theorems, including the laws of large numbers. It is an example of the kind of high-level abstract construction that makes modern mathematics so potent. Why is it so powerful?

Descriptive statistics considers only the possibilities of the constituents. Its conceptual structure is so narrow we have no tool to think about the relations among the parts and hence the structure of the whole. For instance, we cannot ask whether the first of the n flips is a head, because we lack the notion of the order of flips. All we can do is to count the number of heads and tails.

There is much more to a sequence than the numbers of heads and tails it contains. Its essential characters include the ordering of the heads and tails and the independence of successive flips, which require explicit statement. The probability calculus introduces a synthetic conceptual structure that accommodates the relations among the parts and the structure of the whole. The enriched conceptual scheme lays out all possible configurations of a system and enables us to discern new regularities that are obscured if our view is confined to the narrow possibilities of each constituent. With its help we can introduce *macrocharacters* of the sequence such as the jth flip's being a tail or having k heads in a row. For more complicated systems, macropredicates are more numerous and often do not invoke the constituents. In poker, for example, we have pair, straight, full house, and so on. The macropredicates enable us to rise above the canopy intellectually and see the patterns of the forest instead of counting trees on the ground.

In descriptive statistics, a typical question is to ask the relative weight of a particular constituent state: How many heads does the sequence have? Answer: k. This is a bottom-up approach where we sum constituent characters to obtain a system character. In the probability calculus, "having k heads" is not an answer but a question we pose about the system. We ask, What is the relative statistical weight of all those possible microstates of the sequence that conforms to the macrostate of having k heads? We prescribe a macrostate or macrocharacter and ask about the number of microstates that satisfy it. This is the synthetic microanalytic approach where we start from top down.

To answer the question in the probability calculus, we partition the huge microstate space of the sequence by specifying some macrocharacters of the sequences, then find out the relative weights of the parts. For instance, we pick out the part of the microstate space consisting of possible microstates with k heads, or microstates whose jth flip is a head, or microstates whose first three flips are heads. We then compute the statistical weight of the part relative to the weight of the entire microstate space. Since a possible microstate of the sequence is the product of some possible constituent states, its relative weight can be computed from the relative weights of the constituent states. Given the relative weights of the constituent states H and T, we can readily compute the weight of the microstate HTH. Suppose for a bent coin the relative weight of H is 0.7 and that of T is 0.3. For a sequence of three tosses, the relative weight of those microstates with two heads is $(0.3)(0.7)^2 = 0.147$, and there are three such microstates. Thus the relative weight of all the possible microstates with two heads is $3(0.147) = 0.441$. Technically, 0.441 is called the probability of "having two heads in a sequence of three tosses."

In the laws of large numbers, *large number* refers to the number of constituents of the system under study, for instance, the number n in a sequence of n flips of a coin. If the coin is fair, all the 2^n possible microstates of the sequence have equal statistical weight, so that the relative weight of a group of possible microstates is just the ratio of the number of microstates in the group to the total number of microstates. The strong law of large numbers asserts that the ratio of the number of microstates containing half heads to the total number of possible microstates approaches 1 as n approaches infinity. Note that we are again doing descriptive statistics, but this time the variable is not the constituent state but a macrostate of the system, namely, the number of heads it contains. We study the distribution of possible sequence microstates over the number of heads and find that as n becomes large, the distribution is sharply peaked around $n/2$ heads.

The Kinetic Theory of Gases and Statistical Mechanics[8]

I consider only structureless classical particles. Quantum mechanics introduces different ways of counting but does not alter the general statistical framework. Let us start with a single particle, which is analogous to a single flip of a coin. The state of the particle is determined by six numbers, the three components of its position and the three components of its momentum. Thus its state space can be represented by a six-dimensional Euclidean space spanned by the position and momentum variables, each point in which represents a possible state of the particle. The six-dimensional state space of a particle is akin to the two-point state space of a single flip of a coin.

Consider a system of N classical particles of the same kind. The kinetic theory of gases borrows the state space of a constituent particle and calls it the μ-space. It describes the particles by a *distribution function* over the μ-space. The distribution function, analogous to the histogram for coin flips, gives the number of particles with states in the infinitesimal region around each value of position and momentum (Fig. 3.1a).

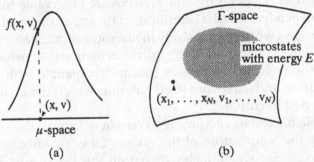

(a)   (b)

FIGURE 3.1 **(a)** The μ-space, depicted by the horizontal line, is the state space of a single particle. Each point in it specifies a possible position $\mathbf{x}$ and a possible velocity $\mathbf{v}$ of the particle. A distribution for an N-particle system gives the number $f(\mathbf{x}, \mathbf{v})$ of particles with position and velocity around an infinitesimal region around $(\mathbf{x}, \mathbf{v})$. The evolution of the distribution is studied in the kinetic theories. **(b)** The Γ-space is the microstate space of the N-particle system as a whole. It includes all possible microstates of the system, each specifying the positions $\mathbf{x}_1, \mathbf{x}_2, \ldots, \mathbf{x}_N$ and velocities $\mathbf{v}_1, \mathbf{v}_2, \ldots, \mathbf{v}_N$ of the N particles. A microstate in the Γ-space determines a unique distribution over the μ-space, but a distribution corresponds to a large number of microstates. A macrocharacter of the system, such as its total energy E, carves out a group of microstates in the Γ-space, depicted by the shaded area. Statistical mechanics computes the relative magnitudes of various groups conforming to various macroconditions.

For the same N particles, statistical mechanics considers the state not of a single particle but of the system as a whole. The N-particle system is analogous to a sequence of n flips of a coin. A possible *microstate* of the N-particle system, analogous to the microstate HTH or TTH for a sequence of three flips, is determined by $6N$ numbers, the components of the positions and momenta of all particles it contains. *Macrostates* are characterized in terms of macrovariables such as the total energy, temperature, or chemical potential, which are akin to the predicate "jth flip" for a sequence of coin flips.

All possible microstates of the N-particle system constitute its microstate space, called the Γ-space (Fig. 3.1b). It is a $6N$-dimensional Euclidean space spanned by $6N$ position and momentum variables, analogous to the microstate space with 2^n points for a sequence of n flips of a coin. The actual microstate of the N-particle system at a moment is specified by a particular point in the Γ-space. Just as the microstate HTH determines a histogram with two heads and a tail but the histogram does not determine the microstate because it is equally satisfied by HHT and THH, the Γ-space description is much more discriminating than the μ-space description. A single point in the Γ-space completely specifies a possible microstate of the system by specifying the characters of each particle. Hence it determines a distribution over the μ-space. However, a distribution does not completely specify a microstate of the system; it only tells the number of particles with a certain momentum but not which ones. A distribution over the μ-space corresponds not to a point but to a large number of points in the Γ-space.

Ensembles, Averages, and Coarse Graining

Macrovariables such as temperature and pressure are observable by themselves and have independent physical significance; they obey thermodynamic and hydrodynamic laws written exclusively in macroconcepts. Statistical mechanics aims to uncover their underlying micromechanisms. To achieve the task, it picks out the possible microstates that are compatible with certain macroscopic conditions and strikes out irrelevant microscopic details by averaging over the selected microstates.

We select possible microstates according to certain macrovariables such as the total energy or the temperature of the system. Like "two heads" in the probability calculus, the macrovariables carved up the Γ-space by picking out those possible microstates that satisfy certain macroscopic conditions. The group of selected microstates is called the *ensemble* conforming to the macroscopic condition. There are various ensembles corresponding to various sets of macrovariables. An ensemble effectively asks a question in macroconcepts and seeks the answer by screening for relevant microinformation. The microstates in the ensemble serve as the basis for obtaining average quantities as is commonly done in the probability calculus. Irrelevant details are averaged out in the process. Statistical mechanics relates the averages to macrovariables whose physical meanings are explicitly defined.

Conceptually, the most basic ensemble is the *microcanonical ensemble*, which picks out those microstates in which the system has the energy value E. These microstates occupy a subset of the Γ-space called the *constant energy surface*. To proceed further, we must make a postulate that is fundamental to the conceptual structure of statistical mechanics. It is not enough to count the microstates with energy E. Just as a coin can be bent, the system may be biased and preferentially realize some possible microstates to the neglect of others. The rejection of bias is a basic assumption in the definition of the density distribution of the microcanonical ensemble. It is sometimes called the *postulate of equal weight* or the *postulate of equal a priori probability*, which stipulates that all microstates in the constant energy surface have equal statistical weight, or that each is equally likely to be realized by the system in equilibrium. The justification of the postulate will be discussed shortly.

With the postulate of equal weight, the statistical weight of a group of microstates is just the number of microstates in the group. Statistical mechanics defines the total number of possible microstates with energy E as the *entropy* of the system. The macrovariable energy carves out a group of possible microstates, whose statistical weight is related to the macrovariable entropy. Thus we establish a microscopically based relation between two macrovariables: the system's entropy as a function of its energy. From this function, we can obtain other thermodynamic quantities and relations by the rules of thermodynamics.

Other ensembles use other macrovariables to pick out other groups of microstates. The *canonical ensemble* picks those microstates kept at a certain temperature by exchanging energy with a heat reservoir. The *grand canonical ensemble* picks those microstates kept at a certain chemical potential by exchanging particles with a material reservoir. In the *thermodynamic limit*, where the volume of the system and the number of particles in it go to infinity as the density stays constant, all three ensembles yield the same result. Thus there are three alternative conceptualizations for a statistical system and physicists choose the one most convenient to the specific problem at hand.

Statistical mechanics is an example of synthetic microanalysis. It is based on an atomistic ontology but it does not try to deduce its way up from microphysics. The question it poses is not "Given a microstate of the system, what are its macrocharacters?" but "Given a macrocharacter of a system, what are the possible microstates?" To address the problem, it posits the microstate space of the system as a whole, defines the macroscopic constraints, and reaches down to seek out the relevant microphysics for the phenomena it wishes to understand. It abstracts from the details of mechanical forces by appropriate statistical methods. The conceptual framework enabling us to grasp the regularities of large numbers is the unique contribution of statistical mechanics that is irreducible to mechanics.

Ergodicity and the Justification of Statistical Mechanics[9]

There are several basic questions regarding the conceptual framework of statistical mechanics. The macroquantities obtained in equilibrium statistical mechanics are the statistical averages taken over all possible microstates of the system compatible with certain macroscopic constraints. What does the average over possible microstates mean, when the system actually realizes only one of the possibilities at any moment? The question is not evaded by interpreting the density distribution as an ensemble of systems realizing all possibilities, for the ensemble is imaginary; in practice we are studying a single system. Then there is the question regarding the postulate of equal statistical weight for all possible microstates. How is the postulate justified? It turns out that these difficulties are related and are partially addressed by the notion of ergodicity.

The word *ergodic* originates in the Greek word for "energy path." The ergodic hypothesis was introduced by Boltzmann in 1871 to justify the procedure of statistical averaging. Boltzmann and Maxwell noted that equilibrium refers to the temporal constancy of macrostates and macroquantities, whereas the microstate of a system changes with every molecular collision. Molecular collisions occur on a time scale much shorter than the time scale characteristic of macroscopic measurements. Since the system actually passes through many microstates during a measurement, measurements of macroquantities yield only time averages over microstates. To argue that the time average over fleetingly realized microstates is equivalent to the statistical average over possible microstates, they advanced the *ergodic hypothesis*, which asserts that an isolated system following its equation of motion will eventually pass through every possible microstate compatible with the macroscopic constraint.

The version of the ergodic hypothesis of Boltzmann and Maxwell is false. Yet it stimulated the *ergodic theory*, which gives it a rigorous formulation. The ergodic theory will be presented more fully in §§ 32–33, in the context of dynamic systems. Here I sketch a few key ideas that are useful later. A dynamic system is *ergodic* if it is not confined to a subset of its possible microstates by its equation of motion; that is, its microstate space is not decomposable into invariant subsets such that if it initially realizes a microstate in a subset then it will only realize other microstates in the subset. Generally, a system is not ergodic if it has one or more constants of motion, for its evolution will be confined to those microstates for which the specific quantities are constant. Thus a system with constant energy, which is the topic of the microcanonical ensemble, is strictly speaking not ergodic. However, it is ergodic within the constant energy surface if there is no other constant of motion that further decomposes the energy surface into subsets.

If a system is ergodic, then in the limit of infinite time and for paths originating from almost all microstates, the time average of a macroscopic dynamic variable over a path is equal to the statistical average of the variable over the accessible microstates, all of which are assigned equal statistical weight. Furthermore, over infinite time, the fraction of time in which it realizes a group of microstates is equal to the ratio of the group to all possible microstates of equal statistical weight. This result, which is a sort of generalization of the law of large numbers, provides some justification for statistical averaging and the equal weight postulate. The latter is further supported by the fact that if and only if the system is ergodic, the microcanonical density distribution that is constant on the energy surface is the only distribution that is invariant to temporal evolution. Thus the ergodicity of a system guarantees the existence of a unique equilibrium state.

Despite the advance in mathematics, the ergodic theory fails to provide a fully satisfactory connection between microdynamics and macroquantities. Ergodicity is far from the universal characteristic of mechanical systems. Some of the simplest and most important systems studied by statistical mechanics are not ergodic. An example of nonergodic systems is the ideal gas that is used, among other things, as a model for electrons in a semiconductor. Another example is the multiple harmonic oscillator that can represent the ideal crystal lattice. Such physical counterexamples undercut the strength of abstract arguments based on the ergodic theory.

Even for ergodic systems, the connection between micro- and macrodescriptions is questionable. Results of the ergodic theory are independent of the size of the systems; their significance is most striking for small systems with a few degrees of freedom. Thus they must be supplemented by the stipulation that the topics of statistical mechanics are large systems with many constituents. Indeed many results of statistical mechanics are exact only in the limit of infinite systems, which is clearly an idealization, as real systems are always finite.

Results of the ergodic theory hold only in the limit of infinite time, but observation times, however long, are finite. Thus they do not explain why almost all systems with arbitrary initial conditions relax to their equilibrium states in finite times and spend most of their lifetimes in equilibrium. The explanation will invoke the condition of large systems, which assures that the number of possible states compatible with equilibrium is overwhelmingly larger than the number of possible states associated with out-of-equilibrium conditions (§ 40).

Oliver Penrose said in his review of the foundations of statistical mechanics, "Statistical mechanics is notorious for conceptual problems to which it is difficult to give a convincing answer." These problems have worried Maxwell, Boltzmann, and many physicists and mathematicians after them. Despite much effort and progress, the toughest problems remain after more than a century. "Alas" is perhaps the most frequent remark in Lawrence Sklar's recent comprehensive survey of the foundational research. R. Jancel has said: "Despite the numerous remarkable efforts – especially in the classical domain – it is in fact illusory to attempt to found the methods of statistical mechanics on a purely dynamic basis." Penrose concured: "A proper account of the foundations of statistical mechanics must use both probability theory and the large size of the system."[10] The concepts of probability and large size are not intrinsic to mechanics. Thus the foundational problems of statistical mechanics, our best theory of composition, testify against eliminative microreductionism.

12. The Unit of Evolution and the Unit of Selection

The ideal subject matter of evolutionary biology is the entire biosphere, but it is too extensive and heterogeneous to be tractable. Population genetics, the core of the modern synthetic theory of evolution, confines itself to the evolution of a single species or a population of organisms belonging to the same species. It is narrower than Darwin's theory because it limits its topic to the microevolutionary processes within a species, disregarding the formation and extinction of species and macroevolutionary trends.

There is much conceptual confusion in population genetics, starting from the articulation of its subject matter. Biologists generally agree that the topic, the unit of evolution, is a species. Controversy breaks out when it is realized that the formulation demands that the species be treated as a composite system made up of individual organisms. How does the treatment reconcile with the usual notion of species as natural kinds? There are also quarrels over whether a gene, an organism, or a group of organisms is the dominant unit of natural selection. Traditionally the answer is the organism; thirty years ago the group was a popular rival; today the gene is so fashionable it has acquired the private label of *genic selectionism*. Much conceptual muddle stems from the statistical nature of population genetics. Statistical descriptioms are better conveyed in mathematical symbols or graphs; in words they readily lead to confusion over the subject of discourse.

This section presents the backbone conceptual structure of population genetics, with emphasis on its representation of *individuals that evolve* and *individuals that are the nexus of natural selection*. The basic concepts are clarified here, so that we can take on controversies arising from simplification and approximation in later chapters.

The Dual Roles of Species as Individuals and Kinds

Is a species a natural kind or an individual? This is one of two disputes over the notion of biological species.[11] Ever since Aristotle, species of organisms have been the paradigm cases of natural kinds. Evolutionary biologists notice that species as kinds seems to be incompatible with the theory of evolution. A kind is a universal and universals do not change, but as the units of evolution, species bud, evolve, split, and go extinct. Some biologists have further argued that the members of a kind must share a common essence, but organisms of the same species vary and the variation is crucial to evolution. On the basis of these and other considerations, Ernst Mayr denounced species as kinds or classes. Michael Ghiselin argued that in the logical sense, species are not *natural kinds with members* but *individuals with parts*: "Species are, then, the most extensive units in the natural economy such that reproductive competition occurs among their parts." "Some biologists will never say 'John Smith is a *Homo sapiens*,' but insist that the correct usage demands 'John Smith is a specimen of *Homo sapiens*.'" The idea was elaborated by the philosopher David Hull, who argued that as a unit of evolution, a species must be conceived as an individual delineated by "spatiotemporal unity and continuity."[12]

The thesis of Ghiselin, Hull, and Mayr ignited a debate. Their argument for the individuality of species based on spatiotemporal continuity is weak. Organisms devour each other and are thus spatiotemporally continuous, but that does not imply the whole biomass is an individual. The inheritance of genetic material is only a minuscule part of the spatiotemporal relations among organisms; if only it counts, then the essential criterion is not spatiotemporal. Ghiselin compared species to corporations. A corporation acts as a unit and is describable independently of its employees. Can the same be said for species? Ford and General Motors compete by promotion campaigns, pricing strategies, and product designs, but those stratagems are totally different from the rivalry between their chief executive officers. In what sense can we say *Leo leo* and *Leo pardus* compete except that individual big cats struggle for existence?

Biologists study how a certain species of bacterium affects a certain kind of cell and reacts to a certain kind of chemical. Here *species* has the same logical role as "kind." Without species as kinds, the notion of organisms collapses. If *Lion* is only a proper name for a species as an individual and not a common noun, we do not know how to talk about those maned creatures that roar. Most of biology would be devastated, nor would evolution be spared. A tooth, a cell, or a pride is no less a part of the individual Lion. Lion cells reproduce, but they are not lions. Ghiselin's definition of species as individuals in terms of "parts" and "reproductive competition" would make no sense if we have not surreptitiously used the familiar notion of lion as a kind. Most arguments against species as kinds are misplaced.[13]

Ghiselin, Hull, and Mayr rightly observe that species are logically individuals in the theory of evolution. They err in excluding the notion of species as kinds for this reason. Species can assume double roles as individuals and as kinds, as Dobzhansky said: "A species, like a race or a genus or a family, is a group concept and a category of classification. A species, is, however, also something else: a superindividual biological system."[14] The theory of evolution demands the duality. A species as an individual that evolves is a composite system, the constituents of which, organisms, are individuated with the concept of the species as a kind. Thus species as individuals are dependent upon species as kinds. Our task is to find a clear articulation for the dual roles.

The duality of species as kinds and as individuals can be most abstractly represented by the notions of sets as universals and as particulars.[15] Here I will use a formulation that more readily accounts for the evolution of species. I have argued in § 10 that a kind is delimited by a range of possible states represented by the state space of its members. Thus a species as a kind is characterized by the state space of the organisms that are its members. A species as an individual is made up of organisms that are members of the species as a kind. It is represented by a distribution over the state space of its constituents. The temporal change of the distribution represents the evolution of the species as an individual, whereas the constant state space represents the immutability of the species as a kind. The formulation naturally accords with the theoretical structure of population genetics, so that we need not appeal to tenuous spatiotemporal continuity for species as individuals.

Species as Kinds and the State Spaces of Organisms

Allow me to digress and introduce some technical terms. The mechanism of heredity, which explains how organismic characters are passed from parents to offspring, is studied in genetics. A *gene* is a unit of heredity. The entire genetic code of an organism is its *genome*. A gene occupies a particular position in a chromosome; the position, called its *locus*, identifies it. A gene can have several alternative forms, each of which is called an *allele*. Organisms with only one set of chromosomes are *haploid*; those with two sets of chromosomes that form pairs are *diploid*. Most common species, including humans, are diploid. In a diploid organism, the pair of alleles in the same chromosomal locus may be the same or different. If they are the same, the organism is *homozygous*; if they are different, it is *heterozygous*. An organism's *genotype state* is all of the genetic characters it inherits from its parents, and its *genotype* is one aspect of its genetic state. For instance, consider a gene with two alleles, *A* and *a*. For diploid organisms, there are three possible genotypes, *AA*, *Aa*, and *aa*. Organisms with alleles *AA* or *aa* are homozygous; those with alleles *Aa* are heterozygous. An organism's *phenotype state* is all of its manifest morphological and behavioral characters, which result from its genotype state and development in a certain environment. Its *phenotype* is a specific morphological or behavioral aspect, such as fleetness of foot. The genotype and phenotype states of organisms separately account for the mechanisms of inheritance and natural selection. A *population* is a group of organisms of the same species sharing a similar environment. A population is *closed* if it does not exchange constituents with the outside. We will consider only closed populations for brevity. The total of genotypes in a population is the population's *gene pool*.

All possible genotype states for an organism constitute its *genotype state space*, which is shared by all organisms belonging to the same species. Similarly, the *phenotype state space* comprises all possible phenotype states for organisms in the species. The genotype and phenotype state spaces are the basic concepts in population genetics.

Since organisms are such complicated individuals, their state spaces are enormous. A point in the genotype state space designates a possible genotype state by specifying an allele for each locus in the genome. For a modest genome with 1,000 loci each with 3 possible alleles, the genotype state space contains $3^{1000} \approx 10^{477}$ points. Because of the huge number of possibilities, it is very likely that individual organisms are genetically unique, even if they all belong to the same species. Also, only a small fraction of the possible genotype states are realized by organisms existing at a particular time. The phenotype state has infinitely many points, for a phenotype state has many aspects – morphological, anatomical, behavioral – many of which vary continuously.

In the theory of evolution, organisms are the constituents of a composite system, a species or a population. Thus the genotype and phenotype state spaces of organisms are conceptually akin to the two-point state space consisting of H and T for flipping a coin, or to the μ-space in the kinetic theory of gases. The notion of the species as a kind ensures that the same genotype and phenotype state spaces are shared by all organisms belonging to it. This assurance underlies the formulation of theories in population genetics.

Species as Individuals or Composite Systems

Arguments for species as individuals can be divided into two groups. The first belongs to macroevolution, in which a species is not explicitly considered as a composite system but simply as an entity with its own heritable characters. I do not consider this aspect because it is not at all clear what these characters are. Let us concentrate on the arguments based on population genetics.

The theoretical framework of population genetics is *statistical*. Wright said: "The conclusion nevertheless seems warranted by the present status of genetics that any theory of evolution must be based on the properties of Mendelian factors, and beyond this, must be concerned largely with the statistical situation in the species." Fisher made a similar assessment: "The investigation of natural selection may be compared to the analytic treatment of the Theory of Gases, in which it is possible to make the most varied assumptions as to the accidental circumstances, and even the essential nature of the individual molecules, and yet to develop the general laws as to the behavior of gases, leaving but a few fundamental constants to be determined by experiment."[16]

Let me flesh out Fisher's analogy between population genetics and the kinetic theory of gases. Consider a gas of electrons in a positively charged background. For simplicity let us neglect the positions of the electrons, so that the state space of an electron is a three-dimensional space representing its three components of velocity. The velocities of the electrons vary. The gas as a whole is described by a velocity distribution over the electron state space, which gives the percentage of electrons with velocities infinitesimally close to each value of velocity. The velocity distribution is changed or maintained by two kinds of causal mechanism: random collisions between the electrons and external forces. The former is like the mechanisms of reproduction and heredity and the latter like natural selection. When two electrons collide, we can say that they "die" and two new electrons with different velocities are "born."[17] The conservation of energy and momentum in collisions can be seen as a kind of "heredity" that preserves certain characters of the "parents." Suppose the external environment has a constant temperature. The velocity distribution reaches an equilibrium and peaks at a certain velocity; the velocity is said to be "selected." Now introduce an external magnetic field that acts on each electron separately according to its velocity. Since the electrons have different velocities, they accelerate differently although they are in the same field. The velocity distribution of the electrons changes, and "directional selection" is in progress. The evolution of the distribution of the gas is governed by a kinetic equation, which accounts for molecular collisions and external forces.

In population genetics, a species is analogous to the electron gas. The genotype and phenotype state spaces are analogous to the state space of an electron in the kinetic theory of gases. Each actual organism of the species is represented by one point in the genotype state space and one in the phenotype state space. The genotype states for all actual organisms constitute a genotype state distribution of the species, which gives the number or percentage of organisms in each genotype state. A phenotype state distribution is defined similarly. Like the gas represented by its changing velocity distribution, the species is represented by its genotype and phenotype state distributions, the changes of which constitute its evolution.

The Conceptual Framework of Population Genetics

Lewontin gave the most lucid exposition of the conceptual structure of population genetics, which consists of a *genotype state space*, a *phenotype state space*, and a set of *causal rules*.[18] At a particular time, each living organism of a species is represented by a point in each state space. The double description is required because the organism's genotype state is responsible for heredity and its phenotype state for survival and selection. The distributions of points over the genotype and phenotype state spaces represent the species and embody the Darwinian notion of variation. The causal rules account for genetic segregation and recombination, mutation, growth and development under a given environment, migration, feeding, fighting, mating, reproduction, accident, and other processes of living. Causal mechanisms act on individual organisms. Developmental processes connect the genotype and phenotype states of an organism. Survival factors decide how its phenotype state changes over time. Reproduction connects the states of parent organisms with that of their offspring. As organisms prosper and reproduce differently, the composition of the species changes accordingly. Thus the causal mechanisms indirectly drive the evolution of the species. With simplified transformations, the conceptual structure is schematically illustrated in Fig. 3.2. The phenotype state distribution in the figure is a snapshot at a particular stage of development. A more detailed rendition would show several phenotype state spaces corresponding to several stages of maturity, or paths in a phenotype state space traced by the points representing the growth of organisms.

Suppose the initial condition of a species is a set of newly fertilized eggs described by a certain genotype state distribution. The genotype state distribution delimits the range of phenotype states the embryos can possibly grow into. Which phenotype states are actualized by the adults depend on the environment and other contingencies of development. Thus the genetic makeup and the environment jointly determine the phenotype state distribution. Natural selection takes effect, and it continues as adults with various phenotype states cope with the environment differently, reproduce with different efficiencies, and contribute differently to the gene pool of the next generation. During reproduction, the genetic material is scrambled. The scrambling, together with genetic mutation and statistical sampling effects, maintains genetic variety. They all contribute to the genotype state distribution of the next generation. The cycle repeats.

The genotype and phenotype state distributions inevitably fluctuate from generation to generation, but the fluctuation is not evolution. For *evolution* to occur, minute changes must be cumulative so that some kind of long-term trend in the change of the state distributions is discernible above the statistical fluctuation.

Natural selection, which accounts for the effects of the environment, is a major evolutionary mechanism. However, it is neither necessary nor sufficient for evolution. It is not sufficient because even when a selective bias decimates a part of a species in a generation, the species can maintain a constant

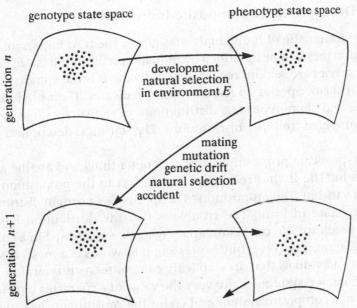

FIGURE 3.2 The genotype (phenotype) state space encompasses all possible genotype (phenotype) states for an organism. The immutable state spaces, shared by all organisms of a species, embody the idea of the species as a natural kind. The species is also represented as an evolving composite system by distributions over the state spaces. At each time, an organism in the species is represented by a point in the genotype state space and a point in the phenotype state space. The collections of points in the genotype (phenotype) state space constitute the genotype (phenotype) state distribution, which characterizes the species at the time. The schematic shows the genotype state distribution at conception and the phenotype state distribution at maturation. Developmental and selective transformations under the given environment map the genotype state of an organism to its phenotype state, if it survives. Reproductive transformations determine the organism's contribution to the gene pool in the next generation. Consequently the genotype and phenotype state distributions change over time, representing the evolution of the species.

genetic distribution across generations because of the scrambling of genetic material during sexual reproduction.[19] It is not necessary because evolution can be driven by other mechanisms.[20] One such mechanism is the *random mutation* of genes. On the basis of the observation of surprisingly large genetic variations in natural populations, the neutral theory argues that most genetic mutations are neither harmful nor beneficial to survival, hence are not discriminated by natural selection. Another mechanism of evolution is *genetic drift*. If the population size is not large enough, statistical errors in the random sampling of the parental gene pool during sexual reproduction lead to fluctuating genotype distributions and eventually the loss of some alleles (§ 37). A third mechanism is the *self-organization* of organisms (§ 24). Various mechanisms are in principle represented by various transformation rules linking the genotype and phenotype states of the organisms. They should all be included in the broad formulation of population genetics.

Statistical Descriptions of Composite Individuals

A general characteristic of individuals that makes them so important in our thoughts is that they can be described by monadic predicates: It is such and so. One reason species are seldom regarded as individuals is the apparent lack of predicates for them. Species do have definite predicates. Their predicates are the genotype and phenotype state distributions, or statistics such as averages or spreads computed from the distributions. The statistical description can be confusing.

In subject–predicate propositions, we pick out a thing and ascribe a predicate to it, as in "The fly has red eyes." The subject in the proposition is the independent variable. The proposition "The species has a certain distribution" is subject–predicate in form. The trouble is that the distribution is such a complicated predicate. No confusion arises when it is expressed in a compact mathematical form, say, $f(v)$, but expressing it in words is a mess. Distributions describe individuals that are explicitly composite. As is typical in statistical descriptions, we systematically vary the value of a character type for the constituents of a composite system and count the constituents bearing various character values. Thus the distribution has its own independent variable, which refers not to things but to character values or predicates of things, such as the redness of eye. To describe a species of fly by a distribution, we say "The species comprises 5 percent yellow-eyed flies, 5 percent white-eyed flies, 90 percent red-eyed flies." The only adjectives in the statement, those denoting eye colors, describe individual flies, although the statement itself describes the species. Consequently it is easy to make the mistake of ascribing attributes of organisms to the species as a whole and to talk about a red-eyed species. The confusion contributed to the once-popular doctrine of group selection. George Williams, who cleared up the confusion, pointed out that the descriptions "an adapted species" and "a fleet population" are incorrect unless they are shorthand for "a species of adapted organisms" or "a population of fleet gazelles."

Units of Selection[21]

The preceding discussion concentrates on the evolution of a species of organisms. Here the organisms are the units of selection because they are the main entities on which the forces of natural selection and other causal mechanisms act. Lewontin, noting the generality of the conceptual structure of population genetics, argued that it is applicable to the evolution of a population of any kind of individual – gene, cell, organism, group of organisms – as long as the individuals have heritable characters and differential survivorship.[22] For instance, it can represent the operation of a mechanism called *meiotic drive*, in which some genes are preferentially passed on during reproduction, leading to evolution on the molecular level. Lewontin's clear and correct observation was soon turned into a prolonged and often polemic dispute over what is *the* unit of selection of organic evolution.

The issue of the debate over the unit of selection is usually formulated as; For whose benefit does natural selection act, the group, the organism, or the gene? The struggle for benefit is so intense that for a moment I thought I was reading economics, in which capitalists fight for profit. The misleading loculation is part of the problem. Natural selection does not act for the benefit of anything. The teleological suggestion is what evolutionary theory is supposed to have debunked.

The unit-of-selection controversy partly concerns the drastic approximations made to obtain definite results, which will be addressed in § 17. More basically, the controversy arises partly from a categorical confusion and a terminological ambiguity. There is a categorical difference between *individuals* and *characters*. An individual such as an organism or a DNA molecule is a particular. A character such as being healthy or having a particular molecular configuration is a universal. A genetic code, being a configuration of genetic molecules, is a character. Individuals and characters are easily confused in genetics because of ambiguity in terminology. A genetic molecule, a kind of genetic molecule, and a genetic code are all called a gene or an allele. In most usages, "a gene" or "an allele" is a character, a molecular configuration or a genetic code, similar to "sodium," which describes a kind of atom. We have no systematic names for the kinds of genetic molecule as we do for kinds of atom because there are too many of them. However, in the question "Is the gene or the organism the unit of selection?" "gene" is a molecule, an individual with the same ontological status as an organism. Thus genes are called replicators. DNA molecules replicate themselves to produce a host bearing the same character or genetic code. A code, being a universal, does not replicate; it is simply instantiated in many particulars.

In evolutionary theory, *what are selected are not individuals but characters*, which may be genetic, morphological, or behavioral. The fittest character in a certain environment spreads, persists, and prevails in a species. Nature has no aims, but in the preferential preservation of characters, its logic is not different from that of breeders who promote characters such as drought resistance or fast growth. Individuals, be they genes, organisms, or groups, are too ephemeral to count in the evolutionary time scale. They are units of selection because they are the bearers of the characters under selection and the locus of selective forces.

The selection of characters is appropriately represented in the statistical formulation of population genetics. Evolving species are described in terms of changing distributions, and the independent variables of the distributions are not individuals but states or characters. If the distribution persistently peaks over certain characters in an environment, these characters are said to be *selected*. A character is favored by natural selection if its relative instantiation, which is the number of individuals bearing it relative to the number of individuals bearing other characters, increases with time. The emphasis on characters instead of individuals is the general characteristic of statistical description, which is a coarse-grain view that ignores fine distinctions among individuals and counts heads under "stereotypes." That is why we often hear people complain that the government worries only about statistics and ignores their individual situations.

In the statistical account of species or populations, the mortality and replacement of the individuals become insignificant, because we simply count heads and do not care whom we are counting. Therefore Richard Dawkins has made a categorical mistake in his argument for genic selectionism. He said that only genes can be the units of selections because "genes are potentially immortal, while bodies and all other higher units are temporary."[23] If the statement compares universals to particulars, it is senseless; if it compares various kinds of particulars, it is false. Genetic molecules do not last forever; they are more ephemeral than bodies and organisms. Genetic codes are universal and atemporal, so are the morphological characters of bodies; to neither does the notion of mortality apply. We can metaphorically say that a morphological character or a genetic code is "extinct" if it was instantiated but is no longer so, but that is logically different from saying an individual is short-lived. The "longevity" or "near immortality" of a genetic pattern or a morphological character is not the condition but the consequence of natural selection. Dawkins's suggestion that nature selects among near-eternal particulars totally distorts the conceptual structure of evolutionary theory.

The *units of selection* are those individuals of which the characters under selection are attributes. They are the constituents of the evolving species described by the distribution and the heads to be counted at various times. They are also the locus of the causal mechanisms that are responsible for the evolution of the distribution.

In the formulation of population genetics in terms of genotype and phenotype state spaces, the unit of selection is obviously an organism; that is the traditional view and is held by most biologists. However, as discussed in § 17, the formulation is too comprehensive to be tractable. Biologists make drastic approximations that sever the genotype and phenotype state spaces and further factorize each into many disjoint characters. Under such approximations, the interpretation of characters becomes ambiguous. An allele or a genetic code can be regarded as the character of a DNA molecule or of an organism, as we say that someone gets a disease because he inherits a bad gene. The controversial questions are the following: To understand the phenomena of organic evolution, what entities should we regard as the bearers of genetic codes? What are the individuals most directly subjected to selective forces? Are they organisms or DNA molecules?

The representation of evolution as changing distributions is general and applicable to aggregates of any entities. The choice of the unit of selection depends mainly on the meaning of the distribution and the causal mechanisms responsible for its change. Natural selection can operate on the molecular level through mechanisms such as the meiotic drive. However, such cases are rare and contribute only slightly to the main trend of evolution. As Maynard Smith said, "If gene selection within individuals (as in meiotic drive and transposition) were the rule rather than the exception, complex organisms would not evolve"[24]. An individual disintegrates if its parts compete rigorously. Usually organisms are the units bearing the storm of life. This fact is tacitly acknowledged by the proponents of genic selectionism in their notion of "a gene for an organismic trait," for example, the gene for a certain disease; the disease affects organisms and not molecules, which do not get sick. Genic selectionism is a bag of rhetorical confusion, as discussed more fully in § 17. I will stick to organisms as units of selection.

13. Economic Individuals as Ideal Optimizers

Microeconomics considers the microstate of an economy, which specifies the state of each of its constituents, households and firms. Most microeconomic models adopt an approximation that factorizes the economy's microstate into the states of individual households and firms, tied together only by the commodity prices that are determined self-consistently with the individual states. Thus the households and firms are by definition situated in the economy, because their states are parameterized by commodity prices, which are characteristics of the economy as a whole. However, the approximation also makes it easy to decouple the individuals from the economy theoretically. All we need to do is to hold the commodity prices constant as we examine the state of individual households and firms. This is what I do in this section. In § 16, we will reintegrate the households and firms into the economy by restoring the variability of prices to study the economy's microstate as represented in the general equilibrium theory for perfectly competitive markets.

Ideal Optimizers in Consumption and Production[25]

We recognize a host of economic units pretheoretically: persons, households, business firms, unions, associations, cartels, government bureaus, nation-states in international trade. They vary greatly in size and organization. Important tasks of theoretical economics are to abstract from the specifics of the diverse units and introduce concepts that capture some general and salient features.

Economic units engage in the activities of promoting material welfare. The physical aspect of economic activities is a poor candidate for generalization because it is so multifarious. A better candidate is the mental aspect; economic activities all involve decisions of some kind. Small or big, simply or hierarchically organized, ideally economic units share the common feature of unity in decision and action. A household has a single budget, Boeing introduces a jet liner, the Organization of Petroleum Exporting Countries (OPEC) decides to curb oil production, Britain goes off the gold standard. A unit may harbor internal conflict, and it disintegrates if the rift becomes too great. However, as long as its constituents hang together so that it announces economic decisions in a single voice and acts on them in unison, it is acknowledged as an individual. The internal organizational structures of the individuals are neglected in a large part of microeconomics, although they are studied in management science and some recent economic theories based on game theories (§ 20). The rationale for the idealization is not unlike that of the neglect of atomic structures in the theory of gases.

Economists postulate that economic individuals are systematic decision makers regarding production, consumption, and exchange. Furthermore, they all make decisions in the same general way: to find the best means to achieve given goals. Abstracting from specifics, all economic units are *optimizers* that systematically survey a given set of possibilities under given constraints and choose the one that maximizes a given objective variable. The general formulation of optimization problems is presented in § 9. We now fill in the substantive possibilities, constraints, and objective variables peculiar to economics.

The set of possibilities open to an individual is its state space. State spaces are the basic postulates of a theory and delimit its topic. What is the topic of economics? According to Mill, "It [political economy] is concerned with him [man] solely as a being who desires to possess wealth, and who is capable of judging the comparative efficacy of means for attaining that end."[26] Thus economics limits the state spaces or possibility sets of its individuals to those items with price tags that can be consumed exclusively by a single individual. The possibility sets exclude priceless elements such as love and honor. They also exclude public goods that must be shared by many individuals, for instance, roads, national defense, clean environment, or television broadcast, although these factors have economic impacts. Except a few scattered topics in welfare economics, the possible performances of the community are omitted from the consideration of individuals. Also expelled are all ethical considerations and personal activities that are not saleable. The limited set of possibilities is the hallmark of *Homo economicus*.

The possible states and objectives of economic individuals are framed in terms of two variables, the *quantities* and *prices* of various *commodities*, or tradable goods. The commodities include the labor and talent, insofar as they are up for sale. A *commodity bundle* consists of various quantities of assorted commodities. Since commodities and prices are characteristics of the economy as a whole, the individuals are intrinsically situated. Strike out the priced commodities and the economic individuals lose meaning.

Economic individuals are classified into two broad types, *firms* and *households*, which respectively decide matters of production and consumption. Firms decide how to transform commodities of some kinds into commodities of other kinds, for instance, iron and other minerals into steel. A firm's possible states are various production plans that detail the quantities of parts, raw material, and finished products. The collection of all possible plans is its *production possibility set*. The range of possible states open to a specific firm is constrained by the *technology* it masters, which is determined by its existing plants and facilities in the short run and by the state-of-the-art technology in the long run. The firm's objective is *profit*, which is the price differential between the output and input commodities. The equilibrium state of the firm is the technologically feasible production plan that rakes in the most profit, or that incurs the least cost if the output product has been decided upon.

Households may receive dividend from firms, but they are not involved in production decisions. They decide what commodities they want to consume. A household's possible states are bundles of commodities, and its actual state is the bundle it possesses; its character is literally its property. Its state space, called its *consumption possibility set*, is the collection of all bundles that the economy can possibly offer. Each commodity bundle has a price. The range of bundles accessible to a specific household is constrained by its *budget*, which is the price of the commodity bundle with which it is initially endowed. Although all households in the economy have the same consumption possibility set, each has access only to the subset delimited by its peculiar budget constraint. It is an equal-opportunity world in principle but not in practice.

A specific household is represented by a *preference order* over the consumption possibility set. The preference orders of different households vary according to their tastes, but they all have the same general form, the basic requirement of which is the consistency of preference. For any two commodity bundles, a household either prefers one to the other or is indifferent between the two, and its preferences are transitive; if it prefers A to B and B to C, then it prefers A to C.

The preference order of households is represented algebraically by a weak ordering. Algebraic representations are general but difficult to analyze. Like other scientists, economists prefer to work with analytic representations. Thus they represent the preference orders by a *utility function*, which is a rule that assigns to each commodity bundle a utility value according to the bundle's desirability. The utility function is the objective function that the household tries to maximize. Given a household's options (the consumption possibility set), constraint (its budget), and objective (its utility function), its equilibrium state is obtained by finding the affordable commodity bundle with the highest utility value.

So far we have considered only decision making under conditions of certainty. The formulation can be expanded to include risk. Decisions in risky situations must take account of a unit's beliefs about the state of the world. A household's beliefs are represented by the probabilities of realization it assigns to various possible states of the world. Usually the beliefs and hence the probability assignments of different households differ.

Accounting for both its desires and beliefs, the possibility set of a household contains not commodity bundles but prospects. Each prospect is like a lottery; it lists several possible commodity bundles and the probability of realization for each. The household establishes a preference order for the prospects. If the set of prospects satisfies certain axiomatic conditions, the preference order can be represented by an *expected utility function*, which is the sum of the products of the utility of each commodity bundle and the probability of its realization. The equilibrium state of the household is obtained by maximizing the expected utility in the standard way.

The economic units illustrate how human beings are conceptualized through the lens of economics. Activities are broadly divided into production and consumption. Production is for profit, and satisfaction is obtained through consumption. Households sell their labor and talent to buy more satisfying commodities. The creative activities enjoyed at leisure are deemed unproductive. The concepts of households and firms are powerful in analyzing our economy. However, they are not the only possible conceptualization of human productivity. Marx, for instance, offered an alternative view in which individuals engage in production not for profit but for the satisfaction of creative activity.

The Question of Rationality

Economists call the defining characteristic of economic individuals, what I have called optimization, *Rationality*. Utility maximization is often called Rational choice theory. It is widely used in the social sciences, and its limitations are widely discussed.[27] I capitalize the word *Rationality* to indicate the technical meaning, because it is narrower than the meaning of *rationality* in ordinary usage or philosophical discourse.[28] In a time when irrationalism is in vogue, the notion of rationality should be treated with double care. A biased or narrow-minded version may be the lemon that drives good cars out of the market.

The technical sense of Rationality has three aspects: unlimited cognitive competence, consistent preferences, and instrumental reason executed by following an algorithm. Unlimited cognitive competence means that an economic unit has complete knowledge of the characteristics and prices of all commodities available in the economy, perfect foresight of future prices and its own future preferences. In risky situations, it can foresee all contingencies that may happen and assign definite probabilities to each. All knowledge is acquired without effort, for the work to collect information contributes to transaction cost, which is assumed to be zero in perfect markets. The cognitive competence of effortlessly knowing all relevant matters is closer to omniscience than rationality. To be rational is to make steady effort to know, not to be in possession of all knowledge.

The consistency of preference is represented by the lineal preference order that excludes dilemmas in choice. It is achieved at the sacrifice of many possibilities. In realistic choice situations, a person's consideration is often multidimensional: personal, familial, civil, religious, emotional, ethical, aesthetical, and so on. It is doubtful that concerns in different dimensions can be meshed into a single variable. The meshing becomes more difficult when regard for others is taken into account. If one alternative is preferable to a second in one dimension but not in another dimension, then the two alternatives cannot be ordered. In these cases not all options can fit into a lineal order. However, that limitation does not rule out the possibility of rational deliberation in the ordinary sense.[29]

Instrumental reason limits itself to finding out the best means to achieve a given goal. It does not exhaust rationality, which calls for openness to all possibilities, the search for alternatives, the formation and assessment of purposes, and the criticism or perhaps the modification of goals in response to experiences. Most of these are spurned by economic individuals, who accept their preference orders and objective functions as given and stick to them without reflection. Within their limited vision, economic individuals are admirably alert to possibilities, partly because they have unlimited cognitive power. An outstanding example of their assessment of possibilities is the concept of opportunity cost, or profit lost from not pursuing some options. By putting opportunity costs on the same footing as out-of-pocket costs, they ensure that no possibility is overlooked. However, they are closed-minded in other ways. They are contented with a restricted set of possibilities that categorically excludes considerations for other individuals, the community, and commodities that can be shared, although such considerations are economically important.

Kenneth Arrow argued that the hypothesis of Rationality is not in principle necessary for economic theories and that utility maximization is not the only way to represent household behaviors. Alternative theories can be framed in terms of, for instance, the formation of and adherence to habits. Routine is a major concept in the evolutionary model of Richard Nelson and Sidney Winter, which depicts economic change without invoking optimizing behaviors. In the "bounded rationality" models of Herbert Simon, the grand optimizing scheme is replaced by a series of partially integrated decisions made under particular situations, with incomplete knowledge and limited reflection. Instead of a given set of options, the models postulate processes for generating options. Individuals do not optimize but try to satisfy certain targets they set for themselves. The results from these models are quite different from the optimization models.[30]

Epistemological Questions Concerning Economic Individuals

As represented in theoretical economics, households and firms are highly abstract and idealistic. Idealization is indispensable in all sciences; for instance, in many mechanics models the sun and the planets are represented as point masses. Idealizations that aid the understanding of complicated phenomena are valuable, provided they are acknowledged as such and placed under constant scrutiny.

Empirical studies have cast doubt on the thesis that people make decisions by maximizing utility. In many experiments, people are found to reverse their preferences consistently in risky conditions, depending on how the questions calling for choice are framed. Discrepancies between empirical data and the fundamental behavioral postulates of microeconomics are widespread and systematic. However, it is not clear which specific axioms of the utility-maximization theory are violated.[31]

The theory of profit-maximizing firms is challenged by the theory of managerial capitalism. Nowadays the bulk of economic activity is conducted by large corporations, control of which is divorced from ownership. Corporations must make profits to survive in the long run. However, it is questionable whether the executive officers aim only to maximize profit. Perhaps they opt for achieving adequate profits and other ends such as power, prestige, sales volume, or control of the market.[32]

The empirical findings on people's decision-making processes would be devastating to economics if it aimed to study human cognition. However, economics is not psychology; its subject matter is the aggregate behaviors of the economy. It is not unusual for theories of aggregates to overlook the detailed behaviors of individual constituents, either because we know little of them or because they have little effect on aggregate behaviors. In some cases, microdetails are filtered out when individual behaviors are aggregated to produce macrophenomena. This happens, for instance, in the critical phenomena in physics, where widely disparate microprocesses produce the same macroscopic result (§ 23). The question is whether similar filtering occurs in economics. More generally, how critically do the behaviors of the economy depend on the optimizing behavior of individuals? How do predictions on the economy change if some postulates about economic individuals are relaxed or replaced? These are pertinent questions for theoretical economics.

Milton Friedman dismissed as irrelevant studies that try to find out whether firms actually make decisions by maximizing profits. He argued that an economic theory is good if it makes reliable predictions for a certain range of phenomena, even if its assumptions, including its axioms on the characters of economic individuals, are unrealistic.[33] I think Friedman makes several valid points, although I reject his positivism. I do not find the assumption that households make complicated optimizing computations so objectionable. We intuitively perform feats such as face recognition that, if formalized, defeat the efforts of supercomputers. A model for human behavior need not reproduce in detail thinking processes, which no one knows anyway. It needs only to capture those gross behaviors relevant to the topic under investigation. This is what Friedman meant when he said economic theories hypothesize that people behave *as if* they always maximize satisfaction.

Although economic units are familiar in our daily activity, they are radically reformulated in scientific theories. The scientific conception is idealistic and far from being a mirror image of reality. Simon Kuznets said that economic individuals are "ideal types" created in theoretical economics "for the purpose of better analysis, explanation, and generalization of the behavior of the whole; and thus depend upon the state of our knowledge, changing in variation with this knowledge."[34] In economics as in other sciences, scientists do not simply take a familiar set of entities and try to combine them. They discern a sphere of phenomena as the topic of inquiry and conceptually crystallize the phenomena by individuating the appropriate units whose characters are compatible with the most reasonable and tractable causal structure of the phenomena. The differentiation of individuals for the sake of studying large systems is what I earlier called synthetic microanalysis.

4

Situated Individuals and the Situation

14. Independent-individual Approximations

The intuitive formulation of a many-body problem is an extrapolation from a few-body problem. We take the familiar character types and relation types of the constituents as they are known from small systems, then add more constituents. Therefore the only difference between small and large systems is the number of constituents, as the character and relation types are common to all. The basic formulation provides a unified view of the world and conveys the idea that exotic systems are made up of ordinary parts related in ordinary ways. Unfortunately, it soon leads to a technical impasse. An n-body problem for $n = 10^{10}$ cannot be solved by extrapolating concepts and methods suitable to $n = 2$, not if the bodies interact. To make headway and get concrete results, we need to recast the problem of large-scale composition into a more tractable form. Fortunately, the theoretical framework in terms of system state spaces presented in the preceding chapter has already enabled us to grasp complex systems as wholes. Now we turn to finding more substantive formulations within the synthetic framework to microanalyze the systems.

Perhaps the most common and successful strategy of microanalysis is *modularization*, which draws on the insight that the whole is the sum of its parts. The saying is wrong generally; it is valid only if the parts are unrelated, and a heap of disjoint parts is hardly worthy of the epithet "*whole*." Yet it does contain a grain of wisdom; the characters of uncorrelated parts are more susceptible to averaging and aggregating, making the characters of the whole much easier to determine. If we are to benefit from the wisdom, we can no longer take as parts the constituents familiar from small systems; if we did, we would be throwing away the relations that cement them into a whole. We have to microanalyze the whole afresh to find suitable parts or modules that are almost noninteracting but cohere automatically. Such modules take a dazzling variety of forms, some of the more complicated of which are discussed in the following chapter.

This chapter examines modules that roughly retain the status of the familiar constituents. They are the products of *independent-individual approximations*, which are the most widely used approximations in many-body problems, and in some complicated cases, the only manageable ones. There are several kinds of independent-individual approximation. Cruder ones brutally sever the relations among the familiar constituents to obtain some result. More sophisticated models define custom-made constituents whose properties internalize most familiar relations, so that they respond independently to a common situation that is determined consistently with their customized properties. This introductory section considers only the sophisticated models. Examples of cruder models are found in the following sections as we examine the sciences separately.

The Snarl of Relations in Large Systems

Relations cement the constituents, ensure the integrity of the composite system, and make the whole more than the sum of its parts. In large-scale composition, relations make possible the endless variety of systems based on a few kinds of constituents. They also make many-body problems difficult. We are so used to talking about the relations among a few individuals we tend to overlook the complication introduced by additional participants. For brevity let us consider only binary relations; multiple relations make the problem worse. Although each binary relation connects only two individuals, each individual can engage in as many binary relations as there are partners, so that a relational system quickly becomes intractable as the number of its constituents increases.

Consider a system comprising n massive bodies interacting gravitationally with each other. The inverse square law of gravity is well known, as is the law of motion for the bodies. If $n = 2$, the motions of the bodies have exact solutions and are solved by students in homework assignments. The difficulty takes a quantum jump with one additional body. Newton could estimate the planetary orbits because he assumed that each planet was attracted only by the sun, thus resolving the solar system into a set of two-body systems. Methods to improve on Newton's approximation by accounting for the gravitational pull among the planets were developed by some of the best mathematicians and physicists in the eighteenth and early nineteenth centuries. These powerful analytic methods enable us to compute the planetary trajectories to great accuracy, but they generally do not provide exact solutions for three-body problems.

For the solar system, the sun's superior mass compared to that of the planets makes Newton's approximation natural. In systems whose constituents are roughly equal in status, there is no justification for singling out a pair and neglecting their relations to other constituents. We must reckon with the network of relations. Perturbing the position of one body in an n-body system changes the gravitational forces it exerts on the other $n - 1$ bodies, each of which responds by changing its position and acceleration, hence changing its forces on the others, including the originally perturbed body. All these changes are interrelated through the network of gravitational interaction and must be determined simultaneously and consistently. One needs little persuasion to be convinced that even an approximate solution is hopeless when n is 1 million.

It can be argued that the problem is actually less complicated because not all of the $n(n-1)/2$ potential binary relations among n bodies are realized or are equally significant. In many systems a constituent is effectively related to only a few others. If the constituents are all related according to a general rule as in gravitation, the interaction strength is negligible except among close neighbors. Truncating relations may or may not simplify the problem. The systematic restriction of relations to constituents within small groups requires its own criteria, which are often difficult to specify. Universal relationships are sometimes easier to handle theoretically. For instance, a basic assumption in economic theories is the existence of all markets so that any economic unit can costlessly trade with any other. The assumption is unrealistic, but its relaxation is tremendously difficult. Furthermore, even if we succeed in formulating models in which relations obtain only among selected constituents, their complexity is still formidable. For example, spins interact only with their immediate neighbors in the Ising model for ferromagnetism, but the model still defies exact solution if the spins form a three-dimensional lattice (§ 23).

Not only do the behaviors of large systems defy solution, individual constituents sink into obscurity. Individuals are clearly defined only if they are relatively independent and the effects of relations on their characters are relatively minor. The effect of a million weak relations acting together is formidable. An individual bound by numerous relations is like a bee trapped in a spider's web; its individuality is strangled and its behavior difficult to predict when it cannot be described without adverting to hundreds of partners that pull the strings that bind it.

The General Idea of Individuals Versus Substantive Descriptions of Individuals

Individuals are important in our thinking because of their analytic advantage. An individual can be characterized by itself, and subject–predicate propositions are much easier to handle than relational propositions. The characters of a collection of unrelated individuals are easy to determine: We simply add up the individual contributions. Relations make matters difficult, but without them the world falls apart. How do we overcome the dilemma?

So far the character and relation types we invoke are the familiar ones extrapolated from small systems. For instance, we simply use the familiar formula for single particle kinetic energy and the inverse square law of gravity between two particles and repeat them to take account of more particles. We stick to the familiar *substantive* concepts and get stuck in a relational web when the system becomes large. An alternative strategy is to follow the *general idea of relatively independent individuals* and to find suitable substantive characterizations that satisfy the general criterion. Noting the analytic advantage of independent individuals, we give up the constituents with familiar characters and relations and reformulate the many-body problem in terms of new constituents that are more independent in their substantive descriptions.

In independent-individual models, we microanalyze the system to find new constituents whose behaviors automatically harmonize with each other so that they naturally fit together without explicit bonds. The new constituents have several general characteristics. First, their monadic character types internalize as much as possible the information originally expressed in the familiar relation types. This condition underlies the saying that the constituents are intrinsically related. Second, with most relations accounted for intrinsically, the constituents can be regarded as externally unrelated to each other in the first approximation without jeopardizing the integrity of the system. When we ignore intrinsic relations hidden in monadic predicates, we simply say they are unrelated. Third, the independent individuals are defined specifically for microanalyzing a certain type of many-body system; thus they do not imply the reducibility of system concepts. The second condition entices us to say causally that the whole is the sum of its parts, but the first and third conditions caution us about the meaning of our saying.

Situated Predicate and the Situation

Logically, "Romeo loves Juliet" expresses a relation; "Romeo is in love with Juliet" attributes a relational character to Romeo. The character is relational because it refers to Juliet. Sometimes, the relational character is implicit because it omits the object, as "Romeo is in love." Implicit relational predicates such as "being in love" carry less information than relations, for they do not specify the relata. For romantic or parental love, whose objects are highly specific, vital information may be lost in going from relations to implicit relational predicates. For diffuse feelings such as philanthropy, the move may not be so deleterious. In fact "loving mankind" is already a semiimplicit predicate; strictly speaking the object of the love is neither an abstract idea nor a conglomerate: The predicate describes a general feeling toward every human being. However, the specification of all the relata is too cumbersome and the effort is wasteful. The idea of mankind summarizes the important information relevant to the feeling. A similar strategy is employed in independent-individual models.

An individual's attributes that internalize at least part of its relations are called its *relational characters*. Its *situated characters* are described by semiimplicit predicates, which describe its responses to the *situation* formed by the others in the system but omit the specifics on the others. Situated characters are customized to specific types of situations.

If we imagine familiar relations as attached strings that impede the movement an individual enjoys in isolation, then independent-individual approximations cut the strings and use situated predicates and a common situation to increase the weight of the individual and account for its sluggishness. The result is an individual that is independent, more significant, and easier to treat theoretically (Fig. 4.1). An independent individual is solitary, for it is not explicitly related to any other individual. However, it is not isolated, for being situated is part of its intrinsic character and its behavior is influenced by the situation that is determined by all individuals in the system.

Two ideas are used by sophisticated independent-individual models to account for the effect of familiar relations. A large part of the relational effect is assimilated into situated predicates that describe the enriched characters of

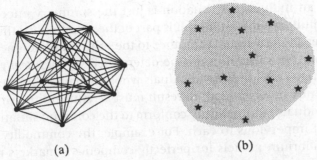

(a)   (b)

FIGURE 4.1 (a) A system with N constituents linked by binary relations is a complicated causal network. (b) In the sophisticated independent-individual approximation, the relations are partly folded into the situated characters of the constituents and partly fused into a common situation determined self-consistently with the situated characters. Whatever residual relations exist are neglected. The result is a system of N modified constituents with no direct mutual relation, each responding independently to the situation jointly created by all.

the constituents. Another part is fused into a situation, which is a rule shared by all individuals and to which alone the individuals respond. The situation is spontaneously generated by the constituents within the system and should not be confused with the external environment that is fixed outside the system and represented by exogenous parameters. Whatever residual relational effect exists is ignored in independent-individual models, but it can be incorporated in other models that take the individuals with situated characters as relata.

In short, independent-individual models replace familiar relations among individuals by the response of each individual to a common situation, or statements of the form "Each individual x_i has character C_i and engages in relation R_{ij} to every individual x_j other than itself" by statements of the form "Each individual x_i has situated character C_i^* and responds to the situation S_i, which is a rule generated by and common to all individuals in the system." The replacement eliminates the double indices in R_{ij}, which signify binary relations and cause most technical difficulties. Note that the situation S is not a constant but a type with varying values S_i for its effects on various individuals x_i. Our main theoretical job is to find the situated character type C^* and the situation S, given the familiar character type C and relation type R. Once the situated character type C^* is known, it is relatively easy to evaluate it for various individuals and aggregate the values to find the behaviors of the system, for there are no longer any entangling relations.

For many-body systems made up of two grossly asymmetric kinds of constituent, the dominant constituents can be approximately treated as a situation that influences the behaviors of the minor constituents but is not affected by them. This happens in solids, where the lattice of heavy ions forms a rigid situation in which electrons roam separately. The situated characters of electrons in various kinds of solids are dramatically different from each other and from the familiar characters of free electrons.

In systems such as a group of households or a sea of electrons, the constituents are all of the same status. It is no longer possible to define a rigid situation, because an individual's situation is just the summary effect of all other individuals similar to it, and it itself is part of the situation for the other individuals. Since each individual is an other to the others and no one is in a dominating position, the situation must be determined *self-consistently*. If we evaluate the collective effect of the individuals whose behaviors in the situation are independently determined, the result must be equal to the situation itself. Thus all individuals automatically conform to the common situation determined by all but impervious to each. For example, the commodity prices in the general equilibrium models for perfectly competitive markets constitute the situation determined self-consistently with the states of individual firms and households. It is the unifying theme of much of microeconomics. Independent-particle models are found in many branches of physics under various names, including self-consistent field theory and Hartree–Fock approximation.

Independent-individual approximations are not only widely used in science but also intuitive in our everyday understanding. They become increasingly relevant as technological developments such as the television and the Internet weaken intimate bonds and draw people individually into a faceless medium.

A colorful example of self-consistent independent-individual models is Leibniz's monadology. Leibniz pushed the reasoning to its extreme and proposed that the world is made up of windowless monads that have no direct relation to each other. The characters of all monads are mutually consistent because they are all situated in the preestablished harmony. Less graphically, Leibniz argued that subject–predicate propositions are sufficient to describe the multifarious world. Relational propositions are not required, for all legitimate relations can be folded into monadic predicates, or situated characters, in our terminology. The compatibility of all the monadic concepts is guaranteed by the preestablished harmony. Since the concept of each monad includes all its relational characters to the others and the world is thoroughly interconnected, the description of a single monad reflects the structure of the whole world. One can see the entire universe in a glass of wine, so to speak. Unfortunately, the monadic concepts are so complicated only God can comprehend them.

The monadic predicates in scientific independent-individual models are comprehensible because we do not demand all relations to be folded into them but go only as far as we can. Thus the models are approximations and far from the final solution that Leibniz envisioned. Explicit correlations such as strategic behaviors must be included to account for more intricate phenomena, as discussed in the following two chapters.

Confusions Arising from Independent-individual Approximations

Most of our ordinary descriptions of personal activities, such as check writing or language use, are situated characters hiding social notions. Perhaps because independent-individual approximations capture some salient features of our impersonal society or perhaps because they are the simplest methods for representing many-body systems, they are frequently employed in social science and political philosophy with minimal explication. Confusion readily arises when people look at the results of independent-individual models without thinking about their logic or derivation.

The self-consistently determined situation is characterized in its own terms, which are qualitatively different from the concepts we use to describe the individuals. Despite its endogenous nature, its direct effect on the independent individuals is similar to that of external forces. Since an individual in a large system is insignificant compared to the aggregate effect of all the rest, the situation constrains individual behaviors and appears to be dominant. For example, although commodity prices are determined self-consistently by the desires of all economic units in the models of perfectly competitive markets, economic units are regarded as passive price takers that respond to the price condition but are powerless to influence it. Since the competitive market model is an equilibrium theory that neglects dynamics, an auctioneer is invented to initiate and control price movements, enhancing the image of exogenous domination. If mathematically rigorous models can spawn such interpretation, imagine the confusion that can arise from less clearly articulated models featuring a host of noninteracting individuals in a dominant situation. The individuals are easily mistaken for isolated beings and the situation is mistaken for an externally imposed institution or norm. Ideological dispute follows.

Methodological holism emphasizes the constraining effect of the situation and advances the notion of systemic constraint. If an additional individual is introduced into a many-body system, its character will be molded by the prevailing situation. Systemic constraint makes sense, provided the endogenous nature of the situation is clarified. Unfortunately it is often not, so that one gets the wrong impression of society as a supraindividual entity imposed externally on all participants.

Methodological individualism, which is much more influential today, rightly argues that society is the spontaneous result of individual actions. In advocating the reductive elimination of social concepts, however, it mistakes situated individuals for bare individuals and overlooks the causal feedback that society has on individuals. It forgets that citizens are not Hobbesian men mushrooming from the earth; even in their most self-centered mode, they have internalized much social relation and conditioning, so that social concepts have been built into in their characterization.[1]

The debate between methodological individualists and holists on the priority of the part or the whole is like the debate on whether the hen or the egg comes first. Carefully formulated scientific models affirm the truism that part and whole arise and fall together. Independent individuals and the situation presuppose each other and are inseparable.

15. Single Particles in the Self-consistent Field

The approximate nature of independent-individual models is most apparent in physics, where the character and relation types of particles in minimal systems are well known, and a precise formulation for large systems of interacting particles is readily available through extrapolation from small systems. This initial formulation, which expressly represents interparticle interaction, serves as the base line of many-body problems and anchors the physical meaning of various terms, but its solution is practically impossible. Its transformation into a more tractable form in which the particles are noninteracting is explicit, revealing all the intermediate reasoning and approximation, which are thereby open to criticism and improvement.

There are many independent-particle models for many-body systems, which hide interparticle interaction with various degrees of refinement. They are also called *single-particle models*. The study of a single representative particle suffices, because the noninteracting particles are of the same kind and have the same situated character type, and scientific theories are mainly concerned with types. In this section we will track the independent-particle models for conduction electrons in a crystalline solid, starting from the crudest, reformulating the problem repeatedly to take account of more factors, and paying special attention to the varying meanings of the electronic predicates in the formulations.

The General Conceptual Structure of Mechanics

Conduction electrons are part of the scope of condensed-matter physics, which studies the micromechanisms underlying macrobehaviors of solids and liquids. It depends as much on quantum mechanics as on the statistical framework discussed in § 11; mechanics provide the substance to the framework. Therefore I begin with a brief description of the general conceptual structure of mechanics. Problems in mechanics are framed in at least three levels of generality: équations of motion, forms of forces, initial and boundary conditions.

On the most general level we have the *equations of motion*: Newton's second law for classical mechanics, Schrödinger's equation for nonrelativistic quantum mechanics, Dirac's equation for quantum fields, and others. The equations cover both dynamic and static situations; in the latter cases the time derivative terms are set to zero. For instance, atomic physics mainly uses the time-independent Schrödinger equation to study atomic structures.

An equation of motion contains a term called the Hamiltonian, which replaces the Newtonian notion of force in the modern formulation of mechanics. The *Hamiltonian* is the form of the total dynamic energy for a kind of system and contains most of its physics. Equations of motion generally leave the Hamiltonians open, and as such they are abstract and lack definite subject matter. The second level of a mechanics problem is to specify the force or the Hamiltonian. For instance, Newton's law says that the product of a particle's mass and acceleration is equal to the force acting on it but does not fix the form of the force. Newton introduced the inverse-square law of the gravitational force separately from the law of motion. Like the inverse-square form of gravitation, the Hamiltonians of more fundamental systems are often called laws. When elementary particle physicists say they are searching for the fundamental laws of nature, they mean they are trying to find the Hamiltonians for the basic building blocks of matter. Not all Hamiltonians have such universal coverage. For instance, the Hamiltonian for harmonic oscillators is applicable only to a small class of systems. Also, there are approximate or effective Hamiltonians, which are introduced in attempts to formulate tractable problems.

The third and most specific level of a problem is to specify the *initial and boundary conditions* for a particular system of the kind represented by the Hamiltonian. Only when the specifics of all three levels are filled can we go on to solve the equation of motion for the behavior of a particular system and perhaps to compare the result to a particular experiment.

The Basic Formulation of a Problem in Condensed-matter Physics

Consider a solid made up atoms of the same kind, say a bar of gold. An atom is made up of a nucleus and many electrons. Most of the electrons form closed shells and are tightly bound to the nucleus. One or two electrons on the outermost shell escape, leaving behind a positively charged ion. The electrons that leave the ions and roam the solid are called *conduction electrons*. They underlie all the electrical and electronic devices on which our everyday life becomes so dependent. Let us see how theoreticians understand their behaviors.

Electrons and ions are quantum mechanical entities. The state of an electron is specified by its wavevector and spin. The wavevector, a quantum-mechanical quantity, is directly proportional to the momentum. An electron can have only two spin states, up or down. Since spin does not play an important role in the processes discussed later, I will ignore it for brevity. The most important character type of electrons is their *energy–wavevector relation* or their *form of kinetic energy*, $E(\mathbf{k})$, which means that if an electron has wavevector $\mathbf{k}$, then its kinetic energy is $E(\mathbf{k})$. The familiar form of energy–wavevector relation for free nonrelativistic electrons is a quadratic relation, where the energy is proportional to the square of the wavevector, $E(\mathbf{k}) \propto \mathbf{k}^2/m$, where m represents the mass of the electron. All electrons have the same form of kinetic energy, but the values of energy and wavevector differ from electron to electron. For simplicity we neglect the internal structure of the ions and regard them as simple entities characterized by a quadratic energy–wavevector relation with a different mass. The ions and electrons are held together by the electrostatic force acting among them; the form of the force is familiar from electromagnetism.

The ions and electrons constitute a mechanical system, the Hamiltonian for which can be readily written down. With the help of indices and the summation symbol, the zillions terms individually characterizing the ions and electrons and their relations can be grouped into five terms in the Hamiltonian. The first includes the kinetic energies of all ions, the second the kinetic energies of all electrons, the third the electrostatic potentials between all pairs of ions, the fourth the electrostatic potentials between all pairs of electrons, and the fifth all the electrostatic potentials between an ion and an electron. External forces such as magnetic fields, if present, are taken into account by additional terms of the Hamiltonian. The Hamiltonian is substituted into the time-independent Schrödinger equation. The basic formulation for the equilibrium structure of solids is completed. The formulation applies to systems of any size; we simply change the upper limit of the summation to account for more electrons and ions.

A small piece of solid consists of more than 10^{20} atoms. Even if we have one of those omnipotent computers advertised in science fiction, we will still be reluctant to hand it the equation and say, "Computer, solve." We anticipate a friendly voice requesting the zillion boundary conditions, which we would be unable to supply. More important, we know that the solution, a matrix with 10^{20} columns and 10^{20} rows of numbers describing the microstate of the solid, would not increase our understanding of solids one bit. The solution is for a specific system. We are interested in the typical behaviors of kinds of system, and we want the information to be so represented that it can be grasped in a few clear concepts.

To gain understanding of solids, physicists make approximations. The models examined in this section all employ some kind of *independent-particle approximation*. The idea behind the approximation is to treat a system of interacting particles as a collection of noninteracting particles with *situated characters*, perhaps moving in a potential generated endogenously within the solid. As in all scientific theories, we are mainly interested in the *form* of the situated character or the situated character *type*, which can take on different *values* for different particles. Once the situated character type is known, we can fill in the specific condition of each particle and obtain a distribution of character values for the particles in the system. By using standard statistical techniques, the distribution can be used in aggregation to yield the desired macroscopic characters of solids.

We will start from the crudest model, which retains the familiar character of free electrons represented by the quadratic energy–wavevector relation $E(\mathbf{k})$. More refined models continue to characterize electrons by their individual energy–wavevector relation $E(\mathbf{k})$, but the relation no longer has the familiar quadratic form. The form of $E(\mathbf{k})$ varies for various kinds of solid, and for the same solid it varies from model to model, depending on what effects of the electrostatic interaction are absorbed into the description of individual electrons.

The Free Electron Model of Conduction[2]

In the crudest approximation of conduction electrons, called the *free-electron model*, the coupling among electrons is simply dropped. We can now consider a representative electron moving in a world populated by the heavy ions that are fixed in their positions. The model further neglects the coupling between the electron and the ions unless they actually collide. Between collisions, the electron is described by the free-electron energy–wavevector relation. It is like a ball in a pinball machine, moving freely and feeling no influence from anything unless it bumps into an ion. When it does, it changes course and moves on free as before. An external electric potential applied to the solid causes the electron to accelerate generally in one direction, similarly to the gravity that causes the ball in the pinball machine to roll downhill generally. The path of the electron is disrupted by the collisions with ions, just as the path of the ball is disrupted by the pins. Despite the disruption, we can estimate the average time interval between collisions and hence the average velocity of the electron.

An electron carries electric charge, and its movement through the solid generates a tiny electric current. The total current through the solid under the applied potential is the sum of the contribution from individual electrons. More current is generated when the electrons suffer fewer collisions and move faster. From the current we can derive the electric conductivity of the solid. Similar reasoning leads to the coefficient of thermal conductivity. This simple model of conduction electrons gives surprisingly good qualitative accounts for many phenomena and the right ballpark numbers for the coefficients of electric and thermal conductivity.

The free-electron model fails in many other aspects. For instance, it cannot explain why some solids are metals and others insulators. Several faulty assumptions underlie the inadequacies of the model. I mention only two. First, the assumption of a pinball-machine environment in which electrons are free except for occasional collisions with ions is wrong. The ionic lattice constitutes an all-pervasive environment that alters the overall behavior of the electrons, not just during collision. Second, the interaction among electrons cannot be neglected generally. Metals contain many conduction electrons and are more like subway stations during rush hours than empty halls. Under such crowded conditions the behaviors of electrons are greatly modified by their mutual electrostatic coupling. We turn to see how these assumptions are ratified in better independent-particle models.

Electrons in the Periodic Potential of a Crystal Lattice[3]

We are surrounded by solids: wood, fabric, plastic, organic bodies. Solid-state physics concentrates on crystalline solids, in which the ions arrange themselves in a *lattice*, or an array of almost perfect regularity. Despite their macroscopic malleability, metals are microscopically as crystalline as diamond and salt. Different kinds of solid have different kinds of lattice structure. Physicists cannot derive the lattice structures from the basic equations of quantum mechanics. However, the configurations of the lattices have been thoroughly studied by X-ray diffraction experiments, the first of which was performed in 1912. Theoreticians simply plunk the experimentally determined lattice configurations into their equations.

The ions in the crystal lattice are positively charged and their electrostatic attraction for the negatively charged electron drops with increasing distance. An electron experiences a greater electrostatic potential when it is near a lattice site where an ion sits than when it is away from a lattice site. Since the lattice sites are regularly spaced, the spatial variation of the electrostatic potential seen by the electron repeats itself periodically. Thus the effect of all the ions constituting the lattice can be represented by a *periodic potential* capturing the essential features of the lattice configuration. The form of the periodic potential is peculiar to the kind of solid, as the lattice configuration varies in different solids. For a specific kind of solid, we introduce an effective Hamiltonian in which the periodic potential replaces the electron–ion interaction. The periodic potential is the situation shared by all electrons. It profoundly alters the electrons' characters.

Many theories for the electronic contribution to the electrical and optical properties of crystalline solids have the following steps: First, electron–electron interaction is neglected. Second, the electron–ion interaction is replaced by a periodic potential in which the electron moves. Third, the electronic equation of motion is solved to yield the situated character type of the electron, called its energy-band structure, which encompasses the possible states of an electron in the specific kind of crystal. Fourth, the distribution of actual electronic states is obtained by accounting for how the possible states are filled by the electrons in the solid. Fifth, the distribution is used to calculate macroscopic quantities such as the conductivity or the optical absorption due to electrons.

The first two steps sever all interparticle relations: electron–electron interaction is neglected and electron–ion interaction replaced by the response of an electron to a background periodic potential. Consequently we can write an equation of motion for a single electron using an effective Hamiltonian featuring the periodic potential. The single-electron equation is still very difficult to solve. Physicists have developed many mathematic methods and approximations to get solutions and many experimental techniques to check the results. We do not go into them but will examine some crude qualitative features of the results.

Figure 4.2a shows the energy–wavevector relation $E(\mathbf{k})$ of a free electron. Figure 4.2b shows the energy–wavevector relation $E^*(\mathbf{k})$ of an electron in silicon. The curves in the figures represent all possible electronic states. Each point in a curve represents two possible states, which have the same energy and wavevector values but different spin orientations. In silicon, there are several possible energy values corresponding to each wavevector value. Each curve is called an *energy band* and $E^*(\mathbf{k})$ the *energy-band structure* of an electron in silicon.

The $E^*(\mathbf{k})$ relation of silicon has a novel feature absent in the free-electron relation $E(\mathbf{k})$. The energy E no longer varies continuously with the wavevector $\mathbf{k}$; there is a pronounced *energy gap*, or a range of energy values that cannot be realized by any electron. The energy bands below the gap are called *valence bands*, those above the gap *conduction bands*. The energy-band structures of various solids differ greatly in details, but they are all messy like that of silicon, and many exhibit energy gaps. The complicated band structures provide more opportunity for us to manipulate electrons. There would be few electronic devices if the band structures of solids are as simple as the free electron $E(\mathbf{k})$. Some important regions of $E^*(\mathbf{k})$ can be approximated in a quadratic form similar to that for free electrons except for a different mass, called the *effective mass m^**, whose values for various kinds of solid are well established and available in tables.

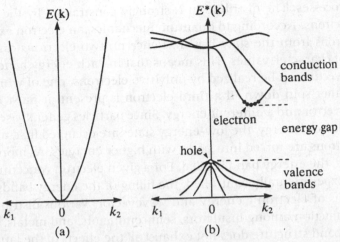

FIGURE 4.2 (a) The energy–wavevector relation $E(\mathbf{k})$ of a free electron. (b) The energy–wavevector relation or energy-band structure $E^*(\mathbf{k})$ of an electron in silicon in two crystal orientations. (Unlike $E(\mathbf{k})$, $E^*(\mathbf{k})$ is different for different directions of $\mathbf{k}$.) Each curve is an energy band, a point of which represents a possible state of the electron. The energy-band structure contains an energy gap with no allowed state. The bands above the gap are called *conduction bands* and those below *valence bands*. Some possible states in the conduction bands are realized by actual electrons, depicted by dots. In most solids, the states in the valence bands are almost all occupied. The few vacancies are called *holes*, depicted by the open circle. The energy gap of silicon is 1.1 eV. Since it is much larger than the thermal energy at room temperature, which is 0.02 eV, electrons in the valence band cannot be thermally excited to the conduction band. The actual concentration of electrons or holes in individual samples can be controlled by introducing impurities that donate or accept electrons. These electrons or holes are responsible for the semiconducting properties of silicon.

The energy-band structure $E^*(\mathbf{k})$ and effective mass m^* represent the *situated character types* of electrons in a specific kind of crystal, in distinction to the character types of free electrons represented by $E(\mathbf{k})$ and m. The situated character types assimilate most of the information of the periodic potential, or of the influence of the ionic lattice on the electron. Given the energy-band structure of electrons in a certain kind of crystal, we can describe their behaviors without mentioning the lattice structure. When engineers design an electronic device using silicon, they treat the electrons as free entities with characters $E^*(\mathbf{k})$ or m^*. Mass is usually regarded as the most basic of the monadic characters, yet the effective mass shows that it can hide much relation. Effective mass is not alone in this respect; we know from the renormalization procedure in quantum field theories that even the free electron mass hides relations (see note 6 in Ch. 2).

The energy-band structure gives the possible states of a crystalline electron and enables us to predict its behavior under specific conditions. To find the macroscopic characters of the solid, we need to sum the behaviors of many electrons. The aggregation requires the knowledge of how the states of actual electrons are distributed, for electrons in different states contribute differently to dynamic processes. The distribution is strongly constrained by the exclusiveness of electrons. According to quantum mechanics, an electron excludes all other electrons from the state it is in, hence no two electrons can have the same set of character values. This means that in each energy band, each value of wavevector can be realized by only two electrons, one of which has spin up, the other spin down. If a third electron is present, it must have a different wavevector and a different energy. Since particles generally seek the state with the lowest energy, the low-energy states are occupied first, and the additional electrons are forced into states with higher energies. As more electrons are added, the energy band is filled. For a given electron concentration, which can be experimentally controlled, the filling of the energy bands gives the distributions of electronic energy and wavevector. Various distributions explain the distinction among insulators, semiconductors, and metals.[4]

The energy-band structure does not exhaust all the effects of the lattice on the electrons. If it did, the solid would have infinite conductivity, because there would be nothing to scatter electrons and prevent them from accelerating to infinite velocity. Conductivity is not infinite, because the crystal lattice is neither perfect nor as rigid as it is represented by the periodic potential. The lattice vibrates at finite temperatures, generating entities called phonons (§ 19). Collisions with phonons, impurities, and dislocations in crystals impede the movement of electrons and produce the finite values of conductivity for solids.

Compare this more sophisticated model of conductivity with the free-electron model discussed earlier. In the free-electron model, conductivity is determined by electrons with familiar characters colliding with ions. In the sophisticated model, electronic motion and collision are still the essential factors in conductivity. However, the electrons involved have situated characters that have taken into account most effects of the ionic lattice, and the scatterers are not ions but phonons, which are modes of motion of the entire lattice. The two models invoke two sets of individuals to account for conductivity. In the sophisticated model, the individuals, situated electrons and phonons, are specialized to the kind of solid under study.

Self-consistent Field Theory for Electron–electron Interaction[5]

Physicists often call the system of crystalline electrons an "electron gas," because the interaction among the electrons is neglected. However, the electron density in a typical metal is thousands of times greater than the density of gases at normal temperatures and pressures. More accurate descriptions of electronic structures of solids take account of the interaction among electrons. The treatment of electron–electron interaction is somewhat similar to that of electron–ion interaction; the electrons are again treated as independent individuals in an effective potential. There is, however, a big difference.

The transition from an electron in an ionic lattice to an electron in an electron sea is like the transition from despotism to democracy. Ions are thousands of times heavier than electrons and they are tightly bound to the lattice sites. To an electron, the periodic potential representing the ions is an externally imposed situation that influences but is not influenced. In an electron sea, the electron is among peers. If the effect of the electron sea is treated as a situation analogous to the periodic potential, then the situation can only be the consensual result of electronic behaviors. Otherwise the electrons will not behave according to it. Consequently the situated character of the independent electron and the effective potential representing the electron sea must be determined self-consistently with each other.

A general method for treating quantum mechanical systems of interacting peers is the *Hartree–Fock approximation*, also called the *self-consistent field theory*. It is an independent-particle approximation; the "field" is the effective potential in which the single particles move. For classical systems, similar ideas are embodied in the *mean field theory* of statistical mechanics and the *molecular field* or the *Weiss field theory* in magnetism.

Developed shortly after the introduction of quantum mechanics in 1925, the Hartree–Fock approximation leaves out many intricate effects, but it is used in many branches of quantum physics. For instance, in some atomic theories, the electrons in an atom are treated as independent particles moving in an average potential determined self-consistently from the motions of all the other electrons and the nucleus. Anderson said: "One may get the impression that modern many-body theory, of which one hears so much, goes far beyond Hartree–Fock, and that therefore we should not bother with such old-fashioned stuff. This is not at all true – in fact, modern many-body theory has mostly just served to show us how, where, and when to use Hartree-Fock theory and how flexible and useful a technique it can be."[6]

Consider a system of N electrons in a uniform background of positive charge that maintains charge neutrality. In the basic quantum mechanical formulation, the N-electron system is represented as a single unit by an N-electron wavefunction that depends on the coordinates of all N electrons. The equation of motion for the system contains a term for electron–electron interaction. The Hartree–Fock approximation makes two major changes. It factorizes the N-electron wavefunction into the product of single-electron wavefunctions; the factorization severs the quantum-mechanical correlation among the electrons. It then replaces the electron–electron interaction term in the equation of motion by an effective potential with the form of an external potential that acts on the electrons individually. With these changes, the equation governing the motion of the N-electron system becomes an equation governing the motion of a single electron under the effective potential.

In the reformulated many-electron problem, the only influence an electron has on the other electrons is through its contribution to the effective potential. In return, it is freed from the myriad forces exerted on it by the others and is influenced only by the effective potential. The effective potential felt by an electron depends only on its own coordinates, for the contributions of all other electrons are averaged over. Since the same consideration applies to all electrons, the effective potential serves as a common situation whose effect varies systematically according to the positions of individual electrons.

If the effective potential is given, the character type of electrons can be obtained by solving the Schrödinger equation. From the character type we can perform statistical aggregation to determine the macroscopic behavior of the electron system. The procedure is similar to the case of crystalline electrons.

The new problem in the self-consistent field theory is to find the effective potential. There are several methods. Iteration, the most practical method, is based on the rationale that the potential is itself a system character determined by aggregating electronic behaviors. We start by guessing an effective potential, put it into the single-electron equation of motion, and solve for the electron wavefunction. We then calculate the effective potential from the wavefunction. The potential so obtained is put back into the single-electron equation, which is solved again to yield a second effective potential. The procedure continues until the potential obtained from the solution of the equation is not substantially different from the potential put into the equation. The effective potential that appears in the single-electron equation constrains the behavior of the electron; the potential calculated from the equation is constituted by all the electrons. The agreement of the two ensures that the electrons spontaneously conform to the situation they together constitute. The effective potential resulting from the iterative process is called the *self-consistent field*.

The situated character type of electrons in the self-consistent field of an N-electron system can again be expressed in terms of an electron energy–wavevector relation $E'(\mathbf{k})$ which, as expected, differs substantially from the $E(\mathbf{k})$ of free electrons. Interaction generally lowers the electronic energy. An electron in an electron sea has lower energy than a free electron with the same wavevector. Consequently, the total energy of a system of N interacting electrons is significantly lower than the sum of the energies of N free electrons.

Quasi Particles and the Principle of Simplicity[7]

The electrostatic interaction among electrons is strong and drops off slowly with increasing distance. For electronic densities typical of metals, the range of the interaction covers many electrons. Why is the independent-electron approximation so successful under such conditions? Part of the answer is that electrons in a crowd screen themselves from interaction and the screening modifies the electrostatic interaction itself.

Imagine an extra positively charged test particle in a sea of electrons. The particle attracts electrons and creates around itself a surplus negative charge, which screens it so that it cannot be "seen" by electrons farther away, thus neutralizing its effect on distant electrons. Screening effectively reduces the strength and range of the electrostatic potential. Similarly, each electron in the electron sea repels other electrons, creates an excess of positive charge around it, and screens itself from the others. A greater electron density produces a stronger screening effect and a greater reduction in the strength and range of the electrostatic coupling. The electron protects its privacy in the crowd and reduces correlation effects, not unlike people who draw into themselves in crowded elevators. This partly explains the success of the independent-electron approximation.

The screened electron in an electron sea is an example of a *quasi particle*, which is a particle plus a certain amount of interactive effect. As a particle moves in an interactive medium, it pushes some particles out of the way and drags others along. The particle and the transient cloud of vacancies and companions it commandeers are integrated into the quasi particle. Unlike the particle that exhibits all its relations, the quasi particle is dressed up and hides most relations under its cloak. The cloud around the particle screens it from further interaction, just as dresses create social distance. Hence the quasi particles are only weakly interacting, and their coupling is neglected in the independent-particle approximation. Since the quasi particle incorporates the cloud, its energy–wavevector relation is quite different from that of the particle itself.

The relation between particles and quasi particles can be explained by what Anderson called the *principle of adiabatic continuity*, which was first developed by Lev Landau. Suppose we intellectually turn off the interaction in a system of interacting particles to get an ideal system of noninteracting particles. We solve for the states of the noninteracting particles. Again intellectually, we turn on the interaction slowly, so that at each instance the system is in equilibrium and the state of the system changes little. By this process we can establish a one-to-one correspondence between the states of the interacting system and the states of its ideal noninteracting counterpart. The states in the noninteracting system belong to *particles*; the states in the interacting system to *quasi particles*, whatever they are. A quasi particle is labeled by a set of quantum numbers, and it corresponds to the particle labeled by the same set of quantum numbers. For instance, the independent electron with wavevector $\mathbf{k}$ corresponds to the free electron with the same $\mathbf{k}$. Quasi particles have substantive characters different from that of the particles, but they inherit the particles' general nature of *not* interacting with each other. Anderson has said the principle of adiabatic continuity, which establishes the correspondence of noninteracting particles and quasi particles, is a basic notion in condensed-matter physics. It "shows that in a very deep and complete sense we can refer back whole classes of problems involving interacting fermions to far simpler and fewer problems involving only noninteracting particles."[8]

The principle of adiabatic continuity illustrates a general research strategy. Noninteracting particles can be described in the simplest form of proposition and their characters aggregated easily. The individuality and simplicity are both strained when interaction is turned on, if the substantive characters of the noninteracting particles continue to dominate. Physicists choose to follow the *general form of thought* with its concept of noninteracting individuals at the cost of abandoning the *substantive character types* of the noninteracting particles. When the interaction among ideal noninteracting particles is turned on, they introduce new characters and new entities that absorb the effect of interactions, thus preserving their general solipsist nature. They continually microanalyze the system, seeking not simple parts but *parts that conform to a simple form of thought*. This strategy goes beyond the models considered in this section, where the quasi particles are still closely linked to the particles. Later we will see that totally new kinds of entity such as phonons and plasmons are introduced because they are almost noninteracting with each other.

Guiding principles in research are neither binding nor universal. The power of independent-individual approximations has limited range, beyond which other forms of thought must take over to account for the effects of interactions. Anderson concluded his lectures on single-particle theories in solid-state physics as follows: "Everything I said during the first term of this course was based on an approximation – an approximation, in fact, which we can easily convince ourselves is a very rough one in most cases. That was the approximation that the electrons in the solid move independently of each other, and of the motions, which we did not even discuss, of the heavy ion cores. At the very best, we have taken into account the interaction of the electrons ... in terms only of its average, i.e., of Hartree–Fock theory. The electrons and ions in a solid do, in fact, interact very strongly, and in what would be expected to be a rapidly fluctuating way."[9] Some effects of interaction that are missed in independent-particle models will be discussed in §§ 19 and 23.

16.   Price Takers in the Perfectly Competitive Market

Theoretical economics does not feature the term "independent-individual approximation." I argue in this section, however, that the general equilibrium theory for perfectly competitive markets, what many call the "keystone of modern neoclassical economics,"[10] is a sophisticated independent-individual model. The signature of the independent-individual approximation is already visible in the definition of households and firms that make up the economy's microstate. As we saw in § 13, economic units are individually defined in terms of commodity prices, which determine the households's budgets and the firms's profits. According to Kenneth Arrow, "The key points in the definition are the parametric role of the prices for each individual and the identity of prices for all individuals."[11] The distinguishing feature of perfect competition is the identity of prices for all individuals without haggling and bargaining, so that none has power to influence price and all are passive price takers. Thus the commodity prices represent the common *situation* analogous to the self-consistent field in physics. Individual states defined in terms of prices are *situated characters* that absorb complicated commercial relations and render economic individuals independent of each other and responsive only to the prices.

The first complete and rigorous general equilibrium models for perfectly competitive markets were developed in the 1950s by Kenneth Arrow, Gerard Debreu, and Lionel McKenzie. These models synthesize most important microeconomic ideas in a rigorous mathematical framework and lay out in exact terms the assumptions underlying the feasibility of a totally decentralized or perfectly competitive economy. With results such as the Invisible Hand Theorem, they substantiate the belief held since Adam Smith that an economy driven by self-interest and guided by the invisible hand of the market can achieve a resource allocation that is more efficient and more satisfying to the participants than other alternatives, *if* certain ideal conditions are met. Thus perfectly competitive market models fit the capitalist ideology well.

The ideal nature of models for perfectly competitive markets is generally acknowledged, but its independent-individual approximation is not. One reason is the lack of a clear and general formulation of interacting economic individuals analogous to the interacting particles in physics, against whose background the approximate nature of independent individuals is unmistakable. The first precise formulation of economic units is the one parameterized by commodity prices, which has already made them independent individuals customized to perfectly competitive markets. Explicit economic interactions represented in game theory come later and usually address small sectors of the economy (§ 20). Consequently the independent individual defined in terms of prices is sometimes regarded as the most "natural" and is then turned around to justify the assumption of the perfectly competitive market.

The following presents a rough outline of a general equilibrium model for perfect competition. Then its picture of the economy is compared to that in the theory of exchange and our commonsense understanding to illustrate its independent-individual approximation.

A Model for Perfectly Competitive Markets without Risk[12]

Microeconomics depicts the *microstates* of the economy. A microstate, called an *allocation*, specifies the state of each constituent of the economy. The constituents are households and firms, whose states are discussed in § 13, where we have decoupled them from the market by holding the commodity prices fixed. Now we will integrate them into the economy by making the prices variables to be determined endogenously by the equilibration of demand and supply.

The states of the households and firms are represented in the two basic variables of the theory: the prices and quantities of various commodities. *Commodities* are infinitely divisible, so they can be traded and consumed in any quantity. They include only private goods and services that can be exclusively consumed. They are classified according to three respects: their physical attributes, spatial locations, and availability at different times, with both space and time divided into finite intervals. Commodities with the same attribute but available at different locations and times are regarded as different kinds and fetch different prices. Thus commodity prices account for the costs of transportation and risk bearing. Except for the risk implicit in future prices, the model contains no factor of risk or uncertainty.

One of the commodities is chosen as a numéraire. The *price* of a commodity is the units of the numéraire that are exchanged for one unit of it. If gold is chosen as the numéraire and ounce the unit, and if a car exchanges for 100 ounces of gold, then the price of the car is 100. Like the exchange values of commodities, prices are relative. Money, as the medium of exchange and the store of value, is a macroeconomic concept that finds no place in microeconomic price theories. Thus microeconomics can be considered as a theory of barter economies minus the restriction of the double coincidence of wants. The restriction, which requires the meeting of two individuals each in possession of a commodity the other wants, is alleviated by the intervention of the market.

A commodity's price is determined by its demand and supply, which are generated by households and firms. The prices can be either positive or zero. Zero prices obtain only for "free goods," goods that are in excess supply, for instance, the air we breathe and the waste no one wants.[13] The prices of all commodities offered in the economy are summarized in a price vector, which specifies the price of each commodity, with the price of the numéraire equal to 1.

Commodities are grouped into *bundles*, each containing a certain quantity of each kind of commodity. Households do not evaluate commodities separately: They evaluate commodity bundles, for the utility of a commodity often depends on the possession of other commodities; chocolate loses its appeal when one is filled with ice cream. When we say we prefer one thing to another, we implicitly assume that everything else is equal. The implicit assumption is made explicit in taking the commodity bundle and not the individual commodity as the variable.

A *household* has two exogenously determined characters: its taste and endowment. The taste is represented by its preference order of commodity bundles, which is in turn represented by a utility function. Taste never changes, so that the preference order is fixed. A household's endowment includes its leisure, talent, labor, and wealth, which it can retain or exchange for other commodities. The household also receives a dividend from some firms. Its budget is the dividend plus the total price of its endowment. The household maximizes its utility under its budget constraint; it tries to get the affordable commodity bundle it finds most satisfying. It is insatiable; no matter how much it gets, it always craves more.

A *firm* is characterized by a production possibility set or a production function, which circumscribes the quantities of commodities it can produce from certain quantities of other commodities and raw material. Production possibilities are constrained by the firm's technological capability and knowledge, which are exogenously given. A firm's production plan specifies the "net-put" of each commodity, or the amount of the commodity it produces minus the amount it uses up in production. Its profit is the total price of all the netputs. The firm maximizes its profit under its technological constraint; it determines the plan of transforming input commodities to output commodities that yields the most profit. Perfectly competitive market models *exclude increasing returns to scale*, which means a larger scale of production is more cost effective. The exclusion precludes the emergence of natural monopolists. It also means that the models are not valid for actual technologies that do favor large-scale production.

All individuals are assumed to have perfect knowledge of the qualities and prices of all available commodities, present and future. Such knowledge comes without cost or effort. However, an individual has no knowledge about the tastes, the production plans, or even the existence of other individuals. There is no direct interference of any kind among the behaviors of the individuals. The independence of the individuals excludes the problem of so-called externalities such as pollution. Models for perfectly competitive markets do not consider problems such as a fishery's stocks being killed by the chemicals discharged by a plant upstream, for the problem involves the direct relation between the production plans of two firms.

A *market* exists for a commodity if the commodity can be transferred from any individual to any other without cost. A market *clears* when, at a certain price, the demand and supply of its commodity are in balance. The price that clears the market is called the *equilibrium price* of the commodity. General equilibrium models posit the *existence* and *clearance of markets for all commodities* available in the present and future. Whatever odd trade the individuals desire, a market must exist for it and the market must clear without friction. This strong assumption rules out the possibility of voluntary unemployment, which means that the labor market does not clear.

Equilibrium takes on two important senses in economics. The broader sense refers to an unchanging state of an economy as discussed in § 9. In the narrower sense, equilibrium prevails when a market clears, and general equilibrium prevails when interconnected markets simultaneously clear. The two senses are interrelated. The economy settles into an unchanging state because the clearance of markets leaves every individual maximally satisfied, so that no one has further incentive for change. Disequilibrium theories relax the notion of market clearing but retain the idea of an unchanging state.

There are various types of equilibrium theory. A *partial equilibrium theory* analyzes a specific market or a particular part of the economy, taking the influence of the rest of the economy as constant. It studies, for example, how the price of oil varies with the prices of all other commodities fixed. Partial equilibrium models are practical in the analysis of realistic economic cases. A *general equilibrium theory* treats the prices and quantities of all commodities as variable to account for the interdependence of markets and the reconciliation of all participants in the economy. The only factors it assumes to be constant are individual preferences, technologies, and gross resources such as population.

Equilibrium models forbid trades unless they occur at the market-clearing prices. Since all individuals have perfect foresight and immutable character, all trades, including those for commodities to be delivered in various periods, can be agreed upon in a one-shot deal at the beginning of time. No further trade is needed because each individual has its optimal choice. Everybody lives happily thereafter, delivering and receiving goods.

An economy is *perfectly competitive* if all individuals are passive price takers that cannot influence prices. There is no limit to the amount an individual can sell, but no matter how much it sells, it does not influence the price. Thus commodity prices are explicitly independent of the behavior of any individual. Monopolists and oligopolists, which can influence prices by limiting production and other methods, are market imperfections disallowed in the perfectly competitive market.[14]

The commodities are the resource of the economy. The resource is allocated through the transformation of commodities in production and the transfer of ownership in exchange. A *resource allocation* is a microstate of the economy specifying the state of each of its constituents, namely, the commodity bundle owned by each household and the production plan adopted by each firm.

A *competitive equilibrium* is a resource allocation in which all markets clear. The states of the households and firms are parameterized by the commodity prices. In competitive equilibrium, a set of equilibrium prices obtains such that (1) each household gets the commodity bundle that maximizes its utility under its budget constraint; (2) each firm engages in the production plan that maximizes its profit under its technological constraint; (3) there is no excess demand: that is, the sum of the optimal bundles for all households is less than or equal to the sum of the initial endowments of all households plus the sum of the optimal netputs from all firms.

The major result of the general equilibrium model for perfectly competitive markets is an existence theorem proving that *under the listed assumptions, competitive equilibria exist for any arbitrary set of initial endowments.* Competitive equilibria are often not unique. The theorem asserts that some competitive equilibria exist but does not tell what they are. It is a separate job to determine the equilibrium prices and the allocation for a specific set of preferences and technologies.

The *Invisible Hand Theorem,* an extension of the perfectly competitive market models, asserts that all competitive equilibria are Pareto-efficient. The theorem is based on the assumption that the individuals are totally self-interested and do not cooperate. Thus it does not rule out the possibility that a Pareto-optimal allocation can be surpassed if the individuals cooperate to achieve a world in which everyone is better off.[15]

An allocation is *Pareto-efficient* or *Pareto-optimal* if any alteration will make at least one individual in the economy worse off; there is no alternative allocation that is preferred by all individuals to the Pareto-optimal allocation, in which each attains its optimal choice. There may be many Pareto-optimal allocations for an economy. An economy in which a single individual owns all available resources is Pareto-optimal because any change in allocation will make the sole owner worse off. Pareto optimality demands that any change for the better must be unanimously approved. As Samuelson remarked, the demand of unanimity tends to immobilize change and maintain the status quo.

Robert Solow said: "The general educated and interested public thinks that economics has 'proved' that 'the free market is efficient,' perhaps even the best of all possible worlds." He pointed to the many assumptions on which the Invisible Hand Theorem rests – omniscient consumers with consistent and immutable preferences, technologies that rule out increasing returns to scale, the existence of markets for all commodities, zero cost for business transaction and information dissemination – and continued: "Not one reader in a thousand of the *Wall Street Journal* has any grasp of the qualifications without which the theorem, as a theorem, is simply false. That would hardly matter if the necessary assumptions were self-evidently true or even genuinely plausible to any open-minded observer. They are neither. Nor do even sophisticated economists have any real grasp of the robustness of the theorem: I do not know of any convincing story, either way, about whether an economy that differs from the abstract economy about as much as ours does comes even close to satisfying the conclusion of the Invisible Hand Theorem."[16]

Monads in the Market-established Harmony

Besides the explicit assumptions underlying the models for perfectly competitive markets, which Solow mentioned, there is a deeper assumption that is even more easily conflated with the ideology of pure self-interest: the independent-individual approximation. Conspicuously absent in the models is any direct interaction among the individuals. A household is neither envious nor benevolent; its taste is preordained and unaffected by what its neighbors possess. A firm neither advertises its product nor takes interest in other firms' products and management; it can always sell what it produces and its technology is fixed. Each economic individual behaves as if the others do not exist; all it sees is the world of commodities with price tags. It is a Leibnizian monad in a harmony maintained by the equilibrium market prices. It has no window through which it can communicate with the others, but it has a television screen through which it acquires perfect knowledge of the commercial cosmos characterized by the list of commodities and the going prices. The knowledge enables it to change its state to suit its peculiar taste. The possibility of a harmony among the states of different individuals is guaranteed by a mathematical theorem. If God, who knows the dispositions of his creatures, endowed each with its optimal commodity bundle and showed the equilibrium prices on the screen, he would be praised for having created the best of all possible worlds in the sense of Pareto optimality.

This picture is far from reality, in which economic units bargain about prices; exchange not only commodities but also ideas that affect their preferences and actions; rival and imitate each other; form coalitions if necessary to gain an edge in realistic competition. Perhaps some of these interactions are peripheral to economics, but surely not to commercial relations. In the perfect-market models, however, individuals never meet to negotiate and trade with each other; they merely respond to the commodity prices. At the equilibrium prices, the initial endowments of the households are miraculously transformed into their optimal bundles. Oliver Hart and Bengt Holmström describe it as "the impersonal Arrow–Debreu market setting where people make trades 'with the market.'"[17]

Perfectly competitive market models, sometimes called price theories, are similar to the self-consistent field models in physics; both belong to self-consistent independent-individual models. Electrons interact, but single-particle models transform them into noninteracting particles that respond independently to an average potential. Economic individuals interact, but perfectly competitive market models depict them as noninteracting individuals that respond independently to commodity prices in the market. An electron responds passively to the potential, which appears in the equation that governs its motion. An economic unit is passive in the competitive market and powerless to influence prices. The average potential is produced by all electrons and determined self-consistently with their behaviors; the noninteracting electrons so behave in the potential that a proper aggregation of their independent behaviors yields the potential. Commodity prices are generated by all economic units and equilibrium prices are determined self-consistently with their individual optimal states; for the equilibrium price of a commodity, the aggregate demand of all individuals in their respective optimal states equals the aggregate supply, and the equilibration of demand and supply determines the equilibrium price.

To substantiate my interpretation of the perfectly competitive market models as independent-individual models, I will discuss three points: the theory of exchange that relates competitive equilibrium to bargain and trade, the alleged process of price adjustment that exemplifies the independence of economic individuals, and the incomparability of utilities for different individuals.

Theory of Exchange[18]

Trade does not enter explicitly in perfectly competitive market models. To find its connection with the models let us examine how the resource allocations of competitive equilibria can be obtained in the *theory of exchange* based on game theory. We begin with the Edgeworth model of exchange, in which two households, holding their endowments, meet, bargain, and finally barter when their marginal rates of substitution of commodities are equal. Since each has several commodity bundles that it equally prefers, there is a whole set of perfect trading results that would maximally satisfy both parties. The game theoretic generalization of the Edgeworth model shows that as the number of households engaging in direct exchange increases indefinitely, the set of perfect trading results shrinks and converges to the competitive allocation of general equilibrium models. Thus in an infinitely large and perfectly competitive market, any two passive price takers end up in one of the many mutually satisfying results that they can achieve by haggling and bargaining with each other.[19]

The individuals in both exchange and general equilibrium models maximize their satisfaction. However, their theoretical representations are different. In exchange models, individuals interact directly with each other. The theory of exchange illustrates that individuals defined in terms of commodity prices are not the only ones capable of achieving optimal resource allocations.

The Auctioneer and the Tâtonnement Process[20]

General equilibrium models describe the terms in which the desires of all economic individuals are compatible, but not how the individuals come to agreement. Economists say that individual behaviors are coordinated through the price mechanism of the market. However, their models show only the prices, not the mechanism. As Arrow remarked, since all parties in the competitive market respond passively to the prices, how prices settle on their equilibrium values becomes a mystery.

In attempting to answer how prices are determined, Walras introduced a fictitious auctioneer that calls out prices; collects bids and offers; adds them up; computes the excess demand for each commodity, raises its price if demand is brisk, and lowers it if demand is slack; and announces a new set of prices. The process of bidding, computation, and price change is repeated until equilibrium prices are hit upon such that excess demands for all commodities vanish. Then everyone breathes a sigh of relief. The process is called *tâtonnement*, meaning "groping" in French. It does not have the same status as the existence theorem for equilibrium.

The tâtonnement process highlights the solipsistic nature of economic individuals. Households and firms do not deal with each other but submit their bids to the auctioneer, the personification of the market, and wait for price announcements. When the equilibrium prices are reached, sellers toss their supply in a pile and buyers take what they demand, trading not with each other but with the market. They are not responsive to each other; their only responses are to the price signals of the market.

Economists acknowledge that the auctioneer is a myth. Perhaps the iterative procedure in the tâtonnement process is better interpreted not as some market mechanism but as merely an analytic tool by which economists compute the equilibrium prices for specific problems. In this interpretation, it is logically similar to the iterative procedure by which physicists determine the self-consistent potential in the Hartree–Fock approximation. As to the question of how equilibrium prices are arrived at, it belongs to dynamic theories that require more concepts than equilibrium theories.

The Incomparability of the Utilities of Different Individuals

A peculiarity of economic individuals as they are defined in perfectly competitive market models is that the utilities of different individuals cannot be compared. Many economists assume that utilities cannot be compared because they represent people's subjective tastes. I think that the incomparability stems not from subjectivity but from the theoretical formulation of economic individuals that excludes interaction or social convention. A similar formulation of individuals in physics leads to similar incomparability of individual characteristics. Comparison is based on relations, and similarity and difference are relations. Thus it is not surprising that when the unrelatedness of individuals is formulated sharply, their characteristics cannot be compared.

The utility function is a definite predicate of the household, and definite predicates always harbor an arbitrary element, which is revealed in precise mathematical definitions. In the precise definition of a plurality of unrelated individuals, be they persons or things, the arbitrary element in predication varies from individual to individual and destroys the basis for comparison. A similar problem occurs in physics, where a similar definition of individual characters leads to a similar result in interacting field theories such as general relativity. To overcome the problem, physicists introduce the means of comparison either by including physical interaction among the individuals, as in interaction field theories, or by using conventions such as global coordinate systems, as in many-body theories.[21] Interaction is absent in the independent-individual assumption of general equilibrium models, as are all social conventions except the commodity price. Consequently the states of different individuals, that is, their individual utility values, cannot be compared. However, this does not imply the absolute impossibility of interindividual utility comparison. Utility may become comparable in models that include social intercourse just as individual characteristics become comparable in theories that include physical interaction.

The rejection of interindividual utility comparison has enormous consequences in welfare and political theories. It almost forces us to rely on Pareto optimality, which considers the condition of each individual but does not require any comparison of well-being or aggregation of utility. Pareto optimality, however, is inadequate for considering such problems as income distribution, economic equality, and social welfare, which inevitably require some kind of utility comparison.[22]

The Limitation of Independent Individuals

Models for perfectly competitive markets are powerful and fruitful, but they are not adequate for the description of many intricate phenomena in modern economies. Systems for which the models hold are quite homogeneous and lack gross patterns and structures and therefore may not be realistic. Important sectors of today's economy are oligopolies, where a few large firms set prices and compete directly and strategically with each other. The farming industry is often cited as a real-life example of perfect competition. However, in many countries, including the United States, the farming industry depends heavily on government intervention such as price support and acreage restriction. Where institutional organizations are important, individuals cannot be assumed to be independent.

The general equilibrium models make extensive use of concepts in topology, notably convexity, separating hyperplanes, and the fixed-point theorem. Compared to their elegance, the mathematics in condensed-matter physics appears wooly. They demonstrate the power of mathematics in the clarification and sharpening of concepts, which contribute greatly to scientific investigation. On the other hand, they also demonstrate the tradeoff between rigor and realism. Of the mathematical elegance of the microeconomic model, Samuelson commented: "But the fine garments sometimes achieved fit only by chopping off some real arms and legs. The theory of cones, polyhedra, and convex sets made possible 'elementary' theorems and lemmas. But they seduced economists away from the phenomena of increasing returns to scale and nonconvex technology that lie at the heart of oligopoly problems and many real-world maximizing assignments."[23]

17. Fitness, Adaptedness, and the Environment

Independent-individual approximations play many important roles in population genetics, for example, in neglecting the interaction among organisms, or in sweeping the effects of organismic interaction into parameters such as "inclusive fitness." Here we consider a particularly pervasive and controversial application, the factorization of organisms into "bundles" of independent characters.

As Lewontin said, most evolutionary models have several hidden assumptions. The first is "the partitioning of organisms into traits." The second assumes that "characters can be isolated in an adaptive analysis and that although there may be interaction among characters, these interactions are secondary and represent constraints on the adaptation of each character separately."[24] These assumptions are forms of independent-individual approximations whose "individuals" are organismic characters. The approximations in evolutionary models are more drastic than those in physics and economics. There is no attempt to modify the characters to internalize interactions among the characters. Constraints on adaptation, which dimly echo the organismic situation in which individual characters function, are sloppily handled, with no regard for self-consistency or the possibility of conflict. Nevertheless, these approximate models have acquired fanciful interpretations that are widely popularized and as widely criticized. We will examine the assumptions behind the approximations, their ramifications, and the controversies they engender.

Factorizing the Organism

As discussed in § 12, the evolutionary dynamics of a species is represented by a distribution over the genotype and phenotype state spaces of individual organisms. The state of an organism includes many characters, which are intimately correlated with each other, as the function of muscles depends on the function of the heart. A genotype state or phenotype state gives a complete description of all genetic or morphological and behavioral characters of an organism. The amount of information it contains is overwhelming, as an organism may have hundreds of characters and thousands of genetic loci. Biologists microanalyze the genotype or phenotype state into various genotype or phenotype aspects. A genotype aspect is a genetic character type such as a single locus or a few linked loci. A phenotype aspect is a morphological or behavioral character type such as wing span, foraging behavior, clutch size, or a combination of a few character types that function together. Each genotype or phenotype aspect has its range of genotypes or phenotype values.

A genotype state distribution or phenotype state distribution, which varies simultaneously with all character types, is intractable in practice. To get a manageable problem, biologists factorize a state distribution into the product of character distributions, *genotype distributions* or *phenotype distributions*. The factorization severs the correlation among various character types so that each can be reckoned with independently. When we concentrate on a particular genotype or phenotype aspect, we sum over or average out the variations in all other genotype or phenotype aspects. The result is a genotype or phenotype distribution.

Sometimes even the genotype distribution is too complicated, for it demands an account of how alleles pair up in a locus. There are $n(n + 1)/2$ genotypes for a locus with n alleles. It is much easier to count alleles singly. Thus the genotype distribution is further factorized into *allele distributions*. Instead of using genotype as a variable with values such as AA, Aa, and aa, an allele distribution uses allele as a variable with values A and a.

Factorization, which effectively breaks up a system into independent parts, is widely used in the sciences because it greatly simplifies problems. For example, physicists use a similar factorization approximation to get tractable kinetic equations for statistical systems (§ 38). Always, the crucial question is whether the approximation is adequate or justified for the problem under study. Is the factorization of organisms justified in evolutionary biology? Is it a good approximation that organic species evolve like bundles of uncorrelated organismic characters? Does the statistical averaging preserve most important causal mechanisms? Is the correlation among various character types of organisms really negligible? To these controversial questions, genic selectionism not only answers with a blanket *yes* but asserts that the allele distribution is *the* correct way to represent evolution. Its opponents counter that although some factorization is inevitable, usually we must retain the correlation among some character types because their functions are biologically inseparable. The indiscriminate use of allele distributions in evolution theory obscures and distorts the causal mechanisms of evolution.[25]

Severing the Genotype and Phenotype Descriptions

Working with character distributions instead of state descriptions poses a special problem because of the dual characterization of organisms. If the theoretical structure of population genetics depicted in Fig. 3.2 is to be preserved, then the factorization of genotype and phenotype states must be performed in such a way that the resultant genotype and phenotype aspects are matched and connectible by proper rules of development. This is usually impossible.

The relation between genotypes and phenotypes is complicated. Usually an allele affects many phenotypes and a phenotype is affected by many alleles. Many phenotype aspects, such as height and weight at birth, vary continuously. Biologists believe that they are controlled by alleles at many, perhaps hundreds of loci; each relevant allele has an additive effect. The optima for these characters usually occur at some intermediate values; it is advantageous to be tall, but not too tall. Thus whether an allele is beneficial depends on how many relevant alleles are present at other loci. As John Emlen said: "It is now generally accepted that genes at virtually no locus act without modifying effects from genes at others, and that virtually all loci are pleiotropic."[26] A *pleiotropic* allele contributes to several phenotypes. To make matters more complicated, an allele can be *antagonistically pleiotropic*, contributing simultaneously to a beneficial phenotype and a deleterious phenotype.

The difficulty of correlating genotype and phenotype distributions forces biologists to cut the link between them. Thus the conceptual structure of population genetics is fragmented into two pieces, one concerned exclusively with changes in genotype or allele distributions, the other with changes in phenotype distributions (Fig. 4.3). As Lewontin said: "One body of theory, what we might call the 'Mendelian corpus,'... seems to be framed entirely in genetic terms. The other theoretical system, the 'biometric,'... appears to be framed in completely phenotypic terms."[27]

Genotypes account for heredity; phenotypes function in the struggle for existence and respond to the force of natural selection. Heredity and selection are both indispensable to evolution. How can they be included in the changes of either the genotype or the phenotype distribution? Good approximate models usually retain some parameters to account for the information suppressed from explicit consideration. More refined theories also provide some connection to models that flesh out the content of the parameters. Such parameters and supportive models are rare in either the Mendelian or the biometric

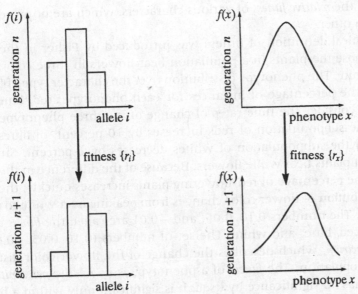

FIGURE 4.3 In practical models, the conceptual structure of population genetics as depicted in Fig. 3.2 is severed into a genetic side and a phenotypic side. Furthermore, the genome is factorized into uncorrelated genes and the phenotype state into disjoint phenotypes. The objects of study become the allele distribution $f(i)$ for a single genetic locus or the phenotype distribution $f(x)$ for a single morphological or behavioral character type. The allele and phenotype distributions are decoupled from each other; the causal mechanisms linking them are mostly neglected. For asexual reproduction, the allele distribution evolves according to the rate equation $df(i)/dt = r_i f(i)$, where the fitness r_i gives the relative rate of growth for the allele i in a fixed environment. Similar interpretation holds for the phenotype distribution.

corpus, partly as a result of the complexity of the genotype–phenotype relation. The transformation rule linking successive genotype distributions contains little indication of how the morphological or behavioral expressions of various alleles fare in life; that is not surprising when it is unclear what these expressions are. The transformation rule linking successive phenotype distributions contains no information on the genetic mechanism underlying the heritability of various phenotypes. Consequently Mendelian models count gene frequencies in successive generations but are unable to explain why the frequencies change. Biometric models explain the adaptedness of various characters but not how the characters are passed on to the next generation. Except in those rare cases where a phenotype is mainly the expression of a single allele, the gulf between Mendelian and biometric models is not bridged.

Fitness and the Charge of Vacuity

In both Mendelian and biometric models, variables that should be determined endogenously become exogenous parameters, or, worse, are neglected altogether. Causal mechanisms and transformations linking genotypes, phenotypes, and environmental conditions are swept wholesale under a set of parameters, the *relative fitness* of various characters, which are now independent of each other.

The technical definition of fitness was introduced by Fisher in the early 1930s. Suppose the plants in a population bear flowers in three colors, red, blue, and white. The phenotype distribution over the character type of flower color gives the percentage of instances for each phenotype: red, blue, and white. Now consider the time rates of change of the three phenotypes: per unit time the subpopulation of reds increases by 10 percent, of blues by 5 percent, and the subpopulation of whites decreases by 1 percent. Suppose initially most plants bear white flowers. Because of the difference in the rates of change, the percentage of red-flowering plants increases quickly; the phenotype distribution of flower color changes from peaking over white to peaking over red. The numbers 0.10, 0.05, and −0.01 are called the *fitness* of the phenotypes red, blue, and white. The set of numbers (0.10, 0.05, −0.01) is the *fitness spectrum*, which describes the change of the flower-color distribution of the population. The *fitness* of a phenotype is its *relative per capita rate of change*. It has no significance by itself; it is significant only within a fitness spectrum, where it is seen relative to the fitness of other phenotypes. The fitness of genotypes and that of alleles are defined similarly.

The technical definition of *fitness* as a relative rate of change in the number of instances of a certain character, which has become a central concept in the modern synthetic theory of evolution, differs from both Darwin's original meaning and ordinary usage.[28] It carries little of the usual meaning of being fit for living. A phenotype or genotype with high fitness need not be thriving; it may be marching to extinction with the population. More important, strictly speaking fitness does not even apply to organisms; it is not an individual but a statistical concept. Since fitness is meaningful only within the fitness spectrum for a distribution and distributions characterize populations, it is part of the characterization of the population as a whole. As Lewontin said: "The probability of survival is an *ensemble* property, not an individual one."[29] Perhaps a less colorful name such as "differential proliferation," "relative growth rate," or even "multiplicative coefficient" would be less confusing.

Fitness includes all contributions to change: natural selection, genetic drift, mutation, and other factors. It is often analyzed into various components. The component that depends on the environment and accounts for natural selection is called *Darwinian fitness*. Population genetics gives adequate explanations for some non-Darwinian mechanisms underlying fitness and growth rates, notably the combinatorics of alleles in sexual reproduction and the sampling problem for small populations. Natural selection fares far worse. In many population-genetic models, the idea of natural selection is entirely encapsulated in Darwinian fitness.[30] This is fine if Darwinian fitness is a function of environmental parameters that inform us, perhaps crudely, of how the relative growth rates depend on the interaction between the environment and organisms bearing various phenotypes or genotypes. Unfortunately, they do not. Aggregate factors such as total resource, population density, or population composition are sometimes included in Darwinian fitness, but specific factors responsible for the difference in the growth rates of various characters are seldom included. In many models, Darwinian fitness are constants empirically determined from vital statistics. The problem of explaining them is passed to other disciplines. Edward Wilson said: "The central formulations of the subject [population genetics] start with the given constant r's [fitness]. The forces that determine the r's fall within the province of ecology."[31]

Any distribution, be it biological, economic, social, or physical, will evolve if its variants grow at different rates. The question an adequate theory of evolution must answer is, What causes the differential proliferation? The mere name "natural selection" will not do. The theory of evolution loses its explanatory power when the causal mechanisms responsible for natural selection are overlooked. Empirically determined Darwinian fitness is suitable only for a specific environment, but the numbers do not tell us what elements in the environment cause the differential proliferation and how. Furthermore, since the numbers are valid only as long as the environment remains static, the idea of evolution driven by a changing environment evaporates. The problem is worse in the evolution of genotype and allele distributions, because alleles are further removed from the environment of life. Without some adequate elaboration of the causal factors underlying Darwinian fitness, population genetics is reduced to what Mayr derided as "beanbag genetics," a statistical scheme for counting genes in successive generations. The common practice of population geneticists of treating Darwinian fitness as given constants throws out the biological content of natural selection and invites the criticism that "survival of the fittest" is vacuous or circular. The weakness is acknowledged by many biologists.[32]

Optimization Models and Their Critics[33]

The adaptation models try to provide some answers to why those with some characters reproduce more successfully in certain environments than those with other characters. Unlike fitness, which is a statistical concept applied to distributions of populations, adaptedness describes an organism's ability to cope in a specific environment. Adaptation, in the sense of a character well suited to the conditions of living, is the conceptual locus of organism–environment interrelation.

Two general approaches to the study of adaptation are the comparative method and the optimization model. The comparative method, based on the observation that similar adaptive characters are found in organisms in a wide range of unrelated groups, is mainly empirical. We will concentrate on the theoretical optimization models. There are two major types of optimization model for studying adaptation: *individual optimization* and *strategic optimization* using game theory. Both conform to the general structure of optimization discussed in § 9. A typical model includes the specification of a set of possible behaviors or characters, a set of constraints including environmental conditions, and the objective to be optimized. In game theoretic models, the possibilities are called strategies and the objectives, payoffs (§§ 20 and 21). The optimization procedure aims to find the behaviors or characters that maximize the objective under the constraints. It is tacitly assumed that the optimal characters or behaviors are those selected, meaning they are the ones in equilibrium with the given environment and the ones most prevalent in the population.

Most causal content of an optimization model lies in the variation of the objective with the possibilities. A ready-made objective for adaptation models seems to be reproductive success, often called fitness, for the result of natural selection is manifested in the number of offspring. However, it is difficult to find the causal relation between reproductive success and the specific character under consideration, unless the character relates directly to reproduction. Thus biologists often use proxy criteria, most of which are economic in nature. A common economic objective is the net caloric gain per unit active time, under the assumption that the best fed organisms are likely to outperform others in reproduction. A model estimates the number of calories produced by each possible behavior or character to determine which one is optimal under the constraints specified. It predicts that the optimal behavior should be most prevalent in nature. The prediction can be checked by field observations.

Optimization models for physiological structure and performance, such as flight pattern, walking gait, skeleton and eggshell structure, eye and wing design, are similar to engineering studies but crude by engineering standards. For instance, wing designs are optimized with respect to gross shape: elliptical, rectangular, triangular. I think few readers would be surprised to learn that many designs found in natural organisms are close to if not the optima. Models of life histories have a stronger biological flavor. Consider a model with the aim of finding the clutch size of birds that maximizes the number of fledglings. The constraints include the supply of food, the length of days for foraging, and the tradeoff between the number of eggs laid and the survival rate of the nestlings. For the great tits in Oxford, the calculated optimal clutch size compared nicely with data collected over twenty-two years.[34]

According to George Oster and Edward Wilson, "[O]ptimization arguments are the foundation upon which a great deal of theoretical biology now rests. Indeed, biologists view natural selection as an optimizing process virtually by definition."[35] Other biologists see the identification of natural selection with optimization as the abusive interpretation of a calculational technique. Gould and Lewontin criticized its degeneration into panadaptationism or the Panglossian paradigm, the "faith in the power of natural selections as an optimizing agent," and the assumption that a behavior shared by most members of a population must be the best, because it is a result of natural selection. They reproached panadaptationists for manipulating data to fit adaptation models and for neglecting competing mechanisms of evolution – allometry, pleiotropy, gene linkage, genetic drift, developmental constraint – many of which are nonadaptive. Since natural opportunities form a continuum in which organisms find various ways to make a living, one can always find some factors for which an arbitrary character is supposed to be the solution. With some ingenuity in framing possibilities, constraints, and objectives, some model can always be found in which a pet character is the optimal. Panadaptationism can be used to justify any behavioral or social status quo in the name of evolution. Maynard Smith agreed in his reply to Lewontin's critique: "The most damaging criticism of optimization theories is that they are untestable. There is a real danger that the search for functional explanations in biology will degenerate into a test of ingenuity."[36]

That natural selection is an optimizing process depends on several dubious assumptions. Optimization depends on the availability of options, but natural selection does not create new characters for choice; it only preferentially preserves existing ones. Thus optimization models implicitly assume that all the options included in their sets of possible characters are already produced by genetic mutation. Critics find this assumption incredible: If *Hamlet* is among the candidates, it will surely be selected as the best play; the question is how it is produced by monkeys playing with typewriters. Adaptationists show computer simulations of the mutation and evolution of systems with a few components and say, Look, it works even for simple systems. The simulation misses the point totally. Even the simplest cells are enormously complex; it is the evolution of large and complex systems that is problematic because of the combinatorial explosion of possibilities. That a monkey can accidentally type out "To be or not to be" does not imply that it can type out all of *Hamlet*. For the availability of evolutionary possibilities, Oster and Wilson frankly admitted: "The assumption, frequently made, that 'sufficient genetic flexibility exists to realize the optimal ecological strategy' is an assertion of pure faith. Rather like the concept of the Holy Trinity, it must be believed to be understood."[37]

The faith in genetic flexibility is not enough. Another leap of faith is required to believe that *equilibrium* models can be the foundation of *evolutionary* biology. Adaptation models are equilibrium models, as the comments of biologists discussed previously in § 9 testify. Both individual and strategic optimization models seek equilibrium or stable-equilibrium behaviors. The models show that certain characters or behaviors are optimal to maintain the organism under certain environments but leave open the question of how the optimal characters are attained through evolution. The adaptedness of organisms was observed long before Darwin and attributed to God's design. A major aim of the evolutionary theory is to find a natural explanation of it. The aim is *not* fulfilled by the optimization models, which articulate clearly *how* the organisms are adapted but not *how come*. The approach to the optimal equilibrium, which is the crux of evolution, is a separate problem whose solution requires *dynamic* theories. We have seen in the previous section that economists face a similar dilemma in asking how optimal allocation is reached. We will see in § 38 that an entire branch of statistical mechanics is required to address the approach to equilibrium. There is nothing remotely comparable to nonequilibrium statistical mechanics in evolutionary biology. The lack of dynamic models is particularly damaging to evolutionary biology, for it shirks the crucial point of contention: What brought about the optimal characters and behaviors?

Most biologists take it for granted that the optima are the results of natural selection; the fittest characters survive. I believe this is mostly true, but the belief is supported not by adaptation models but by historical narratives that present abundant empirical evidence (§ 41). Adaptation models are inept in distinguishing among competing causes; hence they cannot affirm natural selection to be the cause of the optimal behaviors. The behaviors may be the result of learning and social conditioning achieved within a single generation, which is a reasonable interpretation of the Dove–Hawk game discussed in § 21. Or they may be the results of rational choice, as they are in economics; this interpretation cannot be ignored for human behaviors, and animals think too. Or they may be a testimony to God's engineering skill; the perfection of organismic designs was often used as an argument for the existence of God.

Creationism can be rejected on many grounds, but not on the strength of adaptation models, which seem to favor it. Design by optimization is a systematic evaluation of possibilities, which is typical of the rational deliberation of engineers. It stands in sharp contrast to natural selection, which proceeds piecemeal, is short-sighted, and is subject to path-dependent constraints. As Francois Jacob argued, natural selection is not an engineer but a "tinkerer."[38] Thus evolutionists must explain why the optimization models represent the handiwork of a tinkerer and not an engineer. I find no adequate explanation.

Unlike engineering deliberation that can arrive at the optimal design in one fell swoop, evolution depends on the path it takes. The tinkering of natural selection takes only a small step at a time and disallows steps into lower levels of fitness. For such path-dependent processes, there is no guarantee that the result is the best; all kinds of accident can happen that trap the evolution in suboptimal states.[39] A necessary but not sufficient condition for natural selection to produce the characters obtained by engineering optimization is the existence of continuous paths with monotonically increasing fitness that link arbitrary initial characters to the optimum. The existence of evolutionary paths with monotonically increasing fitness is a stringent condition that is seldom satisfied (§ 24).

Independent Characters Become Myths

As we have seen, the factorization of organisms into bundles of independent characters dictates the split of evolutionary theory into a genetic and a phenotypic side. All adaptation models are framed in terms of *phenotypes*, for phenotypes function directly in the business of living. As Maynard Smith said, "Evolutionary game theory is the way of thinking about evolution at the phenotype level." To be models of evolution, the phenotypes must be heritable. To be models of biological evolution, they must be genetically heritable. However, except in the rare cases in which the phenotype is the expression of a single gene, there is no adequate genetic model for the heritability of phenotypes. Maynard Smith explained that the assumption about heredity is usually implicit in the adaptation models, and it is only assumed that "Like begets like." Furthermore, in some cases "'heredity' can be cultural as well as genetic – e.g., for the feeding behavior of oyster-catchers."[40]

The same complicated genotype–phenotype relation that splits the theory of evolution thwarts attempts to explain the genetic heredity of most phenotypes featured in optimization. Maynard Smith, Oster, and Wilson, who developed, used, and defended adaptation models, all expressed skepticism about satisfactorily supporting genetic models. They are not alone. G. A. Parker and P. Hammerstein said in their review article: "The lack of any true genetic framework within evolutionary game theory could present difficulties. On the other hand, the lack of extensive descriptions of the phenotypic interactions is an impediment to classical population genetics theory. It is to be hoped that a marriage of the two disciplines might prove harmonious rather than a mathematical nightmare that stifles any central biological insights; we are not altogether optimistic."[41]

Strangely, these phenotype adaptation models with no adequate genetic explanations become the staple of genic selectionism, a form of microreductionism claiming that evolution is nothing but change in gene frequency and should be expressed solely in genetic terms. Genic selectionists appropriate the phenotype models by a ruse. In Alan Grafen's words, "The phenotype gambit is to examine the evolutionary basis of a character as if the very simplest genetic system controlled it: as if there were a haploid locus at which each distinct strategy was represented by a distinct allele."[42] In short, genic selectionists posit "a gene for a character" such as "a gene for intelligence" and depict the genes fighting for dominance. With the help of a little purple prose, phenotype models without genetic explanations are converted into what George Williams called "selfish-gene theories of adaptation." "Selfish gene" is a phrase introduced by Dawkins to popularize Williams's gene's eye view of evolution.[43] Selfishness is a misleading metaphor for the independence of characters resulting from the approximate factorization of the organism.

The notion of "a gene for a character" gives the adaptation models the appearance of being models of biological evolution. It becomes a central pillar for sociobiology and new social Darwinism, lubricating the rash extrapolation of genetic control into psychology, morality, sociology, epistemology, and cognitive science.[44] Promoters advertise it as a panacea for solving all evolutionary puzzles, although it falls far short. For instance, biologists are striving to explain the evolution of sexual reproduction. Dawkins said the problem can be easily solved: "Sexuality versus non-sexuality will be regarded as an attribute under single-gene control, just like blue eye versus brown eye. A gene 'for' sexuality manipulates all the other genes for its own selfish ends. ... And if sexual, as opposed to asexual, reproduction benefits a gene for sexual reproduction, that is sufficient explanation for the existence of sexual reproduction." Dawkins acknowledged that "this comes perilously close to being a circular argument."[45] He is almost right; his argument is completely empty. We know sexuality is genetically determined and ask how; we know it is selected but ask why. "A gene for sexuality" claims to answer the how and why questions but it does not. Furthermore, it is a muddled concept. To manipulate other genes the "gene for sexuality" must do much more than controlling sexually; it must be capable of scheming and evaluating benefits. It is more like a genie than a gene. Selfish-gene theories are reminiscent of myths in which gods are invented to explain natural phenomena, such as a god for storms or a goddess for spring.

When we look at accounts of "a gene for a character" more closely, we find that "a gene for" can be deleted without diminishing their valid information content. All the valid content of "The gene for sexuality is selected because it is beneficial to life" is contained in "Sexuality is selected because it is beneficial to life." This is hardly surprising if we remember that the scientific models are framed in terms of phenotypes and lack adequate genetic explanations. All that "a gene for" can add is a falsehood. Unlike sexuality, many characters, for instance, aggression and altruism, are not wholly genetically determined. The addition "a gene for" to these characters obscures other possible determining factors, which may be more important than genetic factors.

The expressions of genes in the behaviors of organisms depend heavily on the environments of their development. For organisms dependent on nurture to mature, cultural and social environments may be decisive in shaping certain behaviors. Cultural heritage is already apparent in the behaviors of birds such as oyster-catchers. It is much more important in human behaviors. We may find in a community that speaking English is the most prevalent linguistic behavior and that people who speak English fare better than those who do not. An optimization model based on economic considerations can be developed readily to show that it is indeed optimal to speak English in the community. It is easier for anglophones to gain maximum benefit per unit foraging time, and even to find mates. The optimum is evolutionarily stable; if the great majority stick to English, then a new language cannot invade the community. However, the optimization model does not imply that "a gene for speaking English" is selected or there is "a genetic tendency to speak English." The specific language we speak is socially determined. This obvious counterexample demonstrates how optimization models and the myth of "a gene for a character" can be misleading in more complicated cases.

How does the genetic heritage of a behavior compete with its social or cultural heritage? So far evolutionary models can provide few scientific answers for many behaviors. "A gene for a behavior" suggests the false answer that the genetic factor is all-important. Genic selectionists do mention nongenetic factors and say that "a gene for" means "a genetic contribution to." However, it is not clear whether the qualification is consistent with their extensive accounts of how the genes take the driving wheel and manipulate matters to their benefit.

Genetic determinists say that rejecting "a gene for a character" is equivalent to denying the genetic relevance of the character. This is not true. Many behaviors have some genetic basis; we are only acknowledging that current experiments and theories do not tell us what the bases are or how strongly they contribute to the behaviors. The adaptation models are framed in terms of phenotypes and are not supported by adequate genetic models. The addition of "gene" makes them less and not more scientific; to be scientific is to be honest about what one knows and what one does not know. Darwin is far more scientific than those who posit "a gene for sexuality" to cover up their ignorance of the evolution of sexuality when, pondering the problem of sexual reproduction, he says, "I now see that the whole problem is so intricate that it is safer to leave its solution for the future."[46]

5 Interacting Individuals and Collective Phenomena

18. An Intermediate Layer of Structure and Individuals

There is no ambiguity about the ingredients of solids. A solid is made up of atoms, which are decomposed into ions and electrons, simple and clear. However, a chapter or two into a solid-state text book and one encounters entities such as phonons and plasmons. Not only do they have corpuscular names, they behave like particles, they are treated as particles, and their analogues in elementary particle physics are exactly Particles. What are they and where do they come from? A *phonon* is the concerted motion of many ions, a *plasmon* of many electrons. They are treated as individuals not because they resemble tiny pebbles but because they have distinctive characters and couple weakly to each other. Physicists have microanalyzed the solid afresh to define new entities that emerge from the self-organization of ions and electrons.

Salient structures describable on their own frequently emerge in large composite systems. Systemwide structures will be discussed in the following chapter. Here we examine a class of structure that is microanalyzable into novel individuals, which I call collectives. A *collective* arises from the coherent behavior of a group of strongly interacting constituents. It has strong internal cohesion and weak external coupling, and its characters and causal relations can be conceptualized independently of its participants. Thus it is treated as an individual. Phonons and plasmons are examples of collectives, as are firms and households, which are organized groups of people. Since preferential coupling is the formative force of collectives, to study their internal structures we must explicitly account for the relations among the constituents in *interactive models*.

Collectives constitute an intermediate layer of structure in a many-body system. With their help a many-body problem is decomposed into two simpler and more manageable parts. *Collective analysis* investigates the internal structures of collectives and the relations among them. *System analysis* investigates the contributions of the collectives to the behaviors of the system as a whole. By making certain assumptions, each analysis can proceed on its own. If one of the two is intractable, at least the difficulty is confined and does not prevent us from getting results from the other. This is the strategy of modularization and damage control commonly employed in the theoretical sciences.

The Emergence of Intermediate Structures

In a many-body system that is not perfectly homogeneous, a group of constituents may couple more strongly to each other than to outsiders. The differential interaction may hinder the dispersal of local disturbances, enhance the discriminatory coupling, and draw the group tighter together to form a clump. Other clumps form similarly. In systems comprising large numbers of interacting individuals, the minutest inhomogeneities or fluctuations can lead to pronounced structures.

The universe itself is the best example of structure formation. Before its thousandth birthday, the universe was highly homogeneous and isotropic. Its early homogeneity is evident from the cosmic microwave background radiation, which is the remnant of the ancient time. The fluctuation in the background radiation is less than several parts in a hundred million, according to the recent observation of the satellite *Cosmic Background Explorer*. Such minute fluctuation is the source of all the wonderful structures in the present universe: planets, stars, galaxies, clusters of galaxies, voids tens of millions of light years across.

Prominent structures stemming from small variations in relations are also found on the human scale. The economist Thomas Schelling has given many examples, one of which is the curiosity effect in traffic. An accident in the east-bound side of a divided highway can cause a jam in the west-bound traffic, as people slow at the site of the accident to take a look. Schelling also showed in easy-to-understand models how the slight preference of people to live near others of their own race can lead to self-formed neighborhoods that look like the results of intentional segregation.[1]

Conspicuous structures can form spontaneously without any external influences, as is apparent in the evolution of the universe. They can also be expedited by external forces. The systems we are studying are tiny parts of the universe and are not totally isolated. Sometimes minute inhomogeneities are instigated by external conditions, for instance, the ambient temperature of a solid, the geographic environment of organisms, or the technological advantage of large-scale production. However, the forces that turn the petty inhomogeneities into significant structures are the interactions among the constituents: density fluctuation in solids, social intercourse among organisms, or transaction cost in economies. Therefore the formation of structures is mainly although not exclusively an endogenous process.

Interactions that produce prominent structures are usually rather strong and strong relations are difficult to handle. Thus scientists incessantly try to microanalyze structures into weakly coupled individuals, which are much more tractable. The independent individuals in the preceding chapter are weakly coupled, but they are no longer sufficient when the constituents interact preferentially with each other and situations common to all are not viable. Scientists find new tactics of modularization, the results of which are the collectives: elementary excitations such as phonons in physics, social institutions such as firms in economics, organismic groups such as ant colonies in evolutionary biology.

Collectives

A collective has its own characters, which differ not only in degree but in kind from those of its participants. The individuation of collectives is not merely a grouping of constituents; it is often incomplete and nonspatial, and it does not partition the many-body system. A collective engages many constituents of a system, and a constituent participates in many collectives, as a person works for several firms or the motion of an ion contributes to many phonons. A multinational firm scatters around the globe and a phonon is a mode of a wave that spreads across the crystal lattice. Although these sprawling collectives do not look like entities, which are spatially localized, they are integrated by internal principles such as administrative organization or coherent motion. A collective may have complicated internal structures and perhaps harbor conflicts among its participants, yet externally it appears as an integral unit and interacts as a unit with other collectives and with stray constituents.

Not all collectives are full-blown individuals. Some lack numerical identity and cannot be individually referred to; the characters and causal relations of others are not distinct enough to be carved out. An individual is identified by its spatial position in the world, by which it is distinguished from others with the same character. Entities such as firms that are not localized usually have indirect criteria of identity invoking some spatial designations, for instance, the date and place of registration. Phonons and plasmons are not localized and lack criteria of identity unless they are measured by a spatially localized instrument. They are adequate in physical theories that are concerned not with *the* phonon but only with *a* phonon with a certain character. Groups of organisms suffer the other deficiency in individuality; they are too indistinct to stand out by themselves as the nexus of selective forces. This is a source of the ongoing controversy over group selection in evolutionary biology. The incomplete individuality of the collectives explains why they are generally not miniature many-body systems but constitute an intermediate structure within the systems.

None of the individuals we consider is eternal, but the transience of collectives is most pronounced in our theoretical context. As internal structures of many-body systems, collectives are fluid and dynamic, forming and dissolving as constituents congregate and disperse. Phonons decay, firms fold, organismic groups disintegrate and are promptly replaced by other phonons, firms, and groups. The transience is explicitly represented in many models. Even in equilibrium models, collectives are often qualified with characteristic lifetimes.

Noting the transience and the incomplete individuality of collectives, some people dismiss them as illusions. For instance, some economists argue that the legal person is a legal fiction; a firm is nothing but a nexus of contracts among real persons and should not be treated as an individual. The critics rightly caution against treating collectives as something rigid or something standing above and beyond the constituents, but they overreact in outlawing them. Their argument regards micro- and macroexplanations as mutually exclusive, a mistake common in microreductionism. "Phonon," "colony," and "firm" do represent objective structures. These structures are microanalyzable, but that does not prevent them from having salient characteristics that feature in macroexplanations. As intermediate entities, collectives are the best illustrations of micro- and macroexplanations. Microexplanations account for the characteristics of phonons and their formation from the motion of ions. Macroexplanations use phonon behaviors to calculate the properties of solids such as their specific heat at low temperatures, which is obtained by adding up the contributions of phonons without mentioning ions. The calculation of specific heat is supported instead of invalidated by the microanalysis of phonons. There is tremendous analytic advantage in the conceptualization of phonons and other collectives, for it enables us to break intractable problems into more manageable parts, divide and conquer. We need to qualify carefully the limitations of our descriptions and explanations, not to proscribe them.

Collective Analysis and System Analysis

Models for many-body systems with intermediate layers of structure microanalyzed into collectives have at least four elements: the external environment, the system, the collectives, and the constituents. Causal relations can obtain among the constituents, among the collectives, and among the constituents and collectives. Compositional relations can obtain between the system and the collectives, the system and the constituents, and the collectives and the constituents. The system, collective, and constituent are each characterized by its own concepts. The triple-deckered theoretical structure does not complicate but simplifies many-body problems, because it helps us to make approximations and develop fruitful models.

In *collective analysis*, we try to find out the collective's characters, the behaviors of its participants, and the relation between it and its participants, for instance, the organization of a firm and how it affects the behaviors of its employees. Discriminating interaction is essential to the formation of collectives; hence collective analysis explicitly takes account of the relations among the participants. Even so it is easier than the original many-body problem because the collectives have smaller scale and scope. More important, collective analysis restricts itself to the specific aspects of behaviors and relations that differentiate participants from other constituents. The selected aspects are often more coherent and more susceptible to theoretical representation.

System analysis treats the collectives as individuals in a new many-body problem, in which their contribution to the behaviors of the whole system is determined. The task now is much easier than the original many-body problem, for with most strong coupling among the constituents swept into the internal structures of the collectives, the residual interactions among collectives are weak and scarce. Often the independent-individual approximation suffices. Even when the relations among the collectives are explicitly addressed, the reckoning is easier because there are usually fewer collectives than constituents.

A collective can contribute to many system characters, and a system character can involve several kinds of collective. Hence collective and system analyses are often pursued separately. Ideally, the two are integrated with the system analysis calling on the results of collective analysis when appropriate. This occurs frequently in physics, but it need not be the case generally, for different assumptions may have been made in the collective and system analyses, so what we have are different segments of theories for different phenomena.

Sometimes the internal structures of the collectives are too difficult to determine because of the strong interaction among their participants. Fortunately, often the internal structures of a tightly bound collective affect its contribution to system behaviors only slightly, just as an atom's internal structure slightly affects its contribution to a large molecule. Perhaps the effects of the internal structure can be ignored or summarized by a few parameters to be experimentally determined or intuitively estimated. Many-body theoreticians often introduce additional postulates or cite experimental evidence to sidestep collective analyses and proceed with system analyses. For instance, the ionic lattices of solids are collectives, and their structures are determined not theoretically but experimentally. Sometimes scientists treat the collectives as "black boxes" with only external characters in system analyses. The microeconomic general equilibrium theory for competitive markets is a system analysis that ignores the internal organizations of firms and households.

The following sections mainly consider collective analyses. I will concentrate on models that explicitly address the relations among individuals. Interactive models are more kaleidoscopic than independent-individual models, but unifying methods are not totally absent. Physics has its standard way of treating interaction. Models in economics and evolutionary biology rely heavily on the mathematics of game theory.

19. Collective Excitations and Their Coupling

Waves are the most common form of collectives in physical systems. Despite their corpuscular names, phonons and plasmons are not microscopic marbles but waves, as are Particles in elementary particle physics. (I capitalize *Particle* to emphasize its difference from ordinary particles, which are like tiny pebbles.) *Collective excitations*, the collectives in condensed-matter physics, are the quanta of normal modes of certain types of wave motion. The concept of *normal modes of waves* plays a crucial role in almost all branches of physics. Its importance stems from the fact that under a wide range of conditions, the normal modes have distinct characteristics, do not couple with each other, and can be superimposed to form arbitrary waveforms. Consequently they can be treated as individuals.

This section begins by introducing the general ideas of waves, their normal modes, and their interpretations. The general ideas are then applied to collective excitations to illustrate how a novel scheme of microanalysis can greatly simplify a many-body problem. We will consider the example of *phonons*, the definition and characterization of which are carried out in collective analysis. The system analysis, which evaluates the phonons' contribution to the behavior of the solid, involves both the independent-particle approximation and explicit phonon–electron interactions.

Waves, Normal Modes, and Particles

Ocean waves, in which myriad water molecules move coherently to generate macroscopic patterns, are familiar examples of collective motion. Attributes of waves – amplitude, frequency, wavelength, wavevector – are inapplicable to classical water molecules. Waves propagate and their propagation is characterized by velocity, but the wave velocity is not the velocity of the participating molecules. A boat's wake spreads and travels, but the water molecules constituting the wake do not travel with it. Water molecules behave similarly to floating objects, which move up and down but stay where they are as the wake passes. Waves are not collections of molecules but the *collective motion* of molecules.

Waves are analyzable into normal modes. A *normal mode* is a specific wave that oscillates with a definite frequency and wavelength determined by the boundary conditions and other general characters of the vibrating medium. Consider a violin string with a certain mass, tension, and boundary condition determined by its fixed ends. Its first normal mode has a wavelength that is twice the length of the string and a specific frequency determined by its length, mass, and tension. The wavelength of its second normal mode is equal to its length. Higher normal modes have shorter wavelengths and higher frequencies. The importance of normal modes in physics partly derives from their characteristic of *not interfering with each other* if the amplitude of vibration is not too large, that is, if the vibration is *harmonic*. In harmonic oscillation, the energy fed into one normal mode stays in the mode and does not pass into another mode. Thus each normal mode can be regarded as an individual independent of the other modes.

Another reason for the importance of the normal modes is the *superposition principle*. Normal modes can combine to form complicated waveforms, and an arbitrary small-amplitude wave can be represented as a superposition of appropriate normal modes. For instance, an arbitrary harmonic waveform of the violin string can be produced by superposing various normal modes with various amplitudes. Thus normal modes provide a powerful method for the analysis of waves. Although they are not spatially delineated, they are the "parts" of a harmonic wave in terms in which the wave is the "sum of its parts."

Consider an arbitrary harmonic wave made up of zillions of interacting molecules. All molecules participate in each of the normal modes, just as the entire violin string is involved in any mode of oscillation. With each normal mode, we pick out from the motions of the molecules the specific aspect that produces a specific coherent form. Thus the normal modes do not partition the many-body system. By treating normal modes as individuals, physicists again adhere to the general notion of weakly coupled entities rather than the substantive feature of spatial packaging. Thus the microanalysis of physics differs from mereology and some revisionary philosophies, which are exclusively concerned with spatial partition.

When vibration is more violent and oscillation becomes anharmonic, normal modes do interact. Noninteracting individuals can persist forever, but interaction compromises individualistic characters, for interaction involves energy transfer. When energy leaks from one normal mode to another in interaction, the first mode damps while the second amplifies. Consequently normal modes decay and have a finite lifetime. However, if the anharmonicity is not too strong, the interaction among normal modes is rather weak and can be treated by approximate methods.

So far we are in the realm of classical physics, where the quantity of energy is infinitely divisible. This is not so in quantum mechanics, where energy comes in chunks or quanta, much as in everyday monetary exchange, in which the increment in value is quantized according to the unit of a penny. In quantum mechanics, the energy of a normal mode increases by units or quanta, with the unit of energy proportional to the frequency of the mode. A normal mode with one unit of energy, called a *quantum* of the mode, has many characteristics of a particle. It has a well-defined energy E proportional to its frequency and a well-defined wavevector $\mathbf{k}$ proportional to its wavelength and is represented by an energy–wavevector relation $E(\mathbf{k})$ similar to the energy–wavevector relation of electrons discussed in § 15. It is what a *Particle* in quantum physics usually means; a Particle in elementary particle physics is a quantum of a mode of the wavelike excitation of a quantum field. It is what a *collective excitation* refers to in condensed-matter physics.

A particular violin string supports many normal modes, each with its peculiar frequency and wavelength. All these modes are of the same kind, defined by the general characteristics of the string; a different kind of vibrating medium supports a different kind of mode. Similarly, the Particles and collective excitations are broadly classified into kinds: electron, photon, and quark in elementary particle physics; phonon, plasmon, and magnon in condensed-matter physics. Each kind of Particle is characterized by a general energy–wavevector relation, quantum variables such as the spin orientation, and constant parameters such as the mass, charge, and absolute spin value. The absolute spin value determines the statistics in which the Particles are counted.

With the concept of Particles, the analysis of a wave becomes an accounting of the number of Particles with each value of energy, wavevector, and spin orientation. The interaction among waves or among the normal modes of a wave is represented as the scattering, creation, annihilation of Particles of various kinds, energies, momenta, and spin orientations. Collective excitations play the same role as Particles.

Collective Analysis: Collective Excitations[2]

A many-body system at absolute zero temperature is frozen into its *ground state*, or the state with the lowest energy. The characters of the ground state of solids, which are related to the binding energies of solids and nuclear matter, are difficult to determine. They play a role in low-temperature phenomena, for example, the superfluidity of liquid helium. However, for most interesting condensed-matter phenomena occurring at finite temperatures or in perturbed systems, the ground state contributes little except as a reference point. Many interesting characters are determined by the system's excited states with relatively low excitation energies, which are likely to be occupied at relatively low temperatures or weak external forces. For instance, when we apply an electric potential to a solid to study its conductivity, we are perturbing the solid and probing its excited states. As it turns out, many excited states are quite simple. They have pronounced individuality, are susceptible to rigorous mathematical treatment, and can be studied independently of the details of the ground state. They are called *elementary excitations*.

There are two types of elementary excitation: the *quasi particles* discussed in § 15 and the quanta of collective modes called *collective excitations*. Both depend on the interaction among the constituents of the system. When interaction is turned off, quasi particles become free particles and collective excitations vanish. In both cases, the relevant effects of interactions are incorporated into the characters of the excitations. We concentrate on collective modes here.

A collective mode expresses the correlation among the motions of a large number of interacting electrons or ions. It mainly takes the form of a wave. An example is the elastic waves of crystalline lattices in solids. The ions in a solid are arranged in a nearly perfect array as a result of their mutual interaction. An ion interacts most strongly with its closest neighbors. We can form a crude picture of the lattice by imagining the ions as tiny balls, each connected to its immediate neighbors by tiny springs. At absolute zero temperature the ions are motionless and the lattice is frozen rigid. As the temperature increases, the ions start to agitate, but the amount of energy available for agitation is small and must be economized. It takes quite a bit of energy to displace an ion while keeping its neighbors fixed, because the springs attached to the displaced ion would have to be extended considerably. Therefore a relatively large amount of energy is required to sustain random fluctuations of the ions. Excited states of the lattice with much lower energies can be obtained if the agitations of the ions are so coordinated that the displacement of each ion differs slightly from that of its neighbors, for then each spring is extended only slightly. The most cost-effective way to excite a lattice is to create waves, and waves naturally exist at finite temperatures. A lattice at finite temperatures exhibits periodic compression and rarefaction, with periods that are large compared to the distance between two adjacent ions. The periodic disturbances are reminiscent of sound waves and are called *elastic waves* (Fig. 5.1).

The elastic waves of the crystalline lattice are analyzed into normal modes. As the modes fall within the realm of quantum mechanics, their energies are quantized. The quantum of an elastic wave is called a *phonon*, after the Greek

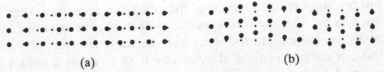

(a)   (b)

FIGURE 5.1 Low-energy excitations of a crystal lattice are not in the form of in-dividual ionic vibration but of collective displacement called an *elastic wave*. The ions are depicted by large dots and their equilibrium positions in the lattice by small dots. **(a)** The concerted compressions and rarefaction of planes of ions generate a longitudinal elastic wave. **(b)** The relative displacements in adjacent planes gener-ate a transverse elastic wave. Elastic waves are analyzable into normal modes each with its characteristic frequency and momentum. The energy of a normal mode increases in discrete chunks or quanta, and a quantum is called a *phonon* with the mode's characteristic frequency and momentum.

"sound." Depending crucially on the concerted motions of ions far way from one another, phonons are genuine collective phenomena. More phonons are created when the crystal is heated up or perturbed in other ways. The phonon is often used as an analogy by elementary particle physicists to explain the concept of Particles. There are optical and acoustic phonons, depending on whether the elastic waves are transverse or longitudinal. Each kind of phonon has its characteristic energy–wavevector relation specifying the possible states that phonons of the kind can assume. The relation depends on the general properties of the lattice such as the mass of the ions and the configuration of their arrangement.

   Another example of collective excitations is the *plasmon*, the quantum of the *plasma wave*, which is the collective oscillation of electrons against a back-ground of positive charges. Yet another example is the *magnon*, the quantum of the *spin wave*, in which the magnetic dipole moments of electrons precess coherently.

   The energy–wavevector relations of phonons, plasmons, magnons, and other collective modes are calculated theoretically from certain effective equa-tions of motion of their participants. The calculation is feasible because it takes into account only the regular and correlated motions of the participants. Such a calculation should not be confused with the purely bottom-up deduction advocated by microreductionism. It is synthetic microanalytic, for its success depends on its being launched from a strategic high ground gained by physi-cal insight into the oscillatory behaviors of the system as a whole. The insight enables physicists to pick out the significant factors worthy of solution and build the simplification into the effective equations to be solved. Without such physical approximations, concepts such as collective excitations would never have fallen out from the brute-force logical manipulations of the basic equations.

   The energy–wavevector relations of various kinds of phonon are also mea-sured in a variety of experiments. In the experiments, probes – beams of light, electrons, neutrons – designed to excite a certain kind of collective mode are sent into the solid. The wavevector spectra of the scattered beams yield the desired energy–wavevector relation of the collective excitation. Other information such as the lifetime of the excitation can also be obtained by more refined analysis of the scattered spectra. Once the characters of the collective excitations are determined, they can be used in many contexts for the investigation of solid-state phenomena.

System Analysis: Specific Heat[3]

A collective excitation contributes to many characters of solids. Let us examine the thermal properties of insulators, so that we need not worry about the effects of electrons. The *specific heat* of a solid is the amount of energy required to raise the temperature of a unit amount of the material by one degree. At and above room temperature, almost all insulators have approximately the same value of specific heat, which is proportional to three times the number of atoms per unit volume and is independent of temperature. That is because the random agitation of each atom has three degrees of freedom, each of which absorbs energy. At low temperatures, the specific heat behaves very differently. Not only is its value lower, it is no longer a constant but varies as the cube of the temperature. The low-temperature behavior of the specific heat cannot be explained by the agitation of individual atoms.

At low temperatures, the energy fed into the solid is insufficient to agitate the ions in the crystal lattice individually; it can only cause the lattice to vibrate collectively. Thus the energy excites phonons with various energies and momenta, generating a distribution of phonons that varies with temperature. Treating the excited phonons as independent particles, we can use standard statistical methods to sum up their energies. The result agrees nicely with the observed temperature dependence of the specific heat.

To calculate the specific heat, we use the independent-particle approximation with respect to the phonons and neglect phonon interactions. Phonons contribute to many other solid-state phenomena in which we must take account of their direct interaction with electrons. For instance, they contribute to the ordinary electric conductivity of metals and semiconductors by acting as scatterers of electrons (§ 15). In some contributions, as in superconductivity (§ 22), the independent-particle approximation is not valid. System analysis too can be complicated and yield surprising results.

20. Strategic Competition and Industrial Organization

The general equilibrium theory for perfectly competitive markets, which dominated microeconomics for many decades, is a system analysis. It relies on the price parameters to coordinate individual behaviors without invoking interaction among individuals. Its individuals are collectives, economic institutions such as firms and households, that it treats as simple units with no internal structure. Oliver Williamson said that such theories belong to the "premicroanalytic" stage of understanding.[4]

Since the 1970s, microeconomists have increasingly turned from system analysis to collective analysis, from independent-individual models to interactive models, in which game theory is heavily used to represent interaction. The interactive models microanalyze institutional structures, investigate explicit relations and realistic competition, and flesh out the details of organization. These models are less unified than the perfectly competitive market models. They spread across information economics, transaction-cost economics, and industrial-organization theories. Compared to that of general equilibrium models that consider the economy as a whole, their vision is comparatively small: focusing on firms and other institutions. On the other hand, they delve deeper into the interrelationships among the constituents of the economy. In this section we examine these collective analyses. The corresponding system analysis, which uses the results of the collective analyses to determine the behaviors of the economy as a whole, remains to be developed.

Interacting Individuals and Governance Structure[5]

The major relation among economic individuals is *contract*, understood in a wide sense that includes the negotiation, the terms of agreement, and the execution of the terms. Exchange is a kind of contract. Spot trades, in which the two sides of a transaction occur simultaneously, are contracts instantaneously negotiated and executed. Even in spot trades, the contractual element is nontrivial if the buyer pays for the purchase with a credit card.

Monadic characters and causal relations complement each other. Thus the introduction of relations simultaneously changes individual characters. Some realistic characters of economic units neglected in the perfectly competitive markets models – their finite cognitive capacity, their reliance on customs and norms, their guile – become important in interactive models.

Ronald Coase complained of perfectly competitive markets models, "The consumer is not a human being but a consistent set of preferences. . . . We have consumers without humanity, firms without organization, and even exchange without markets."[6] Interactive models take a step forward by including some all-too-human factors. They remain within the general ambit of neoclassical economics by retaining the self-interestedness, fixity of goals and preferences, and instrumental reason of individuals. Individuals still care only about themselves, and they still do anything that advances their self-interest, but now anything really means *anything*. Individuals in independent-individual models have no occasion to lie or cheat, for they do not come in contact with each other and the market is deaf. Individuals in interactive models must deal directly with each other and include in their calculation the other's responses to their behaviors. They have plenty of occasions to exercise the full power of their instrumental reason, and they do not waste the opportunities. Consequently they are not only self-interested but self-interested with guile, or *opportunistic*. Opportunism includes the withholding of relevant information, as the seller of a used car refusing to divulge damaging information on the car even when asked.

The presence of others curbs the cognitive capacity of economic individuals. They cannot see through others' minds; sometimes they cannot even fully monitor others' behaviors. Therefore they must face the problems arising from *incomplete information* or *asymmetric information*. As a consequence of specialization and the division of labor, often one party in a relationship is better informed and has an edge over the other party. The asymmetry in information, exacerbated by opportunism, poses problems for the organization of productive effort. Self-interest dictates that the more knowledgeable party advance its benefit at the expense of its partner, resulting in conflict and inefficiency. To mitigate such an outcome, people develop contracts, corporate cultures, and governance structures that minimize the detrimental effects of asymmetric information by aligning the incentives of various parties.

For example, a principal–agent relation occurs when someone, the agent, acts and makes decisions on behalf of another, the principal. Examples are the relations between employer and employee, owner and manager, client and lawyer, patient and doctor. Typically, the principal is less knowledgeable about the job than the agent and is unable to monitor perfectly the agent's behaviors, either directly or indirectly, through observable variables. Thus the principal does not know whether he is being served or being duped; probably it is the latter, judging from the prevalence of opportunism. The problem can be mitigated by designing a contractual relation or governance structure that will align the incentive of the agent to that of the principal, thus inducing the agent to serve the principal's interest even when its behavior cannot be clearly observed. Formally, it is treated as an optimization problem aiming to find a contract that maximizes the expected utility of the principal subjected to the constraint that a compensation scheme is included that induces the agent to accept the contract. Examples of incentive-alignment mechanisms are compensation plans such as commissions, bonuses, and stock options.

Transaction Cost and Industrial Organization[7]

Contracting – the bargaining, drafting, and enforcement of contracts – is costly, sometimes prohibitively so, especially when we account for the opportunity cost of delays caused by negotiation. Contracting costs are major parts of what economists call *transaction costs*, which include search and information costs for finding out with whom and for what one trades, bargaining and decision costs for coming to terms and drawing up contracts, policing and enforcement costs for ensuring that the terms of agreement are fulfilled. Transaction costs, like friction, impede the decentralized market's efficiency in resource allocation and contribute to the appearance of large horizontally or vertically integrated firms with complicated internal managerial structures.[8] The perfectly competitive models are able to exclude such firms only because they ideally assume transaction costs to be zero.

Many commercial contracts involve *transaction-specific investments*, which may be human assets such as a special skill or physical assets such as a special-purpose machine. People make transaction-specific investments because they expect higher returns from the increase in efficiency. They also know the hazard of specialization. Specialists lose their versatility and are valuable only within a small circle of demand; hence they are unlikely to recover the cost of the sunk capital if the special relation collapses.

Williamson pointed out that a contract involving asset specificity effects a "fundamental transformation." To use a physical analogy, two parties engaging in an asset-specific contract are like two particles that form a bound state and henceforth obey a new set of rules different from that for free particles. Before contracting, each party is free to use its capital competitively in the market. After the investment, each specialized party becomes vulnerable to the opportunistic exploitation of the other. The exploitation is often inefficient; as the monopolistic buyer or seller to each other, each party is likely to set prices too high for the optimal volume of trade. Or a party may choose to underinvest, figuring that it will get more benefit even if the total return of the joint venture is diminished. If no provision is made beforehand about how the larger pie expected from specialization is to be divided, the squabble between the partners may reduce efficiency so much that the pie does not get any larger. It is to safeguard against such exploitation and its inefficiency that governance provisions become paramount in contracts.

Comprehensive contracting is impossible under the assumptions of opportunism, bounded rationality, and asset specificity. Because of unforeseen contingencies, bargaining must go on after the contract is signed, but such bargaining is conditioned by the governance structure stipulated by the contract. It is the job of the contract to provide the most cost-effective governance structure. In asset-specific contracts where trades occur infrequently or only once, for instance, the contract for a building, an arbitration scheme may be stipulated. Under other contracts, such as the orders for specialized automobile parts, trades occur frequently and the trading relation needs to be flexible to accommodate frequent design changes. The parties may find it costly to bargain for the changes each time and decide that the most efficient governance structure is unified control. Thus they merge into a single firm that internalizes transactions and economizes on transaction costs. The more highly specialized the asset, the more frequent the trade, the more likely that unified governance within a single firm is more efficient than unstructured bargaining. The efficiency ensures the emergence and flourishing of large and integrated firms. Thus transaction-specific investment, a form of differential interaction, leads to the formation of collectives.

Game Theory[9]

Unlike perfectly competitive firms that forage for profit independently of each other, firms in interactive models engage in hand-to-hand combat. The direct interaction is best represented by game theory. *Game theory* concerns interactive decision making, in which decisions are made in anticipation of the other parties' responses, assuming that all parties use instrumental reason to maximize their payoffs. Parts of game theory were developed early in the twentieth century. Initially it focused on the generalizations of parlor games such as chess or poker. The publication of the *Theory of Games and Economic*

Player B

		β_1	β_2
	α_1	5, 5	0, 7
Player A			
	α_2	7, 0	3, 3

FIGURE 5.2 A two-player game in strategic form. Player A has strategies α_1 and α_2; player B has strategies β_1 and β_2. For each pair of strategies, the first number in the entry represents the payoff to A and the second the payoff to B. This game is usually called the *prisoner's dilemma*, where the strategies α_1 and β_1 are interpreted as "cooperation," and α_2 and β_2 as "defection." Game theory stipulates (α_2, β_2) to be the equilibrium strategies.

Behavior by John von Neumann and Oskar Morgenstern in 1944 secured its impact on the social sciences. It has been further developed in recent decades, as its application to theoretical economics has mushroomed.

There are many kinds of games. In games of *complete information*, the rules of the game and the beliefs and preferences of the players are common knowledge, which every player knows and knows that every player knows. In games of *incomplete information*, the beliefs and preferences of the players may be partially hidden. Games with incomplete information are useful for modeling economic relations with opportunism and asymmetric information.

A game can be cooperative or noncooperative, a distinction descriptive of the rules of the game rather than the attitudes of the players. In *cooperative games*, the commitments, promises, or threats of the players are enforceable. Cooperative game theory has been used to show that the general equilibrium allocation is the limiting result of exchanges among an infinite number of economic units (§ 16). In *noncooperative games*, preplay agreements, even if they are made, are not binding. Thus the only meaningful agreements are self-enforcing, agreements each player finds it in its self-interest to keep, provided the other players act similarly. Noncooperative games constitute the bulk of applications in economics, where economic institutions are regarded as the results of spontaneous strategic interactions among the players. We will consider noncooperative games only.

A game can also be described with the idea of strategies. A *strategy* for a player is a complete plan prescribing the decision for each possible situation in the game. The *strategic* or *normal* form of a game consists of three common-knowledge elements: the set of players in the game; the set of strategies available to each player; the payoff to each player as a function of the strategies chosen by all players. For a two-player game, the strategic form is often depicted as a matrix as in Fig. 5.2. The columns and rows of the matrix are labeled by the particular strategies of the respective players. The entries of the matrix specify the payoffs to the players.

The outcome of a game is an important question in game theory. Unfortunately, there is no unique outcome in many games. Consequently there are many criteria or points of view, usually called *solution concepts*, which try to choose outcomes for classes of games systematically. Equilibrium, which has been an important notion in economics, is increasingly being regarded as a solution concept. The most widely used solution concept in noncooperative games is the Nash equilibrium. A *pure-strategy Nash equilibrium* is a profile of strategies, or a specification of a strategy for each player, such that no single player can increase its payoff by changing its strategy if the rest of the players stick to their strategies.

In the game depicted in Fig. 5.2, the strategy profile (α_2, β_2) is a Nash equilibrium. Player A reasons thus: If B adopts the strategy β_1, then α_2 is the better response; if B adopts β_2, then α_2 is better; therefore α_2 is the sure-win choice. B reasons similarly. So they end up with the payoff $(3, 3)$. The obviously better strategy profile of (α_1, β_1) with payoffs $(5, 5)$ is *not* a Nash equilibrium, for if either player sticks to it, the other can profit by defecting. The game is a model for the *prisoner's dilemma*, a well-known case in the social sciences.

A pure-strategy Nash equilibrium does not always exist. However, at least one mixed-strategy Nash equilibrium exists for any finite noncooperative game. A *mixed strategy* is represented by a probability distribution over the available strategies. An example for mixed-strategy Nash equilibria is the so-called battle of the sexes. A man and a woman engage in a game of deciding whether to attend a ball game or a concert. The man prefers the game, the woman the concert, but each prefers having the company of the other to being alone in the preferred activity. They settle on a Nash equilibrium by flipping a coin to decide where both will go. In more battlelike games, the aim of randomization in mixed strategies is to keep the enemy guessing.

The Nash equilibrium is not a very stringent condition. Often a game has a plethora of equilibria. Thus the identification of an equilibrium in applications is just to find *a* solution, the significance of which should not be overvalued. Furthermore, some of the multiple equilibria may not meet the requirement of self-enforcing agreements. Thus there has been considerable effort to find more restrictive solution concepts, often called the *refinement* or *perfection* of the Nash equilibrium. An example of refinement is the evolutionary stable strategy in biology, which requires the equilibrium to be robust (§ 21).

Game theory is widely applied. In political science, its applications range from voting behavior to strategic deterrence. Economic applications include oligopoly, insurance, antitrust, and bargaining. The applicability of noncooperative games to conflict situations is intuitive. However, noncooperative games also show that mutually beneficial agreements can still emerge spontaneously if the thinking of the purely self-interested players is broadened to include the temporal dimension. Such agreements are crucial to the formation of collectives.

The Possibility of Cooperation: Repeated Games[10]

Let us examine the possibility of cooperation – or collusion, as the public calls it – among oligopolists. In the matured capitalist world, the market structures of the manufacturing sector and to a lesser extent of the mining, finance, and retailing sectors are mainly oligopolistic, with a few large firms capturing the lion's share of the respective market.[11] Oligopolists face a host of strategic decisions that firms in a perfectly competitive market need not worry about. They have to decide on pricing, promotion, new product design, capital investment, research and development expenditure, entrance into or exit from a market, all the while keeping a keen eye on the moves of their rivals.

Adam Smith observed, "People of the same trade seldom meet together, even for merriment or diversion, but the conversation ends in a conspiracy against the public, or in some contrivance to raise prices." In our time, many conspiratorial agreements are not legally enforceable even if they do not invite prosecution. How are they maintained by the parties, each of which is eager to maximize its own profit at the expense of the others?

If there is only one supplier for a product, then the monopolist can charge a price higher than the competitive price by limiting the quantity of supply. The situation changes radically with the entrance of a second supplier, if the products of the suppliers are the same and a firm can always supply the quantity in demand. A market with two oligopolists can be represented as a game similar to that in Fig. 5.2. Suppose the oligopolists conspire to charge high prices. If both stick to the agreement, both make big money. However, each realizes that it can increase its profit further if it cuts its price, wins all customers, and drives the rival out of business. As beings with perfect instrumental reason, they invariably take advantage of the rival's cooperativeness, and the collusion collapses. Thus the theory of instrumental reason predicts that an oligopolist cannot charge anything higher than the competitive price.

Reality is different. Oligopolists usually succeed in cooperating and maintaining prices at above competitive levels. There are many explanations for the oligopolistic phenomenon, some of which are product differentiation, production capacity constraint, and the temporal factor. We will consider only the last explanation.

Oligopolists do no meet only once in the market; they interact repeatedly and know it. Consequently the past and future become important in present decisions. Oligopolists realize that their records convey information on their intentions and will be taken into consideration in their opponents' decisions. Therefore their current behaviors affect not only current but future payoffs. They have to protect their *reputation*, which they now value as an asset. The success of oligopolists in sustaining tacit collusion depends on their ability to signal credible threats of retaliation and to make good on such threats whenever called for. When one airline announces a deep-discount fare, the other major airlines immediately match it. The fare war leaves all airlines bloodied, but the cost is worthwhile; it not only punishes the maverick but builds up the reputation for retaliation and sends a deterring signal to all who are tempted by temporary gain to cut prices. The reputation builds implicit trust and the oligopolists succeed in maintaining monopolistic prices without explicit illegal agreements.

Cooperation arising from repeated interactions of self-interested individuals can be studied by dynamic game theory. A *repeated game* consisting of many rounds of play of the prison-dilemma game in Fig. 5.2 can be made into a model for repeated encounters of two oligopolists. In a repeated game, a player does not consider each component game in isolation. It accumulates information as the game is played out, uses the record to decide its move in the current component game, and aims to maximize its total payoff over the entire time horizon. Depending on how highly the future is valued, future payoffs may be discounted in current decisions, and the discount factors are part of the rules in a repeated game.

A repeated game can be regarded as a composite system in which the basic game is a constituent. Like other composite systems, it often exhibits novel and surprising features. The state space and the set of Nash equilibria of a repeated game are greatly and interestingly enlarged compared to the single play of a component game. A basic result of repeated games, called the *folk theorem*, states that as the number of repetitions goes to infinity, any attainable set of payoffs for a component game can be the realized equilibrium set in any period of the repeated game, if the payoffs give each player more than the minimum it could get. Thus in an infinitely repeated prisoner's-dilemma game, the cooperative strategy profile (α_1, β_1) can be the equilibrium in most or even all periods. Temporal considerations enable the players to achieve in a noncooperative game an outcome normally associated with cooperation.

According to the strict logic of instrumental reason, cooperation is not possible if the game is repeated only a finite number of times. Consider a finite repetition of the prisoner's-dilemma game. Instrumental reason dictates that both parties defect in the last period, for there is no future to worry about. But when the result of the last period is a certainty, the penultimate period becomes the last, and defection is again prescribed. Iterating the argument leads to the stipulation of defection for all periods. Similar reasoning does not hold if the games are infinitely repeated because there is no final period. The logic of the finitely repeated game is not borne out in real life, as is apparent in a computer tournament where the prisoner's-dilemma game is repeated 200 times. Most contestants opted for cooperation and did well.[12] The triumph of cooperation in a finite context illustrates the superiority of commonsense rationality to instrumental Rationality with all its logical rigor.

21. Population Structure and the Evolution of Altruism

The social interaction among organisms and the transient fragmentation of a population into groups generate intermediate-level structures in organic populations. However, the structures are murkier than those in economic and physical systems. Unlike economic and physical collectives that have causally efficacious characters of their own, groups of organisms are characterized only in terms of their size and statistical composition. Their lack of clear causally efficacious characters prevents the separation of population genetics into collective and system analyses, which are treated in a single theory. It also lies at the base of the debate over group selection.

This section examines the representation of social interaction by game theory. Differential interaction leads to social groups. The group structures of populations facilitate the evolution of altruistic behaviors, which will surely be eliminated by natural selection in homogeneous populations.

Evolutionary Game Theory and Frequency-dependent Fitness[13]

To account for causal relations among organisms in the statistical framework of population genetics, biologists introduce the notion of association-specific fitness, the value of which depends on the compositional structure of the population. Association-specific fitness is used in game theory to find the stable-equilibrium composition of the population.

Evolutionary game theory uses the same mathematics discussed in the preceding section. A kind of organism is represented by the mathematical concept of a player, an organismic behavior by a strategy, the interaction of organisms displaying various behaviors by the matching of various strategies, and the association-specific fitness of a behavior by the payoff to a strategy when matched against another.

Consider the Dove–Hawk game, in which a pacificist behavior called Dove and an aggressive behavior called Hawk are represented by two strategies. Suppose there is a certain resource that, when always appropriated in full by an organism behaving in a certain way, increases the fitness of the behavior by r. Any two doves split the resource peacefully; thus the fitness of Dove associated with Dove is $r/2$. Any two hawks split the resource after an injurious fight that decreases Hawk's fitness by d; thus the fitness of Hawk associated with Hawk is $r/2 - d$. The fitness of Hawk associated with Dove is r, of Dove associated with Hawk is 0, for doves always withdraw without confrontation, leaving all the resources to hawks. Dove and Hawk are social behaviors. The fitness of a social behavior is association-specific, for it depends on the behavior of the organisms with which its bearers deal. Pacifism fares better when pacifists deal with pacifists than with aggressors.

To generalize the idea behind the Dove–Hawk game, let us denote by w_{ij} the *association-specific fitness* of the behavior i, or the fitness of the behavior i when organisms displaying it interact with organisms displaying the behavior j. A social behavior is effective only if the relevant social intercourse takes place. Intercourse depends on the presence of partners, and the availability of partners depends on the composition of the population. Organisms with behavior i interact with organisms behaving in many different ways, and each association-specific fitness contributes to the overall fitness w_i of the behavior i. The fitness w_i is obtained by averaging w_{ij} over all j, taking into account the probability that a bearer of i interacts with a bearer of j, given the distributions of both kinds of organism in the population. The resultant fitness w_i depends on the frequency of each behavior. *Thus the fitness of a social behavior depends on the composition of the population.*

Given the association-specific fitness and the initial percentages of doves and hawks in a population, and assuming that each organism of the population interacts once with every other organism, the fitness of Hawk and Dove can be readily computed. The fitness depends on the percentages of doves and hawks in the population. For instance, the fitness of Hawk is higher when there are plenty of doves around than when there are none.

The fitness of a behavior is the relative growth rate of the fraction of organisms in the population exhibiting it. Knowing the fitness of various behaviors, we can determine how the composition of the population changes in successive generations. Evolutionary game theory aims to find a stable composition of the population. Maynard Smith introduced the solution concept called *evolutionarily stable strategy* (ESS). An ESS is a Nash equilibrium that is robust or stable in the sense that slight deviations damp out and the population returns to its equilibrium state. A population in ESS equilibrium cannot be evolutionarily invaded by a few deviants behaving differently, for the deviant behavior cannot spread.

Obviously a population of doves is not stable; Hawk will spread rapidly once a few hawks land in it. The stability of a population of hawks depends on the effect of the resource $r/2$ relative to the effect of the fight d. When the damage due to the fight is low, a hawk population is stable. When the damage outweighs the gain from the resource, it is unstable; Dove has a chance of establishing a foothold as the hawks fight themselves to death. The stable population is a mixture comprising a fraction $r/2d$ of hawks and the balance doves. It is mathematically represented by a mixed strategy. Alternatively, the stable population can be interpreted as comprising organisms that behave as Hawk a fraction $r/2d$ of the times.

The Dove-Hawk game is an example of strategic optimization in the adaptation program and shares its shortcomings. Its considerations are more economic than biological. There is no indication that the strategies have genetic bases; it is possible that the equilibrium population with the right mix of doves and hawks is attained within a single generation as the organisms are conditioned by bloody stimuli to behave properly. Therefore its significance for evolution is unclear.

Emergence of Groups and Population Structure[14]

All models we have examined so far, including the game theoretic models, assume that the population is homogeneous. In reality, many organisms live in groups at least during part of their life. Families, whose participants are related through descent, are only one kind of group. In many mammal and bird populations, the social structures can be multitiered, with groups within herds and coalitions within groups, as the gelada baboons in central Ethiopia form large herds for grazing and small groups for nesting.

The assumption of a homogeneous population treats social interactions as on-or-off deals and makes no provision for the varying strength of interaction. Once an interaction turns on, it obtains uniformly among all organisms of a population, so that every organism in the population interacts with every other with equal strength. The assumption is unrealistic. We have many social traits, but how many persons within your community have you interacted with and influenced?

Perhaps it is impossible to characterize the varying strength of complicated social interactions in a systemic way reminiscent of the inverse square law of gravitation. Perhaps it is impractical even to estimate the effective ranges of social interactions. Nevertheless, a conceptual distinction should at least be made between organisms that are and are not likely to interact. The distinction is made with the introduction of organismic groups, which in turn introduces structures into the population.

There are several definitions of organismic groups, which are closely associated with the unending debate on group selection. We will consider only transient groups, the trait groups.[15] The organisms within the effective range of a social behavior form a *trait group*, so that the fitness of that social behavior depends only on the composition of the group and not on the composition of the population at large. Different social behaviors have different effective ranges and hence different group sizes. A *structured population* is one whose constituents form trait groups during at least part of their lifespan, interact with each other within the groups, then disperse and mate within the population. An important feature of structured populations is that the groups can vary in composition. A group can contain a larger percentage of organisms behaving in a certain way than other groups. The structured population is the secret to the evolution of altruism.

The Possibility of Altruism[16]

In evolutionary biology, altruism is understood in a strictly nonintentional and nonmotivational sense. A social behavior is *altruistic* if it is active and lowers its own fitness relative to the fitness of the recipient behavior. Remember that fitness is a *relative* growth rate. It is not enough for a behavior to have a high growth rate; the behavior is doomed to be crowded out if other behaviors have even higher growth rates. Suppose a population comprises organisms with behaviors i and j. Behavior i is active; its bearer contributes a factor c to itself and a factor b to whomever it interacts with. Behavior j is passive; its bearer does nothing except to receive the contribution from bearers of i. Assuming uniform interaction among organisms, the difference in the fitness of the two behaviors is $w_i - w_j = c - b$. The signs of b and c can be positive or negative, depending on whether the behaviors are beneficial or harmful to the recipient. If everything else is equal, then behavior i grows faster and will spread at the expense of behavior j if $c > b$. This happens if i is selfish ($c > 0, b < 0$), mutualistic ($c > b > 0$), or spiteful ($0 > c > b$). Behavior i is *altruistic* if its bearer incurs a relative cost with respect to the recipient of its action ($c < b$), or more stringently if it incurs an absolute cost ($c < 0$ and $b > 0$). Altruistic behaviors are peculiar in that they always grow more slowly than other behaviors. Thus they seem destined to be eliminated by natural selection. They are not. Why?

Apparently altruistic behaviors abound in the natural world: honeybees kill themselves by stinging common enemies, adult birds rear the young of others instead of breeding, prairie dogs expose themselves to danger by giving alarm calls at the approach of predators. How can these behaviors survive the winnowing of natural selection? There are several explanations. The behavior may be mutualistic and not altruistic, although the benefit may be delayed or hidden. Those who give warning calls also benefit from the calls of others in the community. Birds become helpers only if they cannot find appropriate resources to breed themselves and their helping action may enhance their future success in breeding.[17] Here we consider an important explanation, the structure of the population and the formation of groups, including groups of kin.

Consider a structured population partitioned into trait groups so that organisms in a group interact among themselves but not with outsiders. Since the fitness of a social behavior depends not on the composition of the population at large but on the composition of the group, and since different groups have different compositions, the fitness varies from group to group. Therefore we must distinguish a behavior's *local fitness* defined within a group and its *global fitness* defined within the population at large. We classify a social behavior as altruistic or selfish according to its *local* fitness, for the distinction is significant only if the organisms can interact. On the other hand, the evolution of the behavior is determined by its *global* fitness, which gives its relative growth rate accounting for all factors, including the structure of the population.

If all groups are alike in composition, then the global fitness is the same as the local fitness and altruistic behaviors are doomed. Interesting cases occur when the assortment into groups is not random and the compositions of the groups vary. Suppose organisms behaving in the same way tend to congregate. In groups consisting mostly of altruistic organisms, the majority is taken advantage of by the selfish minority, and the local fitness of altruism is less than that of selfishness as the relative abundance of altruistic organisms within the group declines. In groups consisting mostly of selfish organisms, the altruistic minority fares even worse, and the local fitness of altruism is much lower than that of selfishness. Within each group, the conclusion is not different from that of homogeneous populations. Altruism cannot evolve.

A new factor comes into play in structured populations. Evolutionary models usually do not worry about the absolute sizes of groups or populations; only the relative abundance of organisms with various behaviors counts. However, when a population contains groups with different compositions, the *absolute sizes of the groups* become important. Organisms in a mostly altruistic group thrive in the friendly and cooperative atmosphere, leading to a great increase in the absolute size of the group. In contrast, the size of a group of selfish organisms increases only slightly or perhaps even decreases. Altruism has a diminishing share in each pie, but the pies for which it has a lion's share are getting much bigger. When we average over the local fitness weighted by the sizes of the groups, we find that altruism sometimes comes out ahead globally. Altruism has a lower local fitness than selfishness in each group, as its definition demands. However, in certain structured populations, the global fitness of altruism can exceed that of selfishness, thus ensuring the evolution of altruism.

The effect of population structure on natural selection can be seen in the evolution of the myxoma virus after its 1950 introduction in Australia to control the rampant proliferation of European rabbits. The virus is transmitted by mosquitoes, which bite only living rabbits. Within ten years, the virulence of the virus declined significantly. Similar effects were observed in Britain and France. A virus's virulence is ofen attributed to its reproductive rapidity; the virulent strand reproduces faster than the mild strand. The mild strand is altruistic because its birth rate gives it a lower local fitness. If this explanation is correct, then the decline in virulence cannot be explained if the viruses formed a homogeneous population. However, the actual virus population is highly structured; the viruses in each rabbit form a group. Suppose the composition of different strands of virus in each rabbit varies. Virulent viruses that kill their hosts quickly also kill their opportunity of transmission by the fastidious mosquitoes. Viruses in groups with a high percentage of mild viruses were more likely to be transmitted to other rabbits because their hosts live longer. Thus the population structure helped the spread of the mild strand.[18]

Several factors are crucial for the evolution of altruistic behaviors. The trait groups must persist long enough to allow the effect of social interaction to show up in differential group sizes but briefly enough to prevent selfishness from taking over the initially altruistic groups. This can happen only if the constructive effect of the altruistic interaction is sufficiently large. The environment for each group must have enough resources to support its potential population growth. If resource limitation puts a cap on the size of the groups, then the bigger-pie effect responsible for high global fitness of altruism is destroyed. Finally but not least important, there must be significant variation in the compositions of the groups. The larger the intergroup variation, the more readily altruism can evolve. Larger variation implies there are more chances for organisms behaving similarly to congregate so that an altruistic behavior is more likely to benefit an altruist.

In models of structured populations, the trait groups are the collectives in a population. The definition of local and global fitness distinguishes between collective and system analyses. Collective analysis, which differentiates between altruistic and selfish behaviors, studies the changes in the compositions and sizes of the groups. System analysis determines how the characteristics of the groups contribute to the overall growth rates of various behaviors and the evolution of the population as a whole. The two analyses are integrated in a single model that explains the evolution of altruism.

6 Composite Individuals and Emergent Characters

22. Emergent Characters of the System as a Whole

So far we have examined models in which large composite systems are microanalyzed into uncoupled or weakly coupled entities. In some models, the entities are modified constituents that have internalized much of their former interaction; in others, they are collectives engaging the organized behaviors of many constituents. The models focus on finding the characters and behaviors of the entities that, once known, can be readily aggregated to yield system characters, for the weak coupling among the entities can be neglected in a first approximation. Thus we can crudely view a system in these models as the sum of its parts, although the parts are specialized to it and differ from the familiar constituents of its smaller counterparts.

Not all phenomena are microanalyzable by modularization. Phenomena such as freezing and evaporation, in which the entire structure of a system changes, would be totally obscured from the viewpoint of the parts. Here the correlation among the constituents is too strong and coherent to be swept under the cloaks of individual modules, whatever they are. In such thoroughly interconnected systems, the effects of a slight perturbation on a few constituents can propagate unhampered and produce large systemwide changes. Consequently, the behaviors of these systems are more multifarious, unstable, and surprising. They are often called emergent properties and processes. Emergent characters are most interesting and yet most controversial, for they are most difficult to treat theoretically and incur the wrath of some revisionary philosophies.

Emergence and Explicability

The idea of emergence has long root. Plato distinguished a whole that is "the sum of all the parts" and a whole that is "a single entity that arises out of the parts and is different from the aggregate of the parts." Mill distinguished two sorts of compounds produced by the composition of causes. In the first, exemplified by the mechanical combination of forces, the joint effect of the component causes is identical with the sum of their separate effects. In the second, exemplified by a chemical reaction, the effect of the joint cause is heterogeneous to the component causes and demands characterization of its own. He cited the example of oxygen and hydrogen combining to form water: "Not a trace of the properties of hydrogen or of oxygen is observable in those of their compound, water." "No experimentation on oxygen and hydrogen separately, no knowledge of their laws, could have enabled us deductively to infer that they would produce water." The properties of the compounds formed by the two kinds of composition were later called *resultant* and *emergent*.[1] Mill's account suggests three criteria of emergence: First, an emergent character of a whole is not the sum of the characters of its parts; second, an emergent character is of a type totally different from the character types of the constituents; third, emergent characters are not deducible or predictable from the behaviors of the constituents investigated separately.

Most philosophical doctrines focus on the third criterion and regard emergence as the antithesis of microreductionism. They define a character of a composite system to be emergent relative to a theory if its occurrence cannot be deduced from the attributes and relations of the constituents by means of the theory. This logical definition of emergence seems too loose and arbitrary. The purely bottom-up deduction of microreductionism rarely succeeds in theories for realistic systems of any complexity. There would be too many emergent properties if they occur whenever we appeal to the top-down view of synthetic microanalysis or posit additional conditions and approximations. Worse, a character would be emergent in some theories and not emergent in others, for theories differ in power and generality. Some characters would lose their emergent status simply because we have developed powerful tools such as the computer. So fickle a characteristic is hardly worth contention. To make emergence less dependent on specific theories, some people propose that emergent characters are not deducible from the ideally complete knowledge of the constituents in the ideally complete science. This definition is too speculative and impractical, for it depends on the ideal limit of science, which is itself highly controversial. It makes present ascriptions of emergence arbitrary, because we do not know the substantive details of the complete science and attempts to outguess future science are foolhardy.

Some people push the deducibility criterion to the extreme, equate predictability with explicability, and regard emergent characters as inexplicable and mystical. This view was popularized in the 1920s in the doctrine of emergent evolutionism, which attributed to evolution some mysterious creative force that produced genuinely new life forms. It gave emergence a bad reputation that has not been entirely cleansed today. Only those without practical sense would identify explicability with predictability. Any student knows that peeking at the answer at the back of the book helps to solve a problem. Of course the real problem solving lies in working out the interim steps; otherwise the answers would not be there. Nevertheless the answer indicates the direction to take, and the beacon makes all the difference from groping in total darkness. Similarly, it is easier to explain than to predict, for prediction faces the overwhelming possibilities generated by the combinatorial explosion, whereas explanation knows what it is shooting for. Explanation has the advantage of partial hindsight.

The philosophy of science tends to overestimate the power of deduction and give short shrift to experiment, to which it assigns only the job of testing and verifying theoretical hypotheses. Experiments are much more important; they are scientists' way of peeking at the answers. With experiments scientists seek clues from nature to aid and guide theorization. There are pure theoretical innovations such as general relativity, but they are the exceptions rather than the rule. Most theories, including special relativity and quantum mechanics, are developed to explain the puzzling phenomena observed in experiments. Even after these general theories have established powerful frameworks for prediction, physicists have never slackened their experimental effort. Condensed-matter physics, for example, is as experimental as theoretical. Theories help experimentalist to design incisive experiments, and experiments yield results that guide the thinking of theoreticians. They coevolve. The indispensability of experiments in scientific research and the surprises experiments constantly throw up suggest that nature has many emergent characteristics in store.

Because of the complexity of large-scale organization, we are usually unable to predict emergent characters from scratch on the basis of the information of their constituents. But emergent characters are not inexplicable, even if their microexplanations are not yet available. Once they are experimentally discovered and probed, we can strive to find their underlying micromechanisms. The microexplanations will not be easy and may be years or decades in coming. They typically contain approximations and postulates made possible by partial hindsight, extra factors that explain why deductive prediction is ineffectual.

I deemphasize the deducibility criterion of emergence, because it has little weight in the framework of synthetic microanalysis. Focusing on the criteria of additivity and novelty, I try to find some objective distinctions between resultant and emergent characters and examine the theories we do have for their microexplanations.

Emergent Versus Resultant

Emergence is not simply radical difference. Two things can be as different as heaven and hell, but we do not call the character of either emergent. Emergence results from certain *transformations* of things. Traditionally, the transformation considered is composition. The difference between a composite system and its constituents does not always favor the system. The system can have additional characters; it can also lack some characters of the constituents. Macroscopic systems typically lack the phase entanglement of their quantum constituents. However, by convention only the extra characters of the system are candidates for emergence.

We broadly classify the characters of composite systems as resultant or emergent. Crudely speaking, resultant characters are more closely tied to the *material* content of the constituents; they include aggregative quantities such as mass, energy, force, momentum, and quantities defined exclusively in terms of them. Emergent characters mostly belong to the *structural* aspect of systems and stem mainly from the organization of their constituents. Descriptions of structures and configurations can be quite abstract, often calling for concepts that make little reference to material; for that reason emergent characters often appear peculiar. Large systems are more likely to exhibit emergent characters, because the combinatorial explosion generates endless structural varieties, some of which may be amazing.

Emergent characters of composite systems are often said to be novel. Novel compared to what? How is the novelty gauged? I try to answer these questions by pushing the domain of resultant characters as far as possible, so that the club of emergent characters is reasonably exclusive.

Emergence is closely associated with microexplanations of macrophenomena. Ignoring intermediate levels of organization, microexplanations can involve four kinds of description, two for individual constituents and two for the system. Each description uses its own peculiar character types. The *elemental description* describes the constituents in terms of the character types and relation types familiar in small or minimal systems, so that it must list the myriad relations in a large system. The *constituent description* uses specialized predicates to absorb most relations and characterize individual constituents within the many-body system more concisely. The *microdescription* of a composite system is given in terms of character types, each value of which specifies the behaviors and relations of every constituent. These character types can be constructed from the constituent characters straightforwardly. The *macrodescription* introduces character types with simpler values and seals the unity of the composite system. For example, a value of a microcharacter type for a gas contains trillions of trillions of numbers for the momentum of every molecule; a value of a macrocharacter type is a single number for the pressure of the gas.

Emergent characters usually occur in the macrodescription, for we demand that they be describable concisely in their own terms. That is why people often say organisms have emergent characters but clocks do not. The characters of a contraption of interlocking wheels and springs are obviously quite different from those of its components. When asked to describe it, however, we find ourselves in a microdescription reporting the relative motion of the components. The device itself seems to lack characters of its own; it emerges from the rambling microdescription only when we acknowledge it as a clock. A clock's significance derives from its use, which is embedded in the broader context of purposeful action. In the theoretical context that abstracts from uses and purposes, the contraption can hardly be described independently of its components. In contrast, we readily describe an organism's birth and death, growth and reproduction, which are alien to the molecules that constitute it. The independent predicates give the system its own individuality.

The novelty of emergent character types in macrodescriptions is usually judged by comparison with the character types in elemental descriptions, which are common to systems large and small. Let us try to make the criterion of novelty as stringent as possible.

When we say a system's emergent character differs not only quantitatively but qualitatively from the character of its constituents, *quantitative* has a more general meaning than that regarding amount or size. Every attribute has a typical and a definite aspect, called its type and value (§ 5). I will call the differences in value *quantitative* and the differences in type *qualitative*. Red and green are values of color. Things with various colors are quantitatively different. They are qualitatively different from atoms, to which the character type of color does not apply. The ascription of the wrong value to an individual is an error; the ascription of the wrong type becomes an absurdity or a paradox.

A character type is some kind of rule for assigning values and it can cover a wide domain. Since a type is a rule, its values are systematically related to each other. The rule and relation are most apparent when the values are numerical. Even when they are less apparent in other representations, a conceptual relation is tacitly understood to obtain among phenomena with the same character type. A compound's resultant property shares the same character type as its constituents, and the underlying conceptual relation makes the resultant property "expected." In contrast, the character type of an emergent property differs from those of the constituents; hence the emergent property is said to be "novel" and "unexpected."

The scopes of the character types defined in the theoretical sciences are often very broad; the generality of types contributes much to the potency of the sciences. When we consider such types as acidity or oxidation number, we realize that the differences between many chemical compounds are quantitative in our broad sense. For instance, the combination of two hydrogen atoms and an oxygen atom into a water molecule is a quantitative change in chemical properties, which makes the water molecule resultant. (Note that we are talking at the molecular level; the same cannot be said for the transformation of hydrogen and oxygen gases into liquid water, which was what Mill probably had in mind.)

The difference in type is still too loose a criterion for emergence, for it is possible that some character types exhibit a certain affinity, either conceptually or descriptively. Constituents can change their character drastically in large systems, as a glance at Fig. 4.2 shows. However, the conceptual relation between the character types in the elemental and constituent descriptions is straightforward, as demonstrated by the principle of adiabatic continuity in physics (§ 15). Often the two classes of character type involve the same concepts related in different ways. Thus we can further expand the domain of quantitative difference to include them. For example, we can make the energy–wavevector relation in general into a superordinate type, a value of which is a specific energy–wavevector relation. Such superordinate types are not strangers in scientific theories, where the values of many character types are complicated objects with variable internal structures. If the conceptual relations among a group of character types justify the introduction of a superordinate type, then these characters are quantitatively similar. Properties of compounds falling under the same superordinate type as those of their constituents are resultant.

We now turn to the criterion of additivity. Character types in the macro-description that can be defined in terms of the sum or average of constituent behaviors are usually resultant. Most optical, electric conductive, and thermal conductive properties of solids are aggregate characters, if the notion of "constituents" is generalized to include the appropriate collective excitations. For instance, the absorption spectrum of a solid arises from the coupling of the electromagnetic radiation to each electron and phonon. A material is transparent if none of these absorbs radiation in the visible range. In the calculation for the absorption spectrum, the phonons are treated as a kind of independent individual in the system analysis. Hence formally, the optical property of a solid is the sum of the optical properties of its "parts" suitably defined. Ordinary electric conductivity and thermal conductivity of solids are determined similarly. If we ignore whatever emergent characters the phonons and other collectives may have, we can call these optical and conductive characters resultant.

Aggregation and averaging, which yield resultant properties, are possible only if the constituents are weakly coupled. This implies that certain independent-individual models are available in the first approximation, and that the characters of the constituents to be summed have internalized many relations. The independent-individual approximation involves much theoretical maneuvering. Its sophisticated version includes a self-consistently determined situation to which the individuals respond. It is powerful and widely used. Combined with various modifications that take account of the residue interaction among the individuals and enable us to define superordinate types, independent-individual models have greatly expanded the domain of resultant characters. To call the balance of characters emergent is reasonable. Thus I suggest we call emergent only those macrocharacter types with no satisfactory microexplanations in independent-individual models and their modifications.

The idea of emergence can be generalized to cover nonlinear dynamics (§ 29). Nonlinearity stands in contrast to linearity. The motion of a linear system can be analyzed as the sum or superposition of certain standard motions. For instance, the linear (harmonic) vibration of a violin string is analyzable into the superposition of its normal modes (§ 19). Here the normal modes are analogous to uncoupled individual parts, the superposition principle is analogous to the aggregation of parts, and linear systems have resultant behaviors. The superposition principle is the secret of the tractability of linear systems. It breaks down for nonlinear systems. Nonlinearity underlies the chaotic and unpredictable behavior of deterministic systems. These behaviors are emergent according to the preceding criterion.

The independent-individual approximation fails for systems in which the constituent behaviors are highly coherent and correlated. The principle of superposition fails for nonlinear systems. In such systems it is less meaningful to talk about the individual character of a part because it is so tightly bound to the behaviors of the others. A more holistic treatment of the system is required. Holistic contexts are traditionally closely associated with the idea of emergence.

In sum, a macrocharacter of a composite system is *resultant* if it satisfies two criteria: First, it is qualitatively similar to the properties of its constituents, belonging to the same character type or superordinate type as the constituent properties; second, its microexplanation can be given by approximately microanalyzing the system into independent parts with distinctive characters such that it is the sum or average of the characters of the parts, where the microanalysis includes independent-individual models, the superposition principle, and other available means. A character that is not resultant is *emergent*. To give the maximum range to resultant characters, the "part" in the second criterion is understood in the broadest sense as an independently describable unit that need not be spatially individuated. It is the result of microanalysis aimed at finding disjoint or weakly coupled units in a particular kind of system. Its characters usually absorb most elemental relations and differ from the characters in the elemental description, but the two kinds of character are relatable under superordinate character types.

The distinction between emergent and resultant characters based on the feasibility of microanalyzing the system into distinct parts is not sharp. The independent-individual approximation is often used as a first-cut attempt at solving many-body problems, and it is the only tractable method in some complicated cases. However, in these cases the failings of the approximation are also apparent. Scientists usually have a feel for and a rough consensus on the adequacy of a model for certain phenomena, even when they are stuck with it because of technical difficulties. Thus the phenomena for which independent-individual models are commonly judged to be unsatisfactory, even if the models are the only ones presently available, should be regarded as emergent.

The distinction between resultant and emergent applies mainly to characters, although people sometimes call entities that exhibit prominent emergent characters emergent. All composite systems have some resultant characters. Resultant characters often play crucial roles in the microanalysis of emergent characters, because their own micromechanisms are rather apparent and they are causally related to the emergent characters on the macrolevel. In statistical mechanics, for instance, the definition of the ensembles that bridge the micro- and macrostates is based on resultant characters.

Examples: Conductivity Versus Superconductivity

Ordinary electrical conductivity is a resultant phenomenon; superconductivity is emergent. With varying temperatures, a solid can transform from an ordinary conductor to a superconductor, and the transformation is reversible. Thus conductors and superconductors involve the same electrons, their difference lies solely in electronic organizations, illustrating that emergent characters belong to the structural aspects of systems.

An electron carries an electric charge and its motion generates a microscopic electric current. Usually the electrons move in all directions and their currents cancel each other, so that the total macroscopic current is zero. A voltage applied across a conductor herds the electrons in one direction and generates a macroscopic current that is the sum of individual microscopic currents. Even the best conductors have finite resistivity. If the driving force is turned off, the macroscopic electric current decays and finally vanishes, because whenever an electron suffers a collision and changes its course, the total current decreases a bit, and bit by bit it dies. The decay of the macroscopic current is a good manifestation of its resultant nature. Just as conductivity can be built up by adding individual contributions, it can be destroyed by subtracting individual contributions.

Although electrical conductivity is resultant, it is not derivable from the laws of particle motion by the purely bottom-up method. The theory of conductivity involves all the apparatus discussed in §§ 15 and 19, viz., the theories of individual conduction electrons and phonons that scatter electrons. Both depend on the structures of crystal lattices, which have to be experimentally determined. These theoretical complications show how far we have pushed the domain of resultant properties.

Heike Onnes liquefied helium in 1908 and originated the field of low-temperature physics. Three years later, he reported that when mercury is cooled below a critical temperature, it abruptly transforms into a superconducting state with vanishing resistivity. He won a Nobel prize. Many other metals make the transition to the superconducting state at various critical temperatures. A macroscopic electric current established in a superconducting coil persists forever after the driving force is turned off. Zero resistivity is not the only amazing macroscopic character of superconductors. Among other macroscopic behaviors is the expulsion of all magnetic fields, which is often demonstrated by a bar magnet suspended in air above a superconductor.

Superconductivity is an emergent phenomenon. The superconducting current cannot be obtained by aggregating the contributions of individual electrons; similarly it cannot be destroyed by deflecting individual electrons. The microscopic mechanism underlying the observed superconductivity baffled physicists for many years, and was finally explained in 1957 by the Nobel prize-winning BCS theory developed by John Bardeen, Leon Cooper, and Robert Schrieffer. An electron momentarily deforms the crystal lattice by pulling the ions slightly toward it, creating in its wake a surplus of positive charges, which attract a second electron. Thus the two electrons form a pair, called the Cooper pair. The radius of a Cooper pair, measured by the extent of its wavefunction, is about 0.1 micrometer, which is much larger than the interelectron spacing in metals. Within the extent of a pair fall the centers of millions of pairs. All the Cooper pairs interlock to form a pattern spanning the whole superconductor. Since the interlocking prevents the electrons from being individually scattered, the whole pattern of electrons does not decay but persists forever in moving coherently as a unit. The structure of the interlocked electronic pattern as a whole, not the motion of individual electrons, is the crux of superconductivity.

For a long time most known superconductors were metals. Then in 1986, Georg Bednorz and Alex Müller discovered a new class of superconducting material and immediately received a Nobel prize. The new superconductors, mainly oxides, have higher critical temperatures, but they share most macroscopic behaviors of the old superconductors. Interestingly, they have radically different underlying micromechanisms, which the BCS theory does not explain. Smelling another Nobel-caliber problem, physicists descend on it like a pack of hungry wolves. They make much headway, but after ten years, a totally satisfactory microexplanation was still not available.[2]

Superconductivity highlights several characteristics of emergent properties. No special ingredient is needed for emergence; superconductors involve the same old electrons organized in different ways. The structured organization in superconductivity constrains the scattering of individual electrons and forces them to move with the whole, which can be viewed as a kind of downward causation. There are two types of superconductor with two distinct kinds of micromechanism, showing that an emergent character need not be stuck to a specific kind of material. This point will be elaborated later.

Both kinds of superconductivity were experimentally observed and their macroscopic causal connections understood before their micromechanisms were explained. Without the hints from observations, who would have dreamed that electrons, which electrostatically repulse and quantum-mechanically exclude each other, can form pairs? Perhaps one can argue that superconductivity was observed before physicists had the quantum mechanics required to predict it. The argument fails for the second round. Superconductivity has great potential for application; imagine the saving on energy if long-distance power lines were made up of lossless superconductors. Unfortunately, its application was limited by the low temperatures in which the first kind of superconductivity operates. Obviously there are strong incentives to find high-temperature superconductors. Physicists know all the microscopic laws governing the motions of electrons and ions. They know that superconductivity is possible and possess a theory for one kind of micromechanism for it. They know from numerous other cases that similar macroscopic phenomena can be realized by different micromechanisms. Yet they did not theoretically predict by deducing from quantum-mechanical laws the existence of the second kind of superconductivity, which awaited experimental discovery. The prompt accolades for the second discovery confirm the common saying that emergent phenomena are surprising and unpredictable from the knowledge of the constituents. As usual in science, unpredictability does not imply inexplicability. The micromechanism of high-temperature superconductors is still a mystery, but I am confident it will be fully explained.

Universality and the Supermacrodescription

Superconductivity is not the only emergent property that admits different micromechanisms. Emergent characters are part of the structural aspect of composite systems. A structure can be realized in many kinds of material, as an arch can be built of wood or stone or steel. Similarly, an emergent character can be exhibited by systems with widely disparate constituents and micromechanisms. This is because many microscopic details of the constituents become irrelevant for the salient features of large structures. Physicists have a mathematical theory showing how the microdetails are filtered out in coarse-grained views where emergent characters become prominent. Occasionally, emergent characters are so independent of the specifics of the constituents they are known as *universal*. Phase transitions as different as the freezing of water and the magnetization of iron share universal characters represented by something called critical exponents (§ 23).

The universality of some emergent characters shows that we generalize and define properties independently on the micro- and macrolevels. We define macrocharacter types by how they nest in the network of macroscopic causal relations, as various macroscopic electric and magnetic properties of superconductors. Seldom do we define a macrocharacter type by identification to a microcharacter type. The micro- and macrotypes generally do not match; the connection between them is complicated and contorted. Universality and independent generalization are easily accommodated in the conceptual framework of synthetic microanalysis.

Because we generalize independently on the macro- and microlevels, macrodescriptions are not merely the coarse graining of microdescriptions. They can introduce predicates capable of discriminating between states that are indifferentiable in microdescriptions. Broken symmetry, an important and common effect in physics, offers the best examples. Microscopic laws usually have specific symmetries that forbid certain discriminations. For instance, the parity symmetry of microphysics suppresses the distinction between left and right in microdescriptions of molecules, admitting only the superposition of left-handed and right-handed configurations. Macrodescriptions break the symmetry and differentiate between left-handed and right-handed molecules. The evasion of the laws of microphysics is justified by such considerations as the temporal scale of description, which are contingent factors outside the laws but crucial to our way of theorizing. By incorporating certain contingent factors, macrodescriptions acquire their own power of discernment. Therefore macrocharacter types are generally not supervenient on microcharacter types.[3]

Characters are usually defined together with causal relations. Sometimes emergent characters and relations in the macrodescription form coherent networks with such broad scope they are justifiably called emergent laws, for example, the scaling laws of phase transition or the hydrodynamic equations (§ 38). Macrotheories such as thermodynamics, structural mechanics, or macroeconomics stand on their own. They provide the characters and causal relations for describing the composite systems that become constituents of still larger systems.

The universality of some emergent characters, the discriminating power of macrodescriptions, and the generality of macrolaws call for another level of description besides the four considered earlier. So far our discussion is limited to a single kind of system made up of certain kinds of constituent and relation under a certain kind of external condition. Now we expand our conceptual framework to include the *supermacrodescription* applicable to a superordinate kind covering various kinds of composite system with different kinds of constitution and environment. The supermacroframework enables us to compare different kinds of system, study how their behaviors systematically change with varying constitution and environment, and strengthen the concepts in the macrodescription. If systems with different constitutions exhibit similar macrocharacters, then we know that the details of micromechanisms are insignificant on the macrolevel. It is in the supermacroframework that physicists define universal classes and biologists define ahistorical universals. The supermacrodescription is also useful in dynamic models, for instance in the study of stability and bifurcation (§ 30).

23. Self-organization in Phase Transitions

The most familiar examples of emergent phenomena in physical systems are *phase transitions*, in which matter transforms from one structural state to another, as water to ice. Phase transitions occur not in small systems but only in large or, theoretically speaking, infinite composite systems. This affirms the criterion that systems undergoing phase transitions are not mere sums of parts, for the idea of summation applies to small and large systems alike. Although a few aspects of phase transitions can be crudely explained in the single-particle approximation, explanations of many other aspects go way beyond the approximation.

This section begins by describing the macrofeatures of phase transitions and the macroexplanations of them, with emphasis on a subclass of transition called the *critical phenomenon*. Critical phenomena were observed more than a hundred years ago, but satisfactory microexplanations appeared only in the last three decades. The microexplanations do not diminish but accentuate the notion of emergence, for they articulate clearly the conditions in which macrocharacters appear. The concepts of *broken symmetry* and *broken ergodicity* explain why the laws of microscopic systems cannot be directly extrapolated to macroscopic systems. Macroscopic systems are at once compatible with the laws of microphysics and apparently break the laws with the additional ideas of particularity and historical accident. Experimentalists find that systems with widely disparate micromechanisms behave almost identically on the macrolevel for certain aspects of critical phenomena. The observed *universality* of macrobehaviors is explained by the theory of *renormalization groups*, which successively filters out the irrelevant microdetails.

Order Parameters and Critical Phenomena[4]

There are many kinds of phase transition beside the familiar condensation of gases and the freezing of liquids. An example is the paramagnetic–ferromagnetic transition of magnetic materials such as iron. The phase transition occurs at 771°C for iron. Below 771°C, iron is in the ferromagnetic phase characterized by a net spontaneous magnetization, a macrocharacter by which it can be made into a bar magnet. Above 771°C, iron is in the paramagnetic phase without macroscopic magnetization. Another example is the order–disorder transition in binary alloys such as brass. The copper and zinc atoms of brass occupy alternate sites in a regular lattice in the ordered phase, arbitrary sites in the disordered phase.

In a phase transition, a system passes from a *disordered state* into an *ordered state*, or the converse. To characterize the variation between the macrostates,

Landau introduced the *order parameter*, a macroscopic variable whose value is finite in the ordered phase and zero in the disordered phase. The ordered phase occurs in low temperatures in which the system exhibits a certain macroscopic structure represented by the finite order parameter. The macroscopic structure is destroyed by the random motion of the constituents at elevated temperatures; consequently the order parameter vanishes. For example, in paramagnetic–ferromagnetic transitions, the order parameter is the magnetization, which vanishes in the paramagnetic phase. The order parameter is not always obvious. For the gas–liquid transition, it is the difference in the densities of the gas and the liquid.

Phase transitions can be divided into two broad classes. In *discontinuous* phase transitions, the order parameter jumps at the transition point and the two phases always maintain a finite difference. In *continuous* phase transitions, the order parameter drops continuously to zero so that the difference between the two phases is infinitesimal at the transition point. Continuous phase transitions are also called *critical phenomena*, and the values of the macroscopic variables specifying the conditions in which a transition occurs are summarily called its *critical point*. The transition of some systems can be discontinuous under some conditions and continuous under others. The condensation of gases is a discontinuous phase transition at low pressures and a continuous transition at the critical point. At one atmospheric pressure and 100°C, the densities of steam and water differ by a factor of 1,700. As the pressure increases, the density difference decreases and finally vanishes at the critical point of 217 atmospheres and 374°C. At pressures above the critical point there are no distinct gas and liquid phases. Another example of continuous phase transition is the paramagnetic–ferromagnetic transition. We will only consider critical phenomena here, because they are simpler and better understood than discontinuous transitions.

Broken Symmetry[5]

The universe was born symmetric, but everywhere exhibits order. Order is symmetry broken. Physicists conjectured that the primeval fireball of the Big Bang was rather featureless; it had only one interaction, governed by laws characterized by very high symmetry. More features appeared as the universe cooled and expanded. The universe underwent a series of cosmic phase transitions, in which the primal symmetry was successively broken and the gravitation, strong nuclear, weak nuclear, and electromagnetic interactions successively froze out.

For phase transitions on smaller scales, the breaking of a symmetry is accompanied by the appearance of a structure represented by an order parameter. The ordered phase is less symmetric. For instance, the crystal structures of solids break the transitional and rotational symmetry characteristic of fluids. Not all phase transitions involve broken symmetries; the gas–liquid transition, for example, does not. However, such exceptions are rare. Broken symmetry is a crucial concept for understanding phase transitions.

To grasp the notion of broken symmetry, we must first be clear about the technical meanings of symmetry and the degree of symmetry. A *symmetry* is characterized by a group of transformations that leave certain features of a system unchanged. A system characterized by a larger group of transformations has *higher symmetry*. For instance, there are many more transformations that bring a circle back into itself than there are transformations that leave a square unchanged. We can rotate a circle through an arbitrary angle and cannot distinguish the final configuration from the initial one, but only rotations through multiples of 90° leave a square invariant. Consequently the circle is more symmetric and has fewer features than the square; it lacks corners, which are suppressed by its extra symmetry transformations. Transformations wipe out invariant features; therefore a system of high symmetry is plain. A perfectly homogeneous and isotropic world is just a blank, which is why it is usually associated with empty space. The symmetry of a system is broken when new features emerge that are not invariant under at least some of its transformations, as when corners appear on a circle.

Symmetry is a most important concept in physics, for it is characteristic of universal regularities captured in physical *laws*. Most if not all fundamental physical laws are framed in terms of symmetries. The symmetry of a law encompasses the universal characters of all systems covered by it. The symmetry is *spontaneously broken*, or more accurately *hidden*, if the law allows many possible *ground states*, or states with the same lowest energy, and an actual system settles into one of them in equilibrium. The particularity of the realized equilibrium state hides the symmetry of the universal law without actually violating it.

To see how the symmetry is broken, imagine erecting a knitting needle on the table and pressing vertically on it (Fig. 6.1). The force has a cylindrical

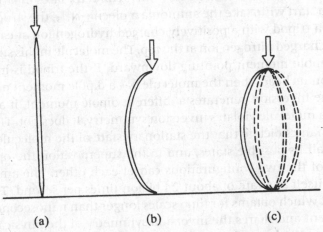

(a) (b) (c)

FIGURE 6.1 (a) Both the needle and the force pushing on it have cylindrical symmetry. **(b)** The buckling of the needle under high pressure breaks the cylindrical symmetry. **(c)** The cylindrical symmetry is restored if we consider all possible configurations of buckling.

symmetry; it is unchanged if rotated by an arbitrary angle about the vertical axis. The symmetry of the force is analogous to the symmetry of physical laws. If the force is small, the needle stands upright; it has a unique ground state that shares the symmetry of the force. As the pressure increases, the needle buckles to relax the stress, and the buckling breaks the cylindrical symmetry. The point at which it first buckles is analogous to a critical point, at which a new feature, the direction of buckling, appears. A needle under high pressure has many possible ground states that are compatible with the cylindrically symmetric force. All these possible ground states are connected to each other by the rotational transformations of the cylindrical symmetry. Thus the symmetry can be recovered if we consider all *possible* ground states. It is broken because the needle, a particular individual, can realize only one of the possibilities. The *realization* of a particular ground state hides the symmetry of the basic law.

Broken Ergodicity and Emergent Characters of Macroscopic Systems[6]

The multiplicity of ground states is not sufficient for the phenomena of broken symmetry. It is possible that the system passes rapidly from one ground state to another, so that we can only observe the superposition of all possible states, which share the symmetry of the law. Broken symmetry obtains only if the system is somehow stuck to its particular state and prevented from going into other possible states. This is where the difference between *microscopic* and *macroscopic* systems becomes important.

Let us consider a specific example, the breaking of the inversion symmetry. In the microscopic world, except some very weak processes, physical laws have *inversion symmetry* that forbids the distinction between right and left or between up and down. Let us examine how this symmetry fares in larger and larger systems. To start with, take the ammonia molecule NH_3, the structure of which looks like a tripod with a positively charged hydrogen ion at each foot and a negatively charged nitrogen ion at the top. The molecule in this structure has an electric dipole moment pointing downward. If the tripod is inverted so that it stands on its head, then the molecule has a dipole moment pointing upward. Since the inversion generates a different dipole moment, it appears that the ammonia molecule violates inversion symmetry. It does not. The laws of quantum mechanics dictate that the stationary state of the molecule is the superposition of all its possible states, and in the superposition the opposite dipole moments of the two configurations cancel each other. The ammonia molecule inverts itself at a rate of about 24 billion times per second. Thus its equilibrium state, which obtains for time scales longer than nanoseconds, has zero dipole moment and shares the inversion symmetry of the physical law.

From ammonia, NH_3, we go to hydrogen phosphide, PH_3, which has a similar tripodlike structure but is about twice as heavy, because phosphorus is heavier than nitrogen. Its inversion rate is ten times slower. Phosphorus trifluoride, PF_3, structurally similar to PH_3 but about sixty times heavier, is theoretically expected to invert, but the time it takes to invert exceeds the limit of human patience, so no one bothers to measure it. For more complex molecules such as sugar with about 40 atoms, the inversion time is so long that stationary states conforming to the inversion symmetry cease to make sense. Instead, they are described in terms of their particular ground states. Thus we have left-handed sugar (*l*-sugar) and right-handed sugar (*d*-sugar), which do not interconvert on human time scales. For biological molecules the inversion symmetry is broken on the evolutionary time scale. The life world observes the left–right distinction; DNA and RNA are composed exclusively of right-handed sugars, whereas proteins contain only left-handed amino acids.

The example shows the emergence of *handedness*, a feature explicitly forbidden by the laws of microphysics but that becomes manifest in large systems in reasonable time scales. It elicits the intuitive idea that qualitative changes occur when the difference in quantity becomes large enough. The blind extrapolation of microscopic laws to the macroscopic level is bound to fail.

The emergence of new features in the symmetry breaking of phase transitions can be viewed in another way. Theoretically, sharp phase transitions occur only in infinite systems, represented by the so-called thermodynamic limit. In this limit, the system stays forever in a particular macrostate and never passes into another. This violates the ergodic hypothesis, which is the basic assumption of statistical mechanics (§ 11). The ergodic hypothesis asserts that a system passes through all its possible microstates, so that we can average over the microstates to obtain macroquantities. The hypothesis is no longer valid after a symmetry is broken, for the system no longer goes through all possible microstates; it limits itself to only those microstates that are compatible with the particular macrostate it settles in. Thus broken symmetry is a case of *broken ergodicity*. With the breaking of ergodicity, the microstate space of the system is radically different from its original microstate space. The difference cannot be deduced from microscopic laws but has to be put in "by hand," bolstering the emergent nature of the macrostate.

The concepts of broken symmetry and broken ergodicity signify that the categorical framework of macroscopic systems has been expanded to include a new general concept, the *particularity* of equilibrium states. Although the system's possibilities are still governed by universal laws, the particular state it realizes becomes preponderant and demands understanding.

A major reason systems in the ordered phases exhibit broken symmetry and broken ergodicity is the strong and long-range correlation among their constituents. Because of the correlation, the parts of a system cannot be changed independently of each other. Everyone is familiar with the phenomenon that a solid rod moves as a whole when one end is disturbed. The disturbance is not transmitted along the rod as a turbulent wave would be. The movement of the two ends of the rod cannot be decoupled without the destruction of a large segment of the molecular arrangement in between. Rigidity is an emergent property of macroscopic bodies that is meaningless for their microscopic constituents. Similar characters, called *generalized rigidity*, are found in other systems in the ordered state. For example, superconductivity and superfluidity are the results when all particles move in a coherently correlated pattern. Because of generalized rigidity, the systems can be altered only as a whole.

Altering the state of a macroscopic system requires large energy that is usually unavailable, and that explains why the system stays put.

An Example: Magnetism[7]

As an example of critical phenomena, let us consider the paramagnetic–ferromagnetic transition. The source of magnetism is the *spin*, or magnetic moment, of atomic electrons. Each atom in a magnetic material carries a spin, which makes it into a tiny magnet that can point in any direction. The idea is aptly captured in illustrations that depict spins as tiny arrows. The microscopic law for the spins has two parts, the kinetic energy of each spin and the interaction among the spins. The kinetic energy is mainly due to thermal agitation, which causes the spins to rotate randomly. The interaction strength between two spins depends on their relative orientations and is largest when they point in the same direction. The interaction range is short; most models include only the interaction between nearest neighbors. In short, interaction favors spin alignment; thermal agitation disrupts alignment.

At temperatures far above the critical temperature, thermal agitation dominates. At any moment, there are as many spins pointing in one direction as in the opposite direction. Consequently the net magnetization vanishes. As the temperature drops below the critical point, thermal agitation yields to spin–spin interaction. Most spins in a macroscopic domain of a ferromagnet are aligned in the same direction, and the alignment becomes more thorough as temperature drops further. The alignment of spins is the order manifested in the macroscopic *magnetization* of ferromagnets. Magnetization, the order parameter in the phase transition, is the thermodynamic average of the spins. It is zero in the paramagnetic phase; its value in the ferromagnetic phase increases with decreasing temperature and levels off at a value for which all spins are aligned.

The microscopic law for the spins has a rotational symmetry; it is unchanged if all spins are rotated through an arbitrary angle. The rotational symmetry is shared by the disordered paramagnetic phase that exhibits no preferred direction. It is broken in the ferromagnetic phase with its particular direction of spin alignment.

According to the basic laws of statistical mechanics, to calculate the magnetization one averages over the microstates of the entire microstate space, that is, over all possible directions of spin alignments. The result is inevitably zero magnetization for ferromagnets. To get finite magnetization, physicists intervene "manually" to restrict the microstate space of the original problem to the realized direction of alignment. The intervention is not part of the microphysical law and is justified by the observation of the macroscopic physical world. It often takes the form of an imaginary external magnetic field. The spin configuration aligned to the imagined field dominates the average, because it has a much lower energy than other configurations. After physicists take the infinite-system limit to lock in the direction of spin alignment along the imagined field, they let the field go to zero to obtain the final equilibrium state in the ferromagnetic phase. The theoretical procedure underscores the *history dependence* of the equilibrium state; the spin orientation of the ordered phase depends on how the state is prepared. In reality with no imaginary magnetic field, the particular orientation of the ordered phase is *accidental*, meaning it is outside the purview of our theories.

By itself, magnetization is a resultant character. Its value can be obtained crudely by an independent-particle approximation called the *mean field theory* or *Weiss molecular field theory*, which is the earliest quantitative analysis of ferromagnetism.[8] The mean field theory is fairly successful in describing certain behaviors of ferromagnets, including the variation of the magnetization with the temperature and the external magnetic field. However, it does not give the correct quantitative behaviors of the system near the critical point. It does not even contain the appropriate concepts to study much that goes on in the phase transition. Here is a clear case where a comprehensive account of the phenomena must go beyond independent-particle models. The extensive body of modern theories on critical phenomena investigates the patterns of fluctuation arising from spin–spin coupling. These fluctuation patterns are emergent.

Fluctuation and Correlation[9]

A question the mean field theory does not answer is that if a ferromagnetic domain contains billions of billions of spins and its size is much larger than the range of spin–spin interaction, how do the spins in a domain accomplish the feat of long-range communication that enables them to line up spontaneously in a specific direction? A solution can be imposed if an external magnetic field is present or if the direction of some boundary spins are somehow fixed. However, these conditions do not generally obtain. A dictatorial solution is impossible, because all spins are equal and none is more powerful than the others. The solution can only be democratic; it can only be the self-organization of the spins via their interaction. How is the self-organization achieved? The question is reminiscent of the economists' question, How does the economy settle on the set of commodity prices that coordinate the behaviors of economic units in the general equilibrium model of perfectly competitive markets? To answer such questions, a completely different approach is required.

In physics, experiments provide a clue. It was observed more than a hundred years ago that liquids at critical points invariably turn milky white, a phenomenon called *critical opalescence*. The phenomenon can be explained by fluctuations in which tiny bubbles of gas intersperse with droplets of liquid. For these bubbles and droplets to scatter light, they must be of a scale comparable to the wavelength of the light, which is thousands of times larger than the atomic spacing in the liquid. Similar violent fluctuations are observed in scattering experiments in other critical transitions. The fluctuations are not merely the constituents going haywire; they contain the correlation necessary to establish the order below the critical temperature.

The order parameter has an important dynamic aspect. Consider magnetism again. The magnetization is not only the average for a ferromagnetic domain. It is also a local variable. When we examine the orientations of spins in a tiny region around a point in the iron, we find that they can deviate from the average magnetization, as long as the energy associated with the deviation is smaller than the characteristic thermal energy. The deviation is called *fluctuation*.

A most important quantity in many-body problems is the *correlation function*, which tells us how much influence the fluctuation at one point has on the fluctuation at another point and how the effect varies with the distance between the two points. Suppose a spin in a magnetic system is forcibly held in the "up" direction. Its immediate neighbors are more likely to point up as a result of spin–spin coupling. Spins farther away are affected to a lesser degree. These behaviors are summarized in the correlation function. The greatest distance over which the influence of a spin can be detected is called the *correlation length*. Spins separated by distances greater than the correlation length are uncorrelated and behave independently of each other. Roughly, the correlation length measures the size of the largest block of spins that rotate in unison. Similarly, it measures the size of the largest droplet in gas–liquid transitions.

Magnetism is perhaps the simplest critical phenomenon, but it is still very complicated. To understand it physicists have developed many ideal models, the simplest of which is the *Ising model*. In the model, the spins are fixed on regular lattice sites and have only two possible orientations, up and down, and each spin interacts only with its nearest neighbors. The one-dimensional Ising model, in which the spins are situated at regular intervals on a line, exhibits no phase transition. The two-dimensional Ising model, in which the spins are arranged on a grid in a plane, exhibits critical phenomena. Critical phenomena are also found in higher-dimensional models. For brevity, we will use the Ising model in our discussion.

At temperatures above the critical point, the numbers of up and down spins are almost equal. However, the correlation length varies with temperature. At temperatures way above the critical temperature, thermal agitation dominates, the correlation length is almost zero, and the system at any moment is a fine-grained mixture of individual spins pointing up or down. As the temperature drops, the relative strength of spin–spin coupling increases, and a few spins start to flip together. As the temperature drops further, sizable blocks of spins flip in unison, and the correlation length increases. This is akin to the formation of droplets in a gas as it cools. Near the critical temperature, the energies characteristic of the ordered and disordered phases are so close the thermal energy permits large blocks of spin to flip together. At precisely the critical temperature where the thermal energy equals the spin coupling, the correlation length becomes infinite, so that each spin is correlated to every other. The long-range correlation coordinates the value of the fluctuation of the magnetization at widely separate points. Below the critical temperature, most spins are aligned in one of the directions. The minority form fluctuating islands in the aligned sea, and the average dimension of the islands gives the correlation length. As temperature drops further, the islands shrink and correlation length decreases.

The Renormalization Group[10]

The structure of a spin system undergoing critical transition is far more complicated than the description in terms of varying correlation length suggests. The infinite correlation length at the critical temperature does not imply the suppression of smaller-scale fluctuations, for statistical mechanics only requires the fluctuation energy to be smaller than or roughly equal to the thermal energy. Since smaller blocks of spins have lower fluctuation energy, they keep on fluctuating near the critical temperature. Thus a snapshot of the system shows a sea of up spins interspersed with land masses of down spins, which contain lakes of up spins, which hold islands of down spins, and so on.

Near the critical point, fluctuations occur simultaneously at many spatial and temporal scales, and the fluctuations in different scales influence each other. The correlation among events with scales ranging from the microscopic to the macroscopic defies the usual averaging techniques and violates the usual understanding that assumes a characteristic scale of length. Large-scale fluctuations dominate the macroscopic phenomena of phase transition, but small-scale fluctuations cannot simply be ignored. A systematic method to screen out the irrelevant details of the small-scale fluctuations is needed. It is provided by the *renormalization group theory* of Kenneth Wilson.

The renormalization group theory has several formulations, the most intuitive of which is the method of decimation or *coarse graining*. Crudely speaking, successive coarse-grained descriptions of the systems are given in terms of new predicates that absorb the salient information of the finer-grained description, until further coarse graining makes no difference. The idea can be roughly illustrated by the block spin method of Leo Kadanoff. Let us start with an Ising spin system. We form blocks, each of which consists of a few spins. The character of a block is obtained by averaging over its constituent spins. It is then streamlined so that a block is represented as a single effective spin coupled to the other effective spins by an effective potential. The distance between the effective spins and the magnitude of effective coupling are rescaled and related to the original physical quantities. Thus the effect of all fluctuations with scales smaller than the block size is neglected, except its contribution to the effective parameters. The new lattice, the basic entities of which are the effective spins, constitutes a coarse-grained representation of the original Ising system. The procedure is repeated; new blocks are formed with the old blocks as "spins." In the process microstructures irrelevant to the macrophenomenon are systematically filtered out.

Physicists have run many computer simulations for the coarse graining of the Ising model. The microscopic snapshots of a two-dimensional system are typically illustrated as arrays of tiny black and white squares representing the up and down spins. The snapshots for the original system contain all the microinformation. They show that the system has an almost equal number of black and white squares at temperatures slightly above and exactly equal to the critical temperature. It has more black squares at temperatures slightly below the critical temperature, but the difference is far from dramatic in this microscopic view. All configurations show patches of black and white and one can hardly tell that a sharp phase transition has occurred. The difference among the configurations is enhanced in the coarse-grained snapshots. After several block-spin iterations, the coarse-grained configurations of the system at slightly above the critical temperature converge to a mixture of salt and pepper. The configurations of the system at slightly below the critical temperature converge to solid black or solid white. The configuration of the system at the critical temperature does not change with coarse graining; it consistently shows patches of black and white. Thus the coarse graining procedure screens out the irrelevant microinformation to reveal the unmistakable difference among the disordered phase, the ordered phase, and transition line separating the two.

Critical Exponents and Universality[11]

In studying the macrobehaviors of critical phenomena, an important question is how the order parameter and other thermodynamic variables vary as the system approaches the critical point. More specifically, we are interested in how they vary with the *reduced temperature*, which is the difference between the critical temperature and the system temperature normalized with respect to the critical temperature. Generally, a thermodynamic variable can be expressed as the product of two terms, a constant magnitude and a variable term, which contains the reduced temperature raised to a certain power, called the *critical exponent*. The critical exponents are peculiar for being neither integers nor simple fractions but jagged numbers such as 0.341. The jaggedness of the critical exponents is typical of a wide class of phenomena, which physicists call *anomalous dimensions* and which are more widely known as *fractals*.[12] Fractals have their own self-similar regularities. Despite their jaggedness, the critical exponents for various thermodynamic variables are related by extremely simple rules; for instance, some of them add up to integers. The relations among the critical exponents are called *scaling laws*.

The critical exponents have another important peculiarity. They are insensitive to microscopic details and depend only on the most general macrocharacters of the system, such as the dimensionality of the system and the dimensionality of its order parameter (the number of components of the parameter). Systems with the same dimensionalities but otherwise very different in structure and microscopic mechanism, such as liquid–gas transitions and the separation of organic chemicals, have the same values of critical exponents and form a *universal class*. The universality would not be so surprising if the exponents were integers; few eyebrows are raised by the fact that the gravitational and electrostatic forces both vary as the distance raised to the power of -2. That the critical exponents converge to jagged values is remarkable and demands explanation.

The universality of the critical exponents and the jaggedness of their values are explained by the renormalization group theory. The critical exponents measure how macroscopic variables change as a function of the reduced temperature when the system approaches the critical temperature. The coarse-graining procedure shows that the variation with respect to the reduced temperature is equivalent to the variation with respect to the scale of coarse graining. Thus the critical exponents reveal the *general structure of fluctuations* at many scales nesting within one another. Microscopic parameters affect the structure of fluctuations only at the smallest scale, and their effects are quickly averaged out in coarse graining. Consequently microstructures do not affect the ways various systems approach their respective critical temperatures, which can be described by a few universal critical exponents and simple scaling laws. Rodlike molecules trying to align themselves in a critical transition fluctuate in the same way as the spins in a magnetic material. They belong to the same universal class and have the same critical exponents. The universality of the critical exponents and the irrelevance of microscopic details in some macrobehaviors justify the use of simple models to study complicated physical systems, for instance, the use of the Ising model to study the behavior of systems of rodlike molecules in phase transition.

Not all aspects of critical phenomena are insensitive to the details of micromechanisms. Differences caused by different microstructures are manifested in the values of the critical temperature, the magnitudes of the order parameter, and the values of other thermodynamic variables. These values vary widely among systems and depend on their specific microstructures.

The scaling laws and the universal classes of critical exponent operate in the supermacrodescription of critical phenomena. By discerning macrocharacters that do not depend on the microscopic specifics of constituents and interactions, they generalize over not merely one kind of system but many kinds. The generalization is properly emergent. Their microfoundation lies not in the aggregation of constituents but in the general patterns of fluctuation extending over wide domains or even the entire system.

The theories of broken symmetry and renormalization groups demonstrate that emergent macrophenomena are not merely the logical consequence of microphysical laws. Suitable conjectures and parameters must be introduced, some of which overshadow but do not overthrow the underlying microphysical laws. They show us clearly how some effects of microphysical laws are systematically filtered out in coarse graining; how symmetry, an important aspect of microphysical laws, is broken in the characters of actual macroscopic systems. Yet the synthetic microanalysis of the emergent characters also demonstrates that they can have adequate microexplanations. Note, however, the complexity of the microexplanation and the unconventionality of renormalization group theory. Physical systems are simple compared to social and biological systems. The experience of physics is a warning to other sciences to beware of deceptive simplistic connections between the micro- and macrodescriptions.

24. Adaptive Organization of Biological Systems

A controversial question in the theory of evolution is whether the history of life contains emergent evolutionary changes or evolutionary novelties that are not merely the accumulation of small adaptive steps. What are the underlying mechanisms of evolutionary novelties if there are any?

Emergent evolutionary changes are frowned on in orthodox population genetics, according to which evolutionary changes must be resultant, with microevolutionary changes adding up to macroevolutionary phenomena. Mutation tinkers with existing organisms and makes minor modifications one at a time, natural selection preserves some modifications and eliminates others, and the preserved modifications gradually accumulate into trends without drastic jumps characteristic of emergent novelties. The tinkering view of evolution and the resultant change it entails are amenable to beanbag genetics that factorizes organisms into uncorrelated characters or genes that evolve separately (§ 17). If organisms are merely bags of genes, organic evolution can be nothing but the replacement of the genes one by one.

Tinkering can go a long way, but it can go only so far. The textile and steel technologies underlying the First Industrial Revolution are mainly due to tinkering; the chemical and electrical technologies driving the Second Industrial Revolution are not, for they depend on the systematic results of scientific research. Can such high-tech products as organisms be the mere results of tinkering? How does the tinkering view fit the historical data of organic evolution?

Puzzles in Macroevolution

Many historical evolutionary phenomena are difficult to explain as the sum of small adaptive steps, because many intermediate steps are implausible. Natural selection is as myopic as those corporate executives who cannot see beyond the quarterly report. It eliminates any mutation with no immediate advantage, even if the mutation is the foundation of a potentially highly advantageous character that can evolve in the future. Thus biologists have great difficulty in explaining phenomena such as the evolution of sexual reproduction. Sexuality is advantageous for providing greater variability in the long run, but in the short run its fitness is only half of the fitness of asexual reproduction. Why was it not immediately eliminated by natural selection? Or consider the bird's wing. It is marvelous only when it is sufficiently well developed and powered by adequate muscles supported by adequate bones and respiratory capacity to take to the air. A half-formed wing with no ancillary structures would be a burden, for to grow it and carry it around require energy that can be better used. How did sexuality and wings evolve? The orthodox answer to such questions is preadaptation: Novel characters originally evolved to serve different adaptive functions before they were converted to their present uses, as flight feathers of birds originally evolved from reptilian scales to keep the body warm. But why flight feathers, which are highly complicated and specialized whereas down is equally good for warmth? Many preadaptive explanations, which lack the support of empirical evidence, are the kind of just-so story criticized by Gould and Lewontin (§ 41).

The doctrine of continuous and gradual change has a hard time in explaining the observed stability of many life forms and the disparate rates of evolution. Comparative morphologists find that myriad species of animals share a small number of *baupläne*, or basic anatomical designs that distinguish various animal phyla, the largest taxonomic grouping unit after the five kingdoms. All baupläne appeared abruptly in the Cambrian Explosion about 550 millions years ago, shortly after the first appearance of multicellular organisms. No new baupläne have appeared since. Why are the basic body plans so stable? Why do they become more inflexible in the course of evolution? Why have new body plans failed to appear in 550 million years?[13]

Many vestiges of the past persist long after their adaptive functions are lost. For instance, embryonic birds and mammals still have gill arches, which have been useless for 400 million years. Why are the vestiges not eliminated by natural selection?

The usual answer is that the baupläne and the vestiges are developed early in the embryo and hence are more difficult to modify than features that develop later. The genetic programs controlling embryonic development were formed in the early days and have been frozen ever since. The argument of developmental constraints is persuasive, but it is double-edged: If developmental constraints are so strong, would they become a factor in evolution able to counteract natural selection?[14] If a once-adaptive but now maladaptive character can be frozen, why cannot a newly appearing maladaptive character be frozen? How can the idea of developmental constraints, or more generally of patterns of frozen structures, be compatible with the orthodox view that each character or gene "selfishly" plays the evolutionary game by itself?

The idea of the accumulation of minor changes has difficulty in explaining why evolutionary processes occur such different rates. The banana was introduced in the Hawaiian Islands about a thousand years ago, but several species of moths feeding exclusively on bananas have already evolved. On the other hand, living fossils such as the alligator, snapping turtle, lungfish, and horseshoe crab have not changed since they came on stage hundreds of millions years ago. How are the widely divergent tempos and modes of evolution explained?[15]

In 1972, Niles Eldredge and Stephen Gould observed that the fossil record does not agree with gradualism. Instead of showing morphological characters changing gradually, the record shows them remaining stagnant for a long time, then changing drastically within a short period, usually when new species are formed. The proposal, known as *punctuated equilibrium*, is hotly debated but enjoys widening support from thorough studies of the fossil record.[16]

All these problems in macroevolution remain unresolved. Everyone agrees that macroevolutionary processes must be compatible with the principles of genetics. However, Mayr said: "Indeed, up to the present time, everything we know about macroevolution has been learned by the study of phenotypes; evolutionary genetics has made virtually no direct contribution so far. In this respect, indeed, macroevolution as a field of study is completely decoupled from microevolution." The population geneticist Francisco Ayala concurred: "Macroevolutionary studies require the formulation of autonomous hypotheses and models (which must be tested using macroevolutionary evidences)."[17]

The Spontaneous Order in Organisms

The combined weight of puzzles strains the credibility of the doctrine that macroevolution must be the resultant of microevolutionary modifications. If emergent changes occur during the formation of species, what can their mechanisms be? Attempts to answer this question once led to the postulation of mysterious creative evolutionary forces that tainted the idea of emergence. We want better answers.

An interesting answer has been proposed by Stuart Kauffman, who argues that organisms are not ad hoc contraptions cobbled together by evolution, as they are in the orthodox view. Once we treat organisms as unitary systems with integral structures, we realize that natural selection is not the sole determinant of organismic structures. An equally important source of structure is *spontaneous order*. Generally, complex systems can spontaneously organize themselves into coherent patterns. Self-organizing phenomena abound in the physical world; the transformation of matter from the disordered to the ordered phase is an example. There is no evidence that similar processes cannot occur in the evolution of organic systems. If they occur, then the internal structures of organisms are as important to the course of evolution as the vicissitudes of external environments. Kauffman said Darwin rightly pointed out the importance of natural selection but neglected the question, What kinds of complex system are best poised to adapt by mutation and natural selection? To answer this question, the conceptual structure of the theory of evolution must be expanded to include properly the integrity of organisms.[18]

To respect organismic integrity, we can no longer employ the independent-individual approximation to factor the organism into uncorrelated genes or characters. The wholeness of organisms becomes a source of emergent evolutionary changes. This section presents the theories of Kauffman and others on the evolutionary effects of *adaptive organization* or self-organization of complex systems responsive to external vicissitudes. I start by introducing the idea of *conflicting constraints* that takes account of the interfering effects of correlated genes. Conflicting constraints lead to the *rugged fitness landscape* and the problem of *global optimization*, in contrast to the local optimization problem addressed in the tinkering models. Two models, the *NK* model and the *NK Boolean network* model, are presented. They illustrate Kauffman's contention that the systems most conducive to evolutionary changes and innovations are those poised at the edge of chaos. They are not unlike systems just above the critical temperature ready to precipitate a phase transition.

Conflicting Constraints

The major assumption in adaptation models, which dictates the resultant nature of evolution, is that organisms can be factorized into discrete character types. Each character type is optimized independently of the others, whose influences are at best indirectly represented as the constraints of optimization. It is assumed that the constraints on various character types never conflict, so that the optimal design of the organism results from cobbling together the optimal values for the types. The optimal whole is nothing but the sum of the optimal parts.

Life is not so simple. A character can change via many developmental paths, some of which can generate countervailing changes in other characters. *Antagonistic pleiotropy*, which occurs when some of the characters controlled by an allele are beneficial and others controlled by the same allele are harmful, is not uncommon. Often an allele advantageous under one environment is deleterious under another; that dual nature can generate conflict because organisms face different environments during their lifetime. We are lucky that the alleles that make bacteria resistant to antibiotics often have strong detrimental effects in a drug-free environment; otherwise natural selection would foil our battle against infectious disease. In short, the constraints on the optimization of various characters can and do conflict with each other. When *conflicting constraints* become important, adaptation models in which each character is optimized individually are no longer justifiable.[19]

Conflicting constraints are found not only in biology but also in economics and physics. There is a well-known "voter paradox." Consider a community with three members facing three social options, A, B, and C. The citizens try to determine a social ordering of the options by voting, so that one option ranks above another if the majority prefers the former to the latter. It happens that the first citizen prefers A to B and B to C, the second prefers B to C and C to A, and the third prefers C to A and A to B. In this situation, it is impossible to find a consistent social ordering. The voter paradox has been generalized by Arrow to the *impossibility theorem*, which proves that under very general conditions, it is impossible to derive from individual preferences a consistent social-choice function that is neither dictatorial nor imposed. The theorem has great ramifications in welfare economics.[20]

In physics, a class of material with frozen disorder called *spin glass* has recently attracted much research attention. Spin glasses are peculiar because they are *frustrated* in the sense that it is impossible to find an equilibrium state in which every particle attains its state of lowest possible energy. It is not unlike the impossibility of finding a social welfare function that makes everyone maximally happy or an organism each of whose characters is at its independent optimum. Fortunately energy is a summable quantity, unlike economists' utility, which cannot even be compared. Physicists can use the "utilitarian principle," or the principle that the spin glass system as a whole settles in a state with the lowest total energy. They are experts in minimizing energies, but this time they find themselves in rough terrain.[21]

Most physical systems have a unique ground state. Systems with broken symmetry have many ground states, but these are systematically related by symmetry transformations. Such ground states are readily found by the standard calculus of energy optimization. Frustrated systems are far more frustrating. A graph of the energy of a frustrated system versus its microstates resembles a rugged landscape, with peaks and valleys of irregular shapes similar to the schematic illustration of Fig. 6.2b. To find the global minima or microstates with the truly lowest energy is a difficult task demanding novel methods. The rugged energy function has many local minima with energy

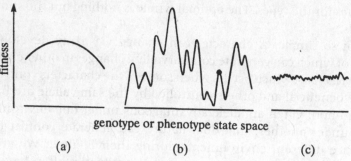

genotype or phenotype state space

(a)   (b)   (c)

FIGURE 6.2 The fitness landscape of a species. The abscissa stands for the entire genotype or phenotype state space of an organism in the species. Each point in it represents a possible genotype or phenotype state, whose fitness is represented by the point in the curve above it. (a) When the parts of the organism are totally uncorrelated, the fitness landscape contains a single broad peak, so that states around the peak have almost the same fitness. (b) When only a few parts of the organism are correlated, the fitness landscape is rugged and contains many peaks and valleys. A species can be stuck in a metastable state whose fitness is a local maximum, as depicted by the dot. (c) When the parts are all correlated, the number of peaks and valleys increases but their heights and depths recede toward the mean fitness value. Thus many states have similar fitness.

higher than the global minima. These are the metastable states. The system may be stuck in one of the metastable states for a long time, making the notion of broken ergodicity truly significant. The frustration of randomly frozen spins leads to many interesting physical phenomena.

The rugged energy function has a close analogy in biology. The difference is that physical systems, preferring lower energy, seek valleys, whereas biological systems, opting for higher fitness, climb hills.

Rugged Fitness Landscapes

To consider an organism as an integral unit, we return to its genotype state space and phenotype state space before they are factorized into alleles and characters (§ 12). Wright introduced the notion of *fitness landscape* for a species in a certain environment. The landscape assigns a fitness value to each point in the genotype (phenotype) state space representing a possible genotype (phenotype) state of the organisms in the species. It generally exhibits peaks and valleys; peaks representing high fitness values are situated on highly adapted organismic states. The actual species is represented by a group of points on the fitness landscape, each of which represents a living organism. Mutation disperses the points and natural selection herds them uphill to higher fitness. Like the genotype and phenotype state spaces, the fitness landscape is typical of all organisms in the species. It contains information about the species's capacity to adapt, evolve, and maintain realized characters. It is used extensively by biologists to make intuitive arguments.

Kauffman used the fitness landscape in several novel ways. First, he adopted the supermacro viewpoint. Instead of looking at the fitness landscape for a particular species, he considered the landscapes for many species and tried to discern and classify their general features. Second, he tried to correlated the general features to the degree of self-organization and conflict in constraints. Third, he did not talk metaphorically but constructed mathematical models that, although unrealistic as representations of organisms, nevertheless allow us to see the basic ideas clearly.

From a supermacro perspective, Kauffman asked, if evolution is to be repeated many times, can we discern any regularities in the results? Are there law-governed processes underlying such regularities? Imagine a set of identical round-top hills under rain. Each hill develops its unique system of rivulets, but all the drainage systems exhibit branching and converging patterns that can be described statistically. We can hope to develop a statistical theory of drainage patterns although not for the rivulets of a particular hill. Thus he turned from the descriptive mentality that focuses on the particular history of life on earth toward a general theory underlying histories of self-organizing organisms.

Substantive definitions for the overall fitness of phenotype states are almost impossible to formulate. Like other biologists, Kauffman used some definite properties as surrogates, for instance, the catalytic rate of proteins. In picking the surrogate, an approximation has been made to restrict attention to certain aspects of the organism. The approximation is nevertheless holistic because it does not further factorize the phenotype space.

The NK Model[22]

Kauffman argued that the ruggedness of the fitness landscape – the number of peaks and valleys, their heights and separations – is determined by the *internal organization* of organisms. To study how it is determined, he introduced an idealized model. The *NK* model abstractly represents an organism as a composite system with *N* parts. Each part contributes to the fitness of the phenotype states and the contribution directly depends on *K* other parts. Together the size N and interconnectedness *K* provide a measure for the complexity of the organism. A specific pair of values for *N* and *K* determines a "landscape" of fitness for various phenotypes. Living organisms are represented as a cluster of points on the fitness landscape. The flow of the points, driven by mutation and selection but restricted by the fitness peaks and valleys, represents the evolution of the species.

The *NK* model is similar to the spin glasses discussed earlier. The fitness landscape is analogous to the energy function of spin glasses, with fitness peaks corresponding to energy valleys. The behavior of a species on a fitness landscape is analogous to the behavior of a spin glass at a fixed temperature. The mutation of a part of the organism is similar to the flipping of a spin in the spin glass; both carry the system into an adjacent state in the state space. Mutation plays the role of thermal agitation and natural selection is analogous to the physical law dictating that physical systems seek states of the lowest energy.

When the parts of the organism are totally uncorrelated ($K = 0$), a part can be changed by mutation without affecting the fitness contributions of the rest. For organisms with many uncorrelated parts, the effect of changing a single part is small. Thus adjacent states have almost the same fitness and the fitness landscape is like a broad hump (Fig. 6.2a). If mutation is strong, then natural selection, which can hardly discriminate between the similar fitnesses of adjacent states, may be unable to hold the species at the top of the broad fitness peak.

As interconnectedness increases, other peaks appear. When the parts are totally correlated ($K = N - 1$), changing a part alters the fitness contributions of all the rest and the fitness values of adjacent states vary independently. What is best for one part is seldom best for all of its correlated partners. The conflicting constraints caused by correlation decrease the total fitness. As the organism increases in complexity, the number of peaks increases and their heights dwindle. When N becomes very large, all peaks recede toward the mean fitness value (Fig. 6.2c).

Both extremes of correlation and no correlation are unfavorable to evolution by natural selection, because both generate a large number of possible phenotype states with approximately the same fitness. The species most conducive to evolve by mutation and selection are those of modular organisms, each part of which is correlated with a few other parts. Here the fitness landscape is rugged, exhibiting many pronounced peaks and valleys (Fig. 6.2b).

The rugged fitness landscape of modular organisms helps to explain some puzzles in macroevolution. Suppose that mutation affects one part at a time so that the species can only "walk" to an adjacent state. Selection decrees that the destination must have higher fitness. In a multipeaked landscape, a species soon finds itself stuck on a local peak. It can no longer evolve, for selection forbids it to go downhill. Thus it sits on an ant hill, congratulating itself on being the fittest. This is the fate of evolution by tinkering. Tinkering can only improve a system so much.

To evolve further, the species must find ways to get off a local peak. It may be helped by a changing environment that lifts some fitness valleys and flattens some fitness peaks. In a fixed environment, it can still evolve by genetic drift, as Wright demonstrated. Suppose the species on a local peak splinters into many relatively isolated small populations. Inbreeding may decrease the adaptedness of organisms in some populations and force a group of points into the fitness valley so that it can explore and find other, ideally higher, peaks, if it is not eliminated by natural selection before it finds them. This mechanism has been used by many biologists to explain the formation of new species.

Another way to escape the fate of tinkering is to "jump" to a distant state via frame mutation or recombination that simultaneously changes many parts of the organism. Unfortunately, it can be proved as a kind of universal law that for such processes, the rate of finding a more adapted state slows very rapidly. If the species starts in states with low fitness, it has an appreciable chance of jumping to a distant state with significantly higher fitness. Thus it is more likely to evolve by jumps than by walks that search for better states locally. With each successful jump, the time required to find a better distant state is doubled. After a few jumps, the waiting time becomes so long it is more efficient to improve fitness by climbing local hills. Finally, the species reaches a local peak and becomes a species of living fossil if it is lucky enough to escape extinction but not lucky enough to jump again.

The different time scales in evolution by large and small mutations in rugged fitness landscapes can be used to explain the difference between two historic events if we expand the phenotype state space to include many species. The Cambrian Explosion saw the appearance of more than 100 phyla of animals, far exceeding the 27 phyla existing today. We can guess that the first multicellular animals were poorly adapted. Therefore they could jump to widely separate states in the phenotype state space to establish various phyla. Some landed near the analogue of the Rockies in the rugged fitness landscape, others near the Andes, still others near the Alps. Suppose those in the Alps and the Andes became extinct, as a result of accident or selection. Those in the Rockies climbed various peaks and diversified into various classes, orders, and other lower taxonomic units. After the Permian Extinction 200 million years ago, which wiped out 96 percent of the species, no new phylum appeared. Instead, the number of species and lower taxonomic units increased rapidly. The existent organisms were already moderately adapted and the species could not jump to better states. They were in the Rockies and may have to wait forever to find the Himalayas. Thus they follow the evolutionary strategy of exploring fully the ridges and summits in their neighborhood. Consequently life exhibits a greater diversity of forms with minor variation but fewer major differences. It appears that cultural evolution is following a similar route; we have more channels on the television but fewer kinds of programs.

Random *NK* Boolean Networks[23]

The *random NK Boolean network* describes the interaction among the parts of organisms more definitely. Again an organism or a genome is treated as a composite system with N parts. Each of the N elements has two possible states and is regulated by K other elements. The specific state an element assumes at a particular moment is determined by the specific states of its K regulators at the preceding moment, the effects of which are combined according to a certain logical switching rule. Suppose the two possible states are active and inactive and the switching rule is the logical OR function; then an element will be active if any of its K regulators is active. Having specified the parameters of a class of Boolean networks – the possible states and the switching rule for each element – we can study various classes by varying N and K.

Boolean networks are dynamic. As discussed in § 29, dynamic systems tend to converge on steady-state cycles called *attractors*, the states of which are repeated. Some attractors are stable; systems perturbed by flipping a few elements or even changing a few switching rules recover and revert to the attractor. Some attractors are unstable and can be damaged by slight perturbations.

For networks with large K or high connectivity, the lengths of the steady-state cycles are extremely long, so that a network practically never repeats itself. Furthermore, these networks are very sensitive to initial conditions and perturbations, changing their patterns of activity with the slightest disturbance. Their behaviors are chaotic.

When K drops to two, the behaviors of the Boolean networks change abruptly and exhibit *spontaneous order*. The reason behind the spontaneous order is the following: Generally it is possible for several elements to lock themselves into either the active or the inactive state. Such unchanging interlocked elements form a "frozen core." In chaotic systems, the cores are usually local and contain only a few elements. The cores increase in size as the connectivity of the networks drops. When $K = 2$, the frozen core is large enough to form a mesh that percolates or extends across the system, isolating the changing elements into local islands. Consequently the network exhibits pronounced order and stability.

A similar situation obtains in phase transitions. Chaotic networks are akin to matter in the disordered phase, and networks with percolating frozen cores are akin to matter in the ordered phase. The most interesting dynamic behaviors are exhibited by networks corresponding to matter near the critical point where phase transition occurs. These networks can handle complex computations and yet are controllable. They also have high capacity to evolve; small changes in them can lead to an avalanche of changes in their behaviors.

Suppose the Boolean networks represent genomes made up of genes regulating each other. Then, Kauffman argued, *those genomes with connectivity that puts them at the edge of chaos have the optimal capacity to evolve*. Most mutations have minor effects on these networks, because they are localized in the active islands. However, a few mutations can trigger an avalanche of modifications that cascade throughout the system, resulting in emergent evolutionary changes. Thus networks at the edge of chaos are able to evolve both by accumulation of small changes and by dramatic changes in which evolutionary novelties emerge. The emergent changes may be triggered by a random mutation or by a change in the environment. The conditions under which natural selection is most powerful are also those in which self-organization and historical contingency are most likely.

Kauffman and his colleagues claim that the theories of self-organization and selection aim to uncover the laws and general regularities of evolution, not the particular history of organic evolution on earth, but evolution in general, whenever and wherever it occurs. They call the typical characters of systems poised at the edge between order and chaos "ahistorical universals" and the thesis that natural selection favors and maintains such systems a law.[24]

Kauffman's call for a major overhaul of the evolutionary theory gets a mixed reception from biologists. His holistic view of organisms ruffles the prevailing practice of factoring the organisms into a bag of genes. His abstract computer models are alien to the customary style of biological research and remote from empirical evidence. Kauffman and his colleagues claim that their models give plausible and coherent explanations of a massive amount of macroevolutionary data, including many that befuddle the orthodox doctrine. However, their explanations remain on an abstract and intuitive level and lack more concrete ties to evidence. As one biologist says, "They have a long way to go to persuade mainstream biologists of the relevance [of their work]."[25]

25. Inflation, Unemployment, and Their Microexplanations

Do economies exhibit emergent characters? Emergent characters of composite systems are describable in concise terms of their own, and their microexplanations cannot be given by summing or averaging the behaviors of constituents. Thus to look for emergent characters we must turn to macroeconomics and its microfoundations, which, respectively, study the macrostates of the economy and their relations to the microstates studied in microeconomics.

Olivier Blanchard said that macroeconomics had "suffered from the start from schizophrenia in its relation to microeconomics."[26] In the writings of Smith and Ricardo, the capitalist economy possesses both deep strength and inherent weakness. Microeconomics inherits the "invisible hand" and portrays a utopian economy in which the market coordinates the desires of individuals with perfect efficiency. Macroeconomics inherits the "dismal science" and depicts an economy plagued by periodic recession and inflation. How does the perfect coordination at the microlevel lead to bungled-up macrophenomena? Does the schizophrenic relation between microeconomics and macroeconomics indicate the presence of emergent phenomena?

At first sight, most macroeconomic variables are resultant. Take, for example, unemployment. It seems to be simply the percentage of unemployed workers in the economy, and so it is, if we are only concerned with descriptive statistics. However, in theoretical economics, unemployment is not merely statistics but a structure of the economy, and as such it is more complicated. Ideally, each worker decides for himself whether leisure is preferable to work at the going wage. The ideal is reasonably well satisfied in good times when the economy operates at or near the full employment level, so that unemployment is a resultant character reflecting the aggregation of individual decisions. Unfortunately, times are not always good. Consider, for example, the Great Depression, when the free-market economy not only coordinated itself into a large-scale slump but locked itself in a pattern of prolonged massive unemployment. Such slumps, some nearly as severe, have dotted the history of the capitalist economy (§ 36). Were it not for the ensuing human suffering, scientists would be excited about observing the development and maintenance of spontaneous order. Is depression-related unemployment an emergent structure of the economy? Is unemployment like electrical conductivity, which is normally resultant, but under certain conditions transforms into the emergent phenomenon of superconductivity? The answers to such questions are controversial among economists.

This section first presents briefly the conceptual structure of macroeconomics. We then examine and compare two approaches to the microfoundations problem. The new classicists stick to the perfectly competitive models of microeconomics that use the independent-individual approximation discussed in § 16. Consequently they deem all macroeconomic phenomena resultant and rule out the possibility of structural unemployment. The new Keynesians reject the assumption of perfect wage and price flexibility that is the linchpin of the independent-individual approximation and explain the emergence of economywide phenomena by appealing to the interactive microeconomic models discussed in § 20.

The Conceptual Framework of Macroeconomics[27]

Macroeconomics characterizes the economy with a handful of macrovariables instead of millions of microvariables. It analyzes the problems ordinary citizens find relevant and important: inflation, unemployment, economic growth, balance of trade, and fiscal and monetary policies by which the government tries to stabilize the economy by mitigating recession and overheating. Dynamic models for business cycles will be examined in § 36. Here we focus on equilibrium concepts.

Equilibrium macroeconomics is similar to thermodynamics. Each describes the behaviors of a complex system in terms of *macrovariables* that can be invoked without referring to the behaviors of microconstituents: gross domestic product, quantity of money, level of employment, price level, interest rate in macroeconomics; entropy, energy, number of particles, temperature, chemical potential in thermodynamics. Each is concerned with the *web of causal relations among the macrovariables*, which provides the most direct explanations of the behaviors of the variables. Macroeconomics represents its web of macrorelations by a set of structural equations; thermodynamics represents its by a set of equations of state. Each engages in *comparative statics*, in which scientists compare a system's equilibrium states corresponding to various values of a control variable to study the effect of the control variable on the system, without worrying about the actual temporal evolution of the system. Economists study how the levels of price and employment vary with tax rate or money supply, thus gauging the relative efficacy of fiscal and monetary policies. Physicists study how the volume and temperature of a gas vary with applied pressure. More similarities and differences are found in the notes.[28]

All variables in macroeconomics pertain to the economy as a whole. They are crudely divided into those determined within the working of the economy and those fixed by noneconomic forces. Some *endogenous variables* are consumption, investment, employment, price level, interest rate, demand of money, and production output, which is the sum of consumption, investment, government spending, and balance of trade. Some *exogenous variables* are the population, stock of capital, and factors decided by the government: tax rate, supply of money, and public spending. For open economies, there are various trade and payment variables that take account of the foreign sector. The variables are interrelated; for instance, investment plunges as interest rate soars, and output and consumption rise together as the economy becomes more affluent. The interrelations are mathematically represented by *structural equations*, which can be classified into behavioral relations and equilibrium conditions.

The *behavioral relations* take account of how consumption depends on output, price level, and tax rate (the consumption function); how output depends on employment and stock of capital (the production function); how investment depends on output and interest rate (the investment function); how the

demand of money depends on output and interest rate (the liquidity preference and transaction demand). The specific forms of the behavioral relations
are the meat of macroeconomic models and the topics of intensive research. In
the geometric idiom prevalent in economics, the discussion is often described
in terms of the shape of a function, such as whether its slope is positive or
negative.

Equilibrium obtains when a market clears or when its demand equals its
supply. Macroeconomics analyzes the economy into several large interrelated
markets, notably a *goods market*, a *money market*, and a *labor market*. The three
markets are interrelated, because their demands and supplies are functions of
the same set of variables. Much debate in macroeconomics centers on whether
equilibrium always obtains in all markets, and whether the equilibrium can
be reached sufficiently rapidly.

Two Approaches to Microfoundations

The economic content of a macroeconomic model resides in the specific forms
of its behavioral equations. Some of the forms can be inferred from econometric data; others depend heavily on the assumptions economists make about the
economy. Keynes and his early followers posited the equations directly on the
macrolevel, relying mainly on their understanding of the economy, acute observation, econometric data, and heuristic reasoning, including plausible stories about how people behave. The procedure is not unlike the development
of thermodynamics, where equations of state were determined by macroscopic experiments and heuristic reasoning involving plausible stories about
colliding particles. Since economic systems are so much more complicated and
their variables so thoroughly interrelated, their structural equations are not as
successful as one might hope. After the crisis of the 1970s, many economists
turned to find firmer grounds for the structural equations, especially to search
for microfoundations by connecting macroeconomics and microeconomics.

I compare macroeconomics to thermodynamics, but I do not compare its
microfoundations research to statistical mechanics that links thermodynamics
to mechanics, for microeconomics is not analogous to mechanics. Mechanics
is more fundamental than thermodynamics, for the laws of mechanics are
universal, governing "all-purpose" elements that are not stuck in thermodynamic contexts. Microeconomics is more detailed but not more fundamental than macroeconomics, for it studies only the microstates specialized to
the economic context and framed in terms of a specific approximation, the
independent-individual approximation. Microstates are often less epistemologically secured; we often observe overall patterns of a system but cannot
discern its microscopic structures.

The usual argument for the superiority of microeconomics is its depiction
of individual behaviors, notably their Rationality. In theoretical economics,
Rationality is essentially optimization under constraints, which is an abstract
procedure not specific to economics (§ 9). The substance that turns it into a
characterization of economic individuals lies in the specification of the possibility set, the constraints, and the objective. Since these are always defined in

terms of the ideal assumptions peculiar to specific microeconomic models, microeconomics depicts not universal individual behaviors but only individuals customized for particular idealizations. The most general definition of Rational economic individuals is parameterized by commodity prices (§ 16). Arrow said: "I want to stress that rationality is not a property of the individual alone, although it is usually presented that way. Rather, it gathers not only its force but also its very meaning from the social context in which it is embedded. . . . [W]e need not merely pure but perfect competition before the rationality hypotheses have their full power."[29] Theoretically, perfect competition, in which all individuals are price takers, depends on the independent-individual approximation. The approximation has only limited validity, even for physical systems that are simpler than economic systems. Empirically, there is scant evidence that realistic markets are perfectly competitive. Thus a Rationality that depends on perfect competition can hardly serve as the foundation of economics.

There are many microeconomic models, each with its peculiar assumptions about the economy and economic individuals (§§ 16, 20). Which ones provide the relevant individual behaviors to certain macroeconomic phenomena? The general choice criterion differentiates two approaches to microfoundations. According to B. Greenwald and J. Stiglitz: "There were two ways in which the two sub-disciplines [micro- and macroeconomics] could be reconnected. Macrotheory could be adapted to microtheory; and the converse. New Classical Economics took the first approach. . . . The other approach seeks to adapt microtheory to macrotheory. For the want of a better term, one can refer to it as the New Keynesian Economics."[30] In my terminology, the approach of new classical economics is microreductionist, that of new Keynesian economics synthetic microanalytic. Let us examine how they address the phenomenon of unemployment.

Microexplanations of Unemployment

In physics, there are spin glasses or other frustrated systems, where the system has no equilibrium state in which all constituents separately attain their individual optimal states (states of the lowest energy). Can similar phenomena of frustration occur in the economy? Is depression-generated unemployment such a phenomenon?

New classical economics, which appeared in the 1970s, specifies a particular microeconomic model, namely, the model for perfectly competitive markets; enshrines its assumption of perfect wage and price flexibility; generalizes its assumption on individual Rationality in the Rational expectations hypothesis; and argues that all macroeconomic phenomena should be derived from them. Its leader Robert Lucas argued that the perfect-market model is "the only 'engine for the discovery of truth' that we have in economics." "The most interesting recent developments in macroeconomic theory seem to me describable as the reincorporation of aggregative problems such as inflation and the business cycle within the general framework of 'microeconomics' theory. If these developments succeed, the term 'macroeconomic' will simply disappear from use and the modifier 'micro' will become superfluous."[31]

Perfect competition or perfect market coordination means perfect wage and price flexibility. A perpetual question in economics is whether wages and prices always adjust themselves to clear all markets, not in the long run when we are all dead, but in a sufficiently short time. The economy is constantly subjected to shocks such as technological change and political unrest. What happens if it is displaced from its equilibrium path by a shock? Can wage and price adjustment rapidly return it to equilibrium with a cleared labor market?

New classical economics answers with a blanket yes and explains unemployment by denying it. Robert Lucas and Thomas Sargent said: "One essential feature of equilibrium models is that all markets clear.... If, therefore, one takes as a basic 'fact' that labor markets do not clear, one arrives immediately at a contradiction between theory and fact. The facts we actually have, however, are simply the available time series on employment and wage rates plus the responses to our unemployment surveys. Cleared market is simply a principle, not verifiable by direct observation, which may or may not be useful in constructing successful hypotheses about the behavior of these series."[32]

Suppose an exogenous shock generates some unemployment; say it floods the economy with veterans when industry demobilizes from war production. In new classical economics, the jobless promptly bid down wages, firms find it more profitable to hire more at the lowered wages, and unemployment disappears in no time. The oft-cited example of the process is the hiring of day laborers by the corporate farms of California. Since the labor market always clears, people are out of work by choice, perhaps choice based on faulty judgment. They may underestimate the real going wage and hence choose ill-afforded leisure, or they may underestimate the real interest rate and invest in unprofitable searches for better offers. However, being Rational, they quickly correct their mistakes and return the economy to full employment. Critics of new classical economics, such as Franco Modigliani, jeer at the denial of involuntary joblessness: "In other words what happened to the United States in the 1930's was a severe attack of contagious laziness."[33]

In the 1980s, new Keynesian economics emerged to find alternative microfoundations for macroeconomics. It argues that the assumptions of perfect Rationality and instantaneous market clearing are inadequate to explain many economic phenomena. Microeconomists themselves have turned to develop models that include information asymmetry, missing market, transaction cost, institutional structure, and other market imperfections (§ 20). Noting the diversity of microeconomic models, new Keynesian economics does not posit beforehand a specific model or a specific form of individual behavior. Rather, it aims to find those individual behaviors that are appropriate to explain certain macrophenomena.

New Keynesians take seriously the fact that massive unemployment occurs occasionally. They argue that the vast majority of workers in the industrialized world are not day laborers who find work in auctions. For steady jobs, wages are not so flexible. In bad times, firms lay off workers but do not cut the wages of those still on the payroll, for good reasons and with the implicit consent of the employees. Firms find it costly to put together teams of experienced and cooperative workers with predictable performance; thus they are reluctant to jeopardize well-trained teams by cutting wages. They also find that the quality of service often depends on the wages; workers cooperate better, shirk less, and are less likely to leave if their pay is above the market-clearing level. Workers feel more secure if employers adjust to fluctuating demands not by varying wages but by varying the work force, effectively instituting some kind of job rationing according to certain established conventions such as the seniority rule, in which they can work toward better positions. To account for these and other behaviors, new Keynesians developed many models. In these models workers optimize their options, but their options are limited by realistic conditions that are neglected in the perfect-market models. Workers may know the real wage and interest rate and they are eager to work. Their problem is that the jobs for which they are qualified are not available in the market.[34]

The eclectic approach of new Keynesian economics enables it to see the emergent characters of the economy barred from the view of new classical economics by its insistence on perfect competition. Models of perfect competition presuppose that individuals interact only implicitly through the wage and price parameters in which they are defined. The presupposition facilitates the aggregation of individual behaviors but precludes the formation of salient structures. When individuals interact directly and differentially, slight inhomogeneity can lead to the emergence of segmentation and other large-scale structures. Solow said, "Among the reasons why market-clearing wage rates do not establish themselves easily and adjust quickly to changing conditions are some that could be described as social conventions, or principles of appropriate behavior, whose source is not entirely individualistic."[35] Causal factors that invalidate the independent-individual approximation explain wage and price stickiness and subsequent emergent structures.

An Oversimplified Microfoundation

A major task of microfoundations research is to link the structural equations of aggregate economic variables to the behaviors of individual households and firms in particular microeconomic models. The households are represented by their utility functions. Thus the aggregation of utility is crucial to microfoundations. Unfortunately, economic theories emphasize the diversity of individual taste but provide no way to compare the utilities of different individuals (§ 16). The incomparability of utilities is a formidable obstacle to aggregation, for without some standard of measure it is impossible to represent the distribution of tastes.

To evade this obstacle, many microfoundations models resort to a drastic approximation: the *homogeneity of individuals*. They assume that all households are identical and have the same utility function; all firms are identical and have the same production function. The only variation is the size of the households and firms, which makes no difference under the assumption of constant returns to scale. With the homogeneity of individuals, macroeconomic relations are closer to an extrapolation than an aggregation of individual behaviors, and the economy becomes nothing but the interplay between two giants, Household and Firm. James Tobin aptly observed, "The myth of macroeconomics is that relations among aggregates are enlarged analogies of relations among corresponding variables for individual households, firms, industries, markets."[36]

In reality, households and firms differ, and the diversity is the major driving force of the economy. Arrow said: "The homogeneity assumption seems to me to be especially dangerous. It denies the fundamental assumption of the economy, that it is built on gains from trade arising from individual differences. Further, it takes attention away from a very important aspect of the economy, namely, the effects of the distribution of income and of other individual characteristics on the working of the economy."[37] The homogeneity assumption also precludes the possibility of emergent structures of the economy.

Rational Expectation

New classical economics raises the assumption of homogeneous individuals to a new height in the hypothesis of Rational expectation, its most famous unique feature. Expectation has always been a major factor in economic models; individuals make decisions in the light of certain expectations. In most models, expectations are adaptive; individuals learn from their experiences and adjust their expectations according to the errors of their past expectations. In contrast, individuals with Rational expectations need not consult experience, because they can divine the future without systematic error (remember the technical sense of Rationality discussed in § 13).

The Rational expectations hypothesis transplants the assumption of Rationality in microeconomics to macroeconomics. A mutation occurs in the transplantation that makes the assumption much more problematic. In microeconomics, the factors individuals reckon with in their decisions are mainly commodity prices. Prices are factual and close to our daily lives; it does not take much to be familiar with the prices of at least some commodities and in ordinary situations to forecast them. In macroeconomics, the factors individuals reckon with in their decisions are the movements of the economy. Individuals with Rational expectation correctly and effortlessly know the structure and performance of the economy, which have frustrated economic professors. This is high-level theoretical knowledge far removed from our everyday experience and requires far greater mental capacity than the knowledge of some commodity prices.

Individual optimization based on prices captures the basic idea that the structure of the economy is the unintended result of individual behaviors. The hand of market coordination is invisible because people responding to price changes need to know very little about the economy. The whole idea changes when individual optimization is based on the knowledge of the structure of the economy. The Rational expectations hypothesis assumes that everyone is as knowledgeable as expert central economic planners and capable of foreseeing the performance of the economy. Arrow said: "If every agent has a complete model of the economy, the hand running the economy is very visible indeed."[38] The Rational expectations hypothesis has strong consequences, including the futility of government stabilization policies, for the effects of the policies would be fully anticipated and discounted by any Rational individual in the street.

New classical economics extrapolates the form of the independent-individual models in microeconomics to macroeconomics by changing the substantive behaviors of the individuals and the self-consistently determined situation to which they respond.[39] Instead of an environment of commodity prices, individuals are now situated in a macroeconomic structure. In microeconomics, individuals have diverse tastes, but their preference orders have the same lineal form. In Rational expectations models, individuals forecast different values for macroeconomic variables, but they all subscribe to the same model of the economy. The same lineal form of preference order means the individuals are the same kind of being. The same economic model means the individuals have the same opinion about their social context. The difference between the two kinds of uniformity is great.

Some people have suggested that new classical economics is influenced by the Austrian economists who championed individualism against collectivism. If it is, something has gone wrong. Friedrich Hayek distinguished between true and false individualism, both of which are theories of society. True individualism, which he ascribed to John Locke, Adam Smith, and Edmund Burke, asserts that society is not the result of any design but the unintended result of spontaneous individual actions. In this view the rational capacity of individuals is very limited; everyone tries to make the best of very imperfect material. False individualism, which he ascribed to Jean-Jacques Rousseau and the French physiocrats, asserts that social organizations should be designed and controlled by Reason, with a capital *R*. In this view Reason is always fully and equally available to all humans, so that all individuals should think and act in the same Rational way. Hayek argued that false individualism, trying to make everyone equally Rational, leads to "a new form of servitude."[40]

Hayek's true individualism is expressed in the microeconomic price theory. In new classical economics, the Rationality of individuals is exaggerated into the Rational expectation based on comprehensive and uniform knowledge of the economy. Instead of a system composed of diverse individuals, it presents one that is a scaled-up version of one of the homogeneous Rational constituents. It is a case of Hayek's false individualism.

Dynamics

7 The Temporality of Dynamic Systems

26. Temporality and Possibility

"Some are brave out of ignorance; when they stop to think they start to fear. Those who are truly brave are those who best know what is sweet in life and what is terrible, then go out undeterred to meet what is to come."[1] The ancient Greeks are not renowned for their sense of history, but these words – put into the mouth of Pericles for a funeral speech honoring fallen warriors in the first winter of the Peloponnesian War by Thucydides some thirty years later, when their city Athens lay in total defeat – reveal the fundamental temporality of the human being: commitments made in view of past experience and future uncertainty. The awareness that we have a past and a future opens a finite temporal horizon for each moment, frees us from the grip of the immediate present, enables us to push back the frontier at either end, to study history and to develop techniques of prediction. Temporality makes possible genuine action; we know that what we choose to do makes our history and changes our destiny. Despite its obscurity, the future is not a blinding fog bank; here and there we see possibilities illuminated by experience including the knowledge of the sweet and the terrible. The past constrains, but it is not a mere dead weight that conditions the behavior of an organism. It contributes to the future by opening our consciousness to a wider range of possibilities and by shaping the character of the person who chooses the aim of his life and adheres to it through vicissitudes. In *The Brothers Karamazov*, Fyodor Dostoyevsky expressed the value of the past in a farewell speech that Alyosha addressed to the children at the grave of Ilusha: "There is nothing higher and stronger and more wholesome and good for life in the future than some good memory, especially a memory of childhood, of home.... Perhaps that one memory may keep us from great evil and we will reflect and say: 'Yes, I was good and brave and honest then.'" Experience illuminates possibility.

Openness to possibility is essential to a thriving life and is the dominant feature of human temporality. Lives with no idea of possibility or with vanishing possibility are vividly portrayed in William Faulkner's *The Sound and the Fury*, which expresses various forms of temporality through the minds of the brothers Compson: Benjy the idiot, Quentin the Harvard man, and Jason the *Homo economicus*. Benjy has no idea that things can be otherwise than the actual routine to which he is conditioned. He bellows with terror when one day his carriage passes to the left of a monument instead of to the right as it routinely does in its daily drive to town. Perpetually captivated by his immediate impressions, Benjy has no sense of possibility and hence no sense of time; he cannot tell whether the impression of the moment is perception or recollection. To use Kant's words, he cannot distinguish between the temporal sequence of subjective impressions and the temporal order of objective events. Hence we say the idiot is incapable of objective experience. As Benjy lives in a present without time, Quentin lives in a past with no exit. Quentin breaking the hands of his watch is a powerful image reminiscent of Rousseau throwing away his watch after rejecting the stifling culture that subjects everything to the turning of mechanical wheels. "Time is dead as long as it is being clicked off by little wheels; only when the clock stops does time come to life," Quentin recalls his father's echo of Heidegger's remark that man loses his temporality and acquires a clock. Yet to Quentin the remark is a dead memory. Right after he breaks the watch he buys the pair of irons that, at the end of the day, assist his suicide by water. Quentin is obsessed with the past, especially with his sister's loss of virginity. Whatever he does, memory erupts and drowns the significance of the present and the future. Weighted down by what he cannot forget, he sees his possibilities dwindle to one. More accurately, he forsakes the idea of possibility, for throughout the day he never considers the alternative of not killing himself. As Friedrich Nietzsche observed, too strong a sense of history can be lethal.[2]

Possibility and the State Space

Human beings are individuals who are conscious of their own possibilities and concerned with the consequences of their actions, aware of their histories and anxious about the meanings of their lives. Choice and decision depend on the knowledge of options; hence the concept of possibility is fundamental to the being and temporality of humans. Charles Sherover, at the end of a detailed comparison of Kant's cognitive and Heidegger's existential investigations, argued that the notion of possibility is fundamental to both: "If the analyses of cognition and temporality, as of moral reasoning, make any sense, events can only be apprehended by us in terms of possibilities; futurity and thereby all temporal experience is constituted in terms of the possible."[3]

I do not address the full complexity of human temporality with its self-awareness, for we are not investigating full human beings active in the world or the phenomenology of temporal consciousness. The temporal structure of utility-maximizing *Homo economicus* is much simpler, and how to incorporate the essential features of human temporality into economic models is a major problem that economists have not yet solved. We are only interested in the theoretical attitude, in which we partly detach ourselves from the purposive handling of things and look at them as mere presence. As a part of the human mind, theoretical reason bears some marks of human temporality, which manifest themselves in scientific theories, the products of human thinking about the world.

The general concept of possibility, the essence of human temporality, shines in theoretical reason and scientific theories. Change, ubiquitous in the world, is the staple of many sciences. A thing can change only if it is possible for it to be otherwise than it is. Changes occur temporally; since Aristotle if not earlier, philosophers have argued that time is inalienably associated with change, although it itself is not change. Possibility, change, and time are inseparable. Philosophers who banish possibility mostly eject time and change as well. Parmenides, who countenanced only thoughts about what is, opted for a timeless world. Quine, who scorned modal concepts, traded in the world of changing three-dimensional things for a world of four-dimensional "physical objects" in which time is totally spatialized and change vanishes.

The idea of possibility unifies equilibrium and dynamic theories. I argue in § 10 that possibility is constitutive of the general concept of individuals. It is represented in scientific theories by the *state space*, which is the structured collection of all possible momentary states of an individual. The sciences use different state spaces, some of which we have examined: the Hilbert space of quantum mechanics; the μ-space of the kinetic theory; the Γ-space of statistical physics; the phase space of thermodynamics; the genotype and phenotype state spaces underlying the theory of evolution; the consumption and production possibility sets delimiting the options of microeconomic units; the macrostate space spanned by the macrovariables of output, price, employment, and interest rate framing macroeconomic models. The state space is also the foundation of dynamics.

A dynamic process consists of the successive states of a system. Deterministic and stochastic dynamics are both based on the state space. A deterministic process traces a path in the state space connecting the states realized by a system at various times. The synthetic grasp of all possible processes for the system enables us to separate determinateness from predictability and to introduce ideas such as chaos and bifurcation in the supermacrodescription. Probability measures the relative magnitudes of various groups of possibilities in a properly quantified state space.

The possibilities of many historical processes are so numerous and heterogeneous they cannot be circumscribed in state spaces. For such processes neither deterministic nor stochastic theories apply; we must resort to narrative explanations. Most explanations of evolutionary events are narrative. The limited applicability of state space indicates the limit of systematic generalization, not the confine of the general concept of possibility. The idea of possibility is most onerous in the genuine uncertainty of historic processes. Human temporality is historical.

27. Processes and Time

Evolution studies the past; economics plans for the future; physics has no notion of tense. Nevertheless all three sciences depend on the notion of time in their dynamic models. To compare and contrast their dynamic models, I first clarify some general temporal concepts.

Time is among the most obvious ideas in our everyday thinking and the most obscure in philosophy. Most philosophers have pondered it. The answers they give to the question "What is time?" range from the measure of motion in space to the measure of extension in mind. Some say time is not real, others that it is really real. Some identify time with becoming; others find the idea of becoming incoherent. Some confine the study of time to physics; others say fundamental physics is timeless.[4]

Objective but Nonsubstantival Time

An influential doctrine conceives time as a kind of substance that subsists independently of things and events and can be studied independently of them. Substantival time, along with substantival space or substantival spacetime, is implicitly assumed in many philosophies, notably those that define things as the material occupants of spatiotemporal regions. There are two competing models for substantival time. Dynamic time, flowing, passing, responsible for becoming and the passage from the future through the present to the past, is often described figuratively as a river carrying events along. Static time is like a linear frame or a container whose slots are variously occupied by events. The addresses of the slots determine the before and after of their occupants.[5]

Substantival time has many difficulties, which become insurmountable after the arrival of relativistic physics. Many philosophers argue that time is at least partially experiential or ideal. Extreme idealists deny the objectivity of time, arguing that things and events are themselves not temporal and are intelligible without temporal concepts. The most recent idealist doctrines are the causal theories of time, which try to eliminate temporal concepts by reducing them to the causal connectability among nontemporal events. They fail, for their abstract models have no relevance to the objective world.[6]

Although philosophers often argue as if one must choose between substantivalism and reductionism, these are not the only alternatives. I reject both, for neither can give a clear account of things and events. A river of time is graphic, but aside from the rhetoric its meaning is obscure. Instead, I argue that time, like spacetime, is objective but not substantival. Time is objective and irreducible because it is inherent in the endurance and change of things, the happening of events, the progression of processes, and the regularity of causation. To echo a remark of Newton's, when any thing, event, or process is posited, time is posited. However, objective time is not substantival, because it is the inalienable structure of things, events, processes, and the world at large and cannot be materially separated from them as a freestanding substance. Time cannot be torn out of a process just as the structure of a chain cannot be torn out of the chain. In thought, perhaps we can abstract a structure of temporal concepts, but the abstract conceptual structure does not refer to a river or a frame that exists on its own. Objective time rejects three features of substantival time: its subsistence independent of things and events, its flow, and its simple monolithic structure covering the entire world.

In the objective but nonsubstantival view, events and processes are not in time; rather, time is inherent in events, processes, and other physical and empirical entities that it helps to make intelligible. There are torrents of events and streams of consciousness but there is no river of time. "Time lasts" means "Things endure," "Times change" means "Things change," "Past times" means "Past events." Instead of arguing about whether time is finite or infinite, one wonders whether the universe has a beginning or an end. Instead of debating on whether time is linear or circular, one asks whether world history repeats itself or progresses monotonically, or whether a causal chain of events can loop back on itself. Dynamic time has its analogue in the tensed theory of time asserting that pastness, presentness, and futurity are changing properties of events. Static time has its analogue in the tenseless theory asserting that the only objective temporal ascription is the relation of before and after between events.

Objective and nonsubstantival time demands a clear articulation of the concepts of thing, event, process, and causation, in which it is inherent. Usually we say a thing changes when its present state differs from its past state. Thus time is integrated through change into the notion of thing.

Dynamic Systems, Events, Processes

This banana was green on Monday and yellow on Wednesday. It changed. Four ideas are involved in the notion of *change*: an enduring thing, its various possible states, the identification of an initial and a final state by the temporal index, and the characterization of these states. *Time* is the index that individuates and identifies the various states of a changing thing. In "The thing changes from state S_1 at time t_1 to state S_2 at time t_2," t_1 and t_2 pinpoint the states, which are characterized by S_1 and S_2. However, without some idea that binds the state S_1 at t_1 and S_2 at t_2, we can only say that S_1 and S_2 are different, not that a change occurs. The unity is provided by the concept of the thing, that which changes and endures through the changes.

A dynamic system is a *thing*, an enduring three-dimensional individual whose state changes through time. We find in §10 that the general concept of individuals contains two elements, the individual's numerical identity and the kind to which it belongs. The numerical identity distinguishes the individual from a crowd of similar individuals in equilibrium theories. To account for change and dynamics, it must be expanded to signify endurance and temporal unity, so that it unifies the thing's many attributes and enables us to reidentify it after its temporary eclipse from attention.

The characterization of a thing involves a type–value duality (§ 5). The character type covers a range of *possible* values and paves the way for the idea of change. A thing's potential to change is limited by the range of possible states admissible to the kind of which it is a member. If the bound of possibility is overstepped, the thing is destroyed. When Lot's wife is transformed by God into a salt pillar, we say the woman is dead, not that she is immortalized as the pillar that allegedly still stands at the shore of the Dead Sea. The distinction between changes in things and changes of kinds is not always so clear-cut, because many substantive details for the sortal concepts of things are lacking. Things belonging to a kind can undergo dramatic changes, as mosquitoes grow from swimming larvae to flying blood suckers. When people lacked sortal concepts of insects that undergo metamorphosis, larvae and mosquitoes were regarded as different kinds of thing, the latter arising from the demise of the former.

A thing changes substantively when its states at different times have different characteristics. The successive states constitute the thing's *history*. They can also be interpreted as the stages of the *process* the thing undergoes. A thing need not change substantively, but an unchanging thing still has a history and undergoes a process, a stationary process. Thus an unchanging thing is not beyond the general notion of change. Rather, it is a special case of change in which the substantive magnitude of change is zero.[7]

All these commonsense ideas are incorporated in scientific theories. Equilibrium models already represent systems by their state spaces that circumscribe all their possible momentary states. The endurance of a system is embodied in a *path* or *trajectory* in its state space that collects and connects a set of possible states, each of which is identified by a unique temporal index. The path represents the history of the system or the process it undergoes. The temporal parameter of the path establishes a temporal order among the chosen states, and the substantive difference of successive states represents the change of the system. "Path" or "trajectory" is mostly without spatial connotation. Although the states it connects can be spatial, as in the case of classical mechanics, they are more often purely qualitative. The states can describe various temperatures, genetic compositions, or gross domestic products.

The identity of a thing through time has always troubled philosophers. Its difficulty is compounded when the things are composite and their constituents are in flux. Hobbes told the story of the ship of Theseus that was replaced plank by plank while riding the high sea, until not one original plank remained. The replaced planks were stored and then used to construct a second ship with the same design. He imagined Athenian sophists disputing which ship, if any, was numerically identical to the original ship of Theseus. If the second, when did the switch of identity occur? If neither, when did the original ship cease to exist? Is identity decided by matter or the uninterrupted history of service? He conceded that ultimately the decision is a matter of convention, albeit a reasoned one.[8] These questions are not academic. For instance, the lack of clear definitions for the beginning and the end of persons plays no small role in the debate on abortion and the right to die. However, in most scientific models they are pushed aside as dynamic systems are defined with various idealizations and approximations.

Besides things, events such as an explosion and processes such as growth and decay often stand as subjects of propositions. Not all events and processes involve things. A process is *pure* if it is not the change of things. The propagation of light is an example of pure processes. Pure processes dominate fundamental physics, the ontology of which is a set of fields with no things in the ordinary sense.

Most familiar events and processes are the transformations of things. Abrupt changes of things are called *events* and more gradual changes *processes*. An event or a process may involve many things, and a process may consist of many events. In scientific theories, processes are represented in the form $S(t)$, where the state S of a system varies dependently on the variation of the temporal parameter t. We can also interpret $dS(t)/dt$ evaluated at a specific time t_0, the almost instantaneous change of a system, as an event. When we treat t not as a variable but as a specific value of the temporal parameter, we intellectually "stop" the process in a specific stage and examine the momentary state of the system undergoing the process.

Unlike things, events and processes are four-dimensional; they are temporally extended and can be decomposed into *temporal parts* called periods, stages, phases, movements, or the like. Furthermore, since they encompass the temporal dimension and the changes in things, they have exhausted the idea of change within their internal structures, so that as wholes they do not come under the notion of change. For instance, the life of a butterfly is a process made up of a larva stage, a chrysalis stage, and an adult stage. During these stages the insect changes from a caterpillar to a pupa to a butterfly. As the sum of the stages, the process of development as a whole does not change. When we talk about the ups and downs of life, we either refer to its successive phases or to the changing state of the organism living through it. In short, things change over time and endure through the changes; processes have temporal extension and are composed of successive stages with different characters.

A Multitude of Processes and Temporal Scales

As the parameter whose values distinguish and identify various states of an enduring thing or various stages of a process, time is inherent in the general concepts of endurance and processes. Note, however, that the temporal parameter is defined individually for each thing or process. This answers a puzzle that confounds the substantival theory of time. People talk about many kinds of time and attribute contradictory characters to time. How can this be if time is a monolithic substance?

Many people, pointing to the irreversibility of most familiar processes, insist that time is anisotropic. Others, noting that most processes in fundamental physics are reversible, argue that time must be isotropic. The protracted controversy conflates the temporal symmetries of various processes with the symmetry of substantival time. We admit both spatially symmetric and asymmetric configurations; why can't we admit both temporally symmetric and asymmetric processes without introducing a contradiction about the symmetry of time itself?

The major trend of a system is not the only ongoing process; there are a multitude of processes proceeding together, each with its own pace and temporal structure. Instead of ascribing various speeds to the flow of time, we talk about the characteristic times of various processes. Physicists distinguish collision times, lifetimes, relaxation times, and other characteristic times for various processes and are careful to note their magnitudes. Biologists compare the rates of mutation and environmental change, and the rates of evolution during and between speciation. Economists talk about short-term fluctuation, intermediate-term adjustment and cycle, and long-term growth. Historians distinguish a matter of years for political events, tens of years for social and economic developments, and hundreds of years for tradition and other civilization-shaping factors. The magnitudes of the characteristic times and the relative magnitudes for competing processes contain the relevant physics, biology, economics, and history. They inform scientists of which variations dominate in a particular time scale, which processes can be treated as stationary, and which factors can be averaged over. Usually, a dynamic equation implicitly assumes a particular time scale, and it addresses those processes that change in the scale.

Phenomena on different time scales are as distinctive as phenomena on different spatial scales or organizational levels. Processes with different characteristic times or rates do interact with each other, but often they can be studied independently to good approximations. Consequently people sometimes call them different notions of time and laud the initial study of some processes as the "discovery of time." Such sayings confuse general and substantive concepts. Time, as a general concept, cannot be monopolized by any process. The substantive issue is not the notion of time but the temporal characteristic of various processes.

Local Times and the Global Temporal Structure

The multitude of temporal rates underscores that the time parameterizing a process is *local* to the process, a fact corroborated by relativistic physics. We have not yet addressed the problem of synchronizing the clocks of various processes, although we have tacitly assumed some synchronization when we compare various processes. Above all, we have not considered whether it is possible for a single *global* time to parameterize all processes in the universe.

A process is *synchronized* with another if the successive values of its temporal parameter can be identified with the successive values of the other's temporal parameter with proper scaling. For a group of synchronized processes, we have a shared temporal structure consisting of a sequence of simultaneous planes indexed by the values of the shared temporal parameter. Successive planes contain the successive stages of each process. If all processes in the universe can be synchronized, then the global temporal structure is particularly simple. It consists of a sequence of global planes of simultaneity or a sequence of global moments, each of which is the state of the three-dimensional universe at the moment indexed by the global temporal parameter. The global parameter fuels the image of a universal time that flows equably.

Since the triumph of Newtonian mechanics, the idea of a global temporal parameter is so ingrained in our thinking it became a metaphysical principle until it is challenged by relativistic physics. The theories of relativity may seem exotic and unnatural, but we will do well to recall that the Newtonian view is itself quite recent. In ancient histories, chronology is dominated by local times; people in different cities or countries kept different calendars.

Methods of synchronizing local times always call on physical interaction, including separate observations of the same astronomical event. Thus the shared temporal structure covering an ever-expanding domain of events is firmly rooted in things and processes. Most of our daily thinking and scientific theories make use of a shared temporal parameter for a significant range of processes. The three sciences we examine are no exception. However, since the topics of these sciences are regional in the cosmological view, their times, although extensive, are nevertheless local.

Is it possible to synchronize all processes in the universe to obtain a global temporal parameter? To this empirical question, whose answer depends on the processes and interactions in the universe, relativistic physics answers no. The rejection of a global temporal parameter does not imply the rejection of a global temporal structure. It only means that the global temporal structure is more complicated than a sequence of global moments parameterized by a single variable. The global temporal structure must accommodate all processes in all temporal scales. It is obtained not by generalizing the features of some particular processes but by finding the maximum possible agreement among all processes. In a way it is akin to the structure of composite systems, which is obtained not by extrapolating the characters of the constituents but by accounting for their interactions. To account for the complicated global temporal structure, relativistic physics presents a four-dimensional view of the world.

Does a Four-Dimensional Worldview Necessarily Petrify Time?

People often say that time is spatialized and change frozen in four-dimensional worldviews. This is not necessarily so. Some four-dimensional views do petrify time. One example is Quine's world of four-dimensional "physical objects" composed of temporal parts. A more interesting example is the Arrow–Debreu model of perfectly competitive markets, which treats time exactly as it treats space (§ 16). Here both time and space are discrete indices of commodities, specifying when and where they are available. Interest charges, which differentiate the prices of commodities delivered at different times, are the strict counterparts to shipping fees that differentiate the prices of commodities from different places. The economy is a four-dimensional landscape with no change and no uncertainty, so that transactions for all times can be settled in a one-shot deal. It is "a pattern of timeless moments," to borrow T. S. Eliot's words. As Martin Currie and Ian Steedman said: "One of the most persistent criticisms of the [Arrow–Debreu] model has been that it 'reduces' time to space."[9]

In contrast, relativistic physics does not petrify time. It has two notions of time, the time coordinate and the proper time. The *time coordinate*, together with the space coordinates, constitute a set of four variables that picks out and identifies individual events in the world. It has a counterpart in the Arrow–Debreu model. What the Arrow–Debreu model lacks is the analogue to the *proper time*, which, as the index that points out successive states in the world lines of enduring particles, underlies relativistic dynamics. Relativistic physics not only distinguishes between space and time but also relates them through the crucial concept of *velocity*, the analogue of which is again absent in models that spatialize time. Because velocity is a dynamic quantity categorically different from the gradient that relates different dimensions of space, time is not merely another spatial dimension.[10]

Relativistic physics also distinguishes time from space via the concept of *mass* or *energy*. Timelike paths in the four-dimensional world represent processes or world lines of enduring particles and are associated with the conservation of mass or energy. Spacelike paths represent composite entities obtained by the aggregation of mass or energy. They separately account for process and thing, endurance and composition. Physicists integrate over the spatial dimensions but not the temporal dimension to find the total mass of a thing. This difference between space and time is absent from the four-dimensional worlds of revisionary philosophies, where things and processes are undifferentiated and the notion of mass is obscure.

If the world is made up of three-dimensional particles, the changes of which are represented by world lines, why does relativistic physics adopt a four-dimensional worldview? Why not treat the whole world as an evolving three-dimensional object as in Newtonian physics? We cannot, not objectively.

Proper time is *local* and defined with respect to a particular process undergone by a particular particle. There are as many local times as there are particles in the world. To obtain the three-dimensional Newtonian worldview, we must synchronize the clocks carried by individual particles to obtain a global temporal parameter that applies to global planes of simultaneity. We can achieve the synchronization by physical means such as radar echo, but the planes of simultaneity we obtain are peculiar to the particular synchronization process represented by a particular inertial frame. Events simultaneous in one frame can occur at different times in another inertial frame moving with a constant velocity with respect to the first. Even worse, under some conditions it is possible for an event to precede another in one frame and succeed it in a second. There is no way to abstract from the inertial frames and obtain an objective definition of a global temporal parameter as we can for the proper times local to individual particles.

At least two attitudes are possible in the situations. The dictatorial solution imposes a specific synchronization scheme, as in the Newtonian absolute time. Relativistic physics adopts the alternative of instituting a conceptual framework encompassing the results of all synchronization schemes, the rules for transforming among them, and the objective features of the world abstracted from the particularities of the schemes. The complexity of the framework mandates a four-dimensional worldview.

The lack of an objective global time parameterizing global planes of simultaneity in the four-dimensional relativistic world does not imply that nothing happens anymore. It implies only that the Newtonian view in which all processes in the world march in synchronous steps according to a global beat is conventional and too simplistic. The objective happenings and becoming in the world are much more convoluted. To understand them we must grasp all interacting processes in a comprehensive view.

To grasp a process mentally in its entirety is *not* freezing the changes within the process. Ludwig van Beethoven said that when a piece of music matured in his mind, "I see and hear the picture in all its extent and dimensions stand before my mind like a cast." Wolfgang Amadeus Mozart allegedly said that when his music emerged, "the whole, though it be long, stands almost complete and finished in my mind, so that I can survey it, like a fine picture or a beautiful statue, at a glance."[11] Mozart and Beethoven are the last mortals to be suspected of abandoning time and turning music into sculpture. Their works are paragons of dynamic processes that can be and demand to be embraced as wholes even as they unfold temporally. Relativistic physics treats dynamic processes in a similar way. The difference is that instead of individual processes it provides a conceptual framework that accommodates all processes.

28. Past, Present, Future

Aristotle viewed becoming, destruction, and motion as different kinds of change. Motion included both locomotion and qualitative changes of substances; becoming meant coming into existence.[12] Today becoming is generalized to cover all changes, as the sky becomes cloudy or Jones becomes sick. With the conservation of mass or energy, generation and destruction are interpreted as changes in kind, as distinct from the change of a thing belonging to a kind.

Some philosophers argue that becoming is more than the changes discussed in the previous section. The essence of becoming is the passage of time, which brings future events to the present and then pushes them into the past, or, to say it without suggesting a substantival time, future events advance to the present and then recede into the past. The passage underlies the "dynamic time" that flows or flies, as opposed to the "static time" we have been considering, so called because it does not contain the ideas of future and past and hence cannot come to pass.

The ideas of past, present, and future are so important they are built into the grammar of many languages as the tense changes of verbs. Are they objective? Why are they absent in many scientific theories? Are they totally absent? This section briefly examines the questions regarding tense that are relevant to our subsequent investigation of the sciences. *Tense* here includes both verbal tense and terms representing the past, present, and future, such as "ten days ago," "now," and "tomorrow." Such terms are found in languages in which verbs are not tensed, for instance, Chinese.

We distinguish metaphysical and conceptual problems regarding tense. Metaphysics concerns questions such as whether past and future events are real, whether assertions about them are true or false. Interesting as these questions are, we will leave them aside. Conceptually, philosophers in the tensed camp argue that tense concepts are necessary for the objective descriptions of the world. The tenseless camp splits in response. Older tenseless theories of time assert that tensed sentences can be translated into tenseless sentences by assigning dates to all events, including the utterance of sentences. "The event *e* occurs now" uttered at date d_0 is translated into "*e* occurs at d_0," where the tense inflection of the verb is abolished. The tense of verbs is like the gender of nouns in some languages; they and other tense terms can be eliminated from our language and conceptual framework without loss or regret. The eliminatist claim is abandoned by more recent tenseless theories, which concede that tensed sentences cannot be translated into tenseless ones without loss of meaning, because the two kinds of sentence have different kinds of truth condition. The new tenseless theories also acknowledge the need for tense concepts in human experiences, expectations, and timely actions. However, they insist that tense terms are not required for objective theories of the world.[13]

Indexical: Context-sensitive Designator

Do tense concepts have a place in physical theories? They can be spotted at once. Physicists use the present mass density of the universe to predict whether the universe is open or closed. The geological timetable is defined in terms of millions of years ago; for instance, the Jurassic Period lasted from 190 to 136 millions of years ago.

Are the tense notions in existing physical theories defects? Can they be eliminated in better theories? To answer these questions let us put aside all metaphysical interpretations of tense notions and study their logical functions. In natural language, the logical function of *now* and *yesterday* is similar to that of *here* and *there*. They belong to a group of expressions philosophers call *indexicals*, which also includes demonstratives such as *this* and *that* and pronouns such as *I* or *it*. Indexicals are designators, by which we identify particular entities and refer to them directly without requiring the entities to satisfy certain definite descriptions.

The striking characteristic of indexicals is that their referents conspicuously depend on the contexts of their usage; "this table" or "the event occurring now" in different contexts refers to different things or events. The meaning of indexicals has been extensively studied by analytic philosophers. David Kaplan argued convincingly that they belong to those denoting phrases that "should not be considered part of the content of what is said but should rather be thought of as contextual factors which help us to interpret the actual physical utterance as having a certain content."[14] The indexicals are semantic rules that assign definite referents in given contexts; for instance, *I* is the rule that assigns the speaker of the sentence. The rules of other indexicals are often difficult to spell out, but we have tacitly understood and applied them when we distinguish the meanings of say, "I will come tomorrow" and "I will come the next day." Being rules, indexicals are by themselves general. Without the specification of a context, their meaning is indeterminate. Since indexicals have referents only within a given context, the truth values of statements containing them are also sensitive to contexts. Whether "Jones was sick yesterday" is true depends on when the sentence is uttered.

Among the indexicals, *now* is peculiar because its context changes compulsively in most daily usages. Things, observers, and language users are all bound in their respective processes of change. They cannot free themselves from the processes as they can stand still in a place. They cannot jump from one stage of the process to another as they can with spatial locations. The time scales of most processes we encounter in our everyday life are comparable to the time scale characteristic of our cognitive responses. In these time scales, the contexts of statements using temporal indexicals change so fast the events to which *now* refers almost always have different characters, leading to the impression of time as a moving now.

Depending on the time scales characteristic of the processes under consideration, *now* is not necessarily as fleeting as it is in our everyday speech. In cosmological and geological time scales, the present is practically forever to us short-lived mortals. In them, or even in the shorter time scales of organic evolution or archaeology, tense notions are capable of supporting systematic articulations of objective facts. That is why the "present" is accepted without demur in many scientific time tables.

Tense and Date

Can *now* as a context-dependent designator of events, states of things, and stages of processes be eliminated from objective empirical theories? Two questions must be clearly distinguished. The first is whether the particularity of the context in which a designator is used has anything to do with the object. The answer is no; the object is independent of the context. The second question is whether context-dependent designators are dispensable in physical theories in which objective statements are made and empirically tested. I argue that *context-dependent designators are indispensable*, for without them the objective statements are untestable and all we have is an abstract theory without an anchor in the real world, hence without empirical and physical significance. Thus we must carefully separate the contextual and objective elements in our theories and refrain from ascribing the context to the object. Ascribing contexts to objects is precisely what philosophies that opt to outlaw indexicals unwarily do.

Superficially, it seems we can gain our independence from temporal contexts by using dates instead of tenses. Such a solution was proposed by Quine in his regimentation of language. Quine called a sentence that excludes tense and designates all events by dates and the relation of earlier and later "an eternal sentence: a sentence whose truth value stays fixed through time and from speaker to speaker." He argued that all sentences should be eternalized to free language from "dependence on circumstances of utterance." Eternal sentences are part of his prescription for "the inclusive conceptual scheme of science."[15] Quine's regimentation fails on two accounts: Dates are not context-independent, and the regimented language cannot discharge the essential functions of the language it replaces. His prescription for "science" runs blatantly against relativistic physics.

Dates do designate events more efficiently and systematically than indexicals. Their apparent context independence, however, is deceptive. Dates are the temporal coordinates in a coordinate system, and a dating system is a convention, a kind of context. According to relativistic physics, if events A and B are spacelike separated, then A can be earlier than B in one coordinate system and later than B in another coordinate system moving with a constant speed with respect to the first. Thus the truth values of sentences about the temporal ordering of events vary from speakers adopting one dating system to speakers adopting another. The conventionality of dates has been thoroughly exposed by relativistic physics. Only if we close our eyes to the conventionality of our own dating system and banish the idea of rival contenders from our thought can we fall prey to the illusion that the truth values of dated sentences are invariant among speakers.

Indexicals are prevalent in scientific theories as floating independent variables of universal statements. Like *now*, the t in Newton's law $F(t) = ma(t)$ waits for the specification of a context or an initial condition to designate anything definite. When we say a particle has energy $E(t)$ without specifying any context, $E(t) = E_1$ means no more than "The particle has energy E_1 now." Probably this point is noticed by journalists who abandon *now* for the more scientific-sounding "this point in time." When an initial time t_0 is specified, then $E(t_0 + \Delta t) = E_1$ can be interpreted as "The particle has energy E_1 at the date $t_0 + \Delta t$." However, we know that we can change the dating system represented by t_0 at will, so that $E(t_0 + \Delta t) = E_1$ is true for one value of t_0 and false for another.

The context in which *now* acquires its meaning represents a temporal perspective of some human subject, a speaker of a sentence or an observer of an event. Kaplan said Quine's move to dates aspires to achieve "sentences which do not express a perspective from within space-time."[16] Only God can view the world from a perspective outside it and hence transcending its spatiotemporal structure. Human beings are radically situated in the world. A ruthless eradication of human perspectives from the general structure of our thought makes the thought totally abstract. Logic and pure mathematics do not invoke any temporal perspective. Empirical science cannot act similarly, for scientific theories must make contact with the world through experience. The aspiration to eternity is illusory.

Without a proper anchoring in the *now*, a dating system is like a calendar in science fiction. "Starday 4354.27" sounds scientific, but it floats in fantasy because it has no point of contact with reality. The utility of dating systems is based on their ability to synchronize the experiences and actions of individuals, so that each can say, "Today is March 1, 1995," or "Now it is 4 o'clock." Similarly, a theory has no physical significance if its results cannot be correlated with experiments, which are always performed by particular human beings in their specific perspectives. Only by including observational perspectives can a theory account for empirical evidence explicitly. Only by noting the conventional and contextual nature of the perspectives can the theory devise ways to compensate for their arbitrariness and articulate the objective content of observations. Thus the categorical framework of physical theories must include the conceptual means for making experimentally testable *empirical statements* as well as *objective statements* that abstract from the perspectives of experiments. It is incorporating this categorical framework that makes the logical structure of modern physical theories so much more complicated than the structure of classical theories.

Modern physical theories do *not* banish coordinate systems, which they associate with observational conditions. Instead, they institute a broad categorical framework that includes all possible perspectives and the rules to transform among them and to abstract from their peculiarities for objective statements that are coordinate-free. The form of objective statements is nothing like the eternal sentence, which is simply blind to its own conventionality and arbitrariness.

Physical theories make statements about objects in a way meaningful not only to God but also to human beings. The conceptual structure of modern physical theories shows that context-dependent designators, the logical analogue of the tenses, are indispensable to our comprehension of the objective world. Theoretical statements are objective both because they can be abstracted from empirical perspectives and because they can be connected to objects through us the knowers, who are in the presence of the world here and now.

8 Complexity in the Temporal Dimension

29. Deterministic Dynamics: State-space Portraits

Dynamic systems are things undergoing processes. This and the following chapter examine two classes of processes and three mathematical theories applicable to them. Deterministic processes are governed by dynamic rules represented by differential equations (§§ 29–31). Stochastic processes are treated by theories that make use of the probability calculus but do not mention dynamic rules (§§ 35–38). A dynamic process can be characterized deterministically in a fine-grained description and stochastically in a coarse-grained description. Both characterizations are included in the ergodic theory, which employs both dynamic rules and statistical concepts (§§ 32–33). The mathematics that unites deterministic and stochastic concepts in a single dynamic system exposes the irrelevancy of the metaphysical doctrines of determinism and tychism (the dominion of chance) (§§ 34, 39).

A deterministic process follows a *dynamic rule* that specifies a unique successor state for each state of the system undergoing the process. The rule-governed change makes the behaviors of deterministic systems predictable and controllable to a significant extent. In recent decades, high-speed digital computers have enabled scientists to study dynamic systems previously deemed too difficult, notably nonlinear systems. Some of these systems exhibit chaotic behaviors that are unpredictable in the long run, because the slightest inaccuracy in the initial state is amplified exponentially, so that the error eventually overwhelms the result of the dynamic rule.

Stochastic processes are represented by *distribution functions* that give the number of stages in a process having certain characters. A stochastic process is *statistically deterministic* if its future stages are predictable with probability 1 on the basis of the record of its entire past. The distribution function is less informative than the dynamic rule; whereas the behavior of a deterministic system is predictable given its state at only one time, the predictability of a statistically deterministic system depends on all its past states.

Ergodic theory, originated with the effort to connect statistical and Newtonian mechanics, has burgeoned into a thriving branch of mathematics that unites deterministic and stochastic theories. It shows that although deterministic and stochastic theories employ different concepts, determinacy and stochasticity are not intrinsically antagonistic; they can be two kinds of description for the evolution of the same dynamic system. When the dynamics of a system is too complex, the statistical description dominates because the deterministic description is impractical even if a dynamic rule is available.

A *dynamic process* is composed of successive stages, which correspond to the changing states of the dynamic system undergoing the process. If a dynamic process is compared to a many-body system, then its deterministic and stochastic characterizations are analogous to the micro- and macrodescriptions of the many-body system. A deterministic theory presents a *fine-grained description* of the process that specifies the character of each constituent stage and the relation between successive stages. A stochastic theory presents a *coarse-grained description* of the process that overlooks the individuality of the stages and accounts for the statistics of types of stage and the statistical correlation among them. When both descriptions are available, the fine-grained description is often interpreted objectively and the coarse-grained description associated with the results of measurements with finite resolution.

This chapter presents deterministic theories and examines their philosophical implications. It is organized differently from other chapters, where the sciences are considered separately because there is no unifying mathematics. Deterministic dynamic theories in all the sciences share a common mathematical structure, that of the ordinary differential equation, or difference equation if time is treated as discrete. I will follow the mathematics of dynamics and draw examples from the sciences when appropriate.

This section presents the basic conceptual structure of deterministic dynamics in its modern geometric formulation. A fundamental notion of the formulation is the *state space*, which connects dynamic theories to the equilibrium theories presented in Part I. The geometric formulation grasps dynamic processes as wholes and paves the way for holistic concepts such as chaos, bifurcation, and strange attractors, which are discussed in the following sections.

The Geometric View of Dynamics[1]

Much of the earlier research on deterministic dynamics addresses problems in celestial mechanics. In the nineteenth century, a major problem was the stability of the solar system. The inadequacy of traditional methods in solving the problem led to the modern phase of dynamic theory. Henri Poincaré, who pioneered the modern phase, introduced the *global geometric view* in which all possible motions of a dynamic system are grasped as a whole and represented by a portrait in the system's state space. Instead of solving for the path of a system with a specific initial condition, one looks at paths under all possible conditions and studies their general patterns. Novel concepts are introduced to represent and classify the general patterns, and rulelike relations among these concepts are delineated. Poincaré's achievement is an instance of the nineteenth-century revolution of higher abstraction in mathematics, which bears tremendous fruit in twentieth-century science.

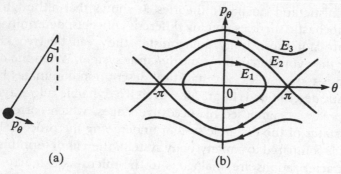

FIGURE 8.1 (a) A state of a pendulum is characterized by two dynamic variables, the angular displacement θ and the angular momentum p_θ. (b) A state-space portrait of the motion of an undamped and undriven pendulum includes all possible paths the pendulum traces in its state space spanned by the variables θ and p_θ. A pendulum with low energy E_1 swings back and forth. With high energy E_3, it rotates about the pivot. The trajectory at the critical energy E_2, which separates the two kinds of motion, is called a *separatrix*. The equilibrium point of the paths occurring at $p_\theta = 0$, $\theta = 0$, is stable; the equilibrium points at $p_\theta = 0, \theta = \pm\pi$, are unstable.

The geometric formulation of dynamic theories does not turn dynamics into statics. With it we mentally grasp temporal processes in their entirety, just as we do in the four-dimensional worldview of relativistic physics. To see how it works, consider a simple system, the undamped and undriven pendulum (Fig. 8.1a). A pendulum is characterized at several levels of fixity. The factors determining the pendulum's natural frequency – the mass of the bob, the length of the ideally weightless rod, and the force of gravity – are fixed once and for all. The total energy of the pendulum, which is encapsulated in an initial condition, is fixed for the duration of an experiment but can vary from experiment to experiment. Finally, the angular displacement θ and the angular momentum p_θ vary with time and characterize the changing state of the pendulum. Classical dynamics is satisfied with finding the temporal evolution of θ and p_θ for a given set of initial values. Modern dynamics considers much more.

The state space of the pendulum includes states for all possible values of θ and p_θ. A pendulum with a small energy E_1 exercises harmonic motion at its natural frequency. If we plot the bob's motion in its state space, we find that it traces out an ellipse (marked E_1 in Fig. 8.1b). Its displacement is maximum when its momentum is zero, zero when its momentum is maximum. The ellipse is repeated again and again as the pendulum swings back and forth. A higher energy yields a larger ellipse, because the energy determines the maximum displacement. Finally, at a critical energy E_2 the pendulum swings so hard it goes over the top. At still higher energy E_3 the behavior of the pendulum changes drastically; instead of going back and forth and stopping at the turning points, it rotates around the pivot and its momentum never vanishes. The critical path E_2, which separates two qualitatively distinct modes of motion, is called a *separatrix*.

The state-space portrait of the pendulum includes all possible paths. A pendulum is capable of three classes of process: swings typified by the path with energy E_1, rotations with E_3, and the separatrix with E_2. It illustrates the emphasis of the modern formulation of dynamics. Dynamic theories still determine the successive states in a path given an initial condition. However, the emphasis is no longer on the solution for specific paths but the delineation of broad classes of path and the articulation of universal features shared by all paths in a class. For instance, the definition of chaotic systems is based on the behavior not of a single path but of a group of paths. The supermacroview of modern dynamics that grasps many classes of paths also paves the way for introducing statistical concepts to study more complicated systems.

The Conceptual Structure of Deterministic Dynamic Theories

The example of the pendulum displays the conceptual elements of a dynamic theory: time; the dynamic rule; exogenous parameters, which include the initial condition; the state of the system at a time; the state space encompassing all possible states; the path in the state space representing the dynamic process the system subjected to a given set of exogenous parameters undergoes; the state-space portrait including paths for all possible exogenous parameters; the classification of the paths according to the exogenous parameters.

Exogenous parameters such as the energy of the pendulum represent those factors that are fixed during the period under consideration but can be changed from period to period. Alternately, they can represent slowly varying factors that can be regarded as constants in the characteristic time scale of the process under investigation. Often they contain information on the influence of the external environment on the system. Since the exogenous parameters are often varied by experimentalists to manipulate the system, they are also called *control parameters*. The variation of the exogenous parameters is important in stability and bifurcation theories.

The *state* of a system contains the information about the system that is relevant to the motion governed by the dynamic rule. It need not be the complete state of the system; the pendulum may be changing in color, but color is excluded from the state as irrelevant to its dynamics. The state is described by one or more *state variables*, which characterize the system at each moment. The minimum number of variables is sometimes called the system's *dimensionality*. The collection of all possible states of the system is its *state space* or *phase space*, the coordinate axes of which are the state variables. The state variables of a pendulum are its angular displacement and momentum. The state variable can be the population or the price level or anything relevant to a dynamic system.

The *dynamic rule*, which may depend on the values of some exogenous parameters, encapsulates the temporal evolution of the system. It is a single-valued function f that transforms the state at a particular time to a unique state at a later time. Given an *initial state*, the dynamic rule generates a *path* in the state space representing the temporal evolution of the system, for instance, the ellipse E_1 in Fig. 8.1b. The path is sometimes called the "motion" of the system. *Motion* here is understood in the Aristotelian sense that is not limited to locomotion; the path is not in "physical" space but in the state space and can represent qualitative changes of the system.

The dynamic rule specifies how a system changes. It is parameterized by the variable *time*, which is the sole independent variable for those systems that do not vary qualitatively spatially. The time variable can be either continuous or discrete. In cases with *continuous time*, the infinitesimal change of the system is represented by a set of coupled ordinary differential equations $dx(t)/t = f(x(t))$, where $x(t)$ is the state of the system at time t and dx/dt the infinitesimal change of the state. For a given initial state, the solution of the equation by integration gives the path of the system. The dynamic rule $f(x)$ is also called a *flow*. The name *flow* aptly captures the image of the state-space diagrams that depict the infinitesimal displacement of each possible state by a tiny arrow, giving the impression of a flowing fluid. It reminds us that generally a dynamic theory accounts for the transformation of every state in the state space.

In cases with *discrete time*, the dynamic rule f is called a *map* and is written as $x_{n+1} = f(x_n)$, where x_n is the state of the system at time t_n and f transforms it to the state at t_{n+1}. The notation shows clearly the indexical function of time. The map is iterated as the temporal index n increases, so that x_n can be obtained from the initial state x_0 by repeating n times the rule f.

Discrete time can be interpreted as a representation of our monitoring the system at regular intervals, for instance, every hour on the hour. A more general interpretation of discrete time is given by the *Poincaré surface section*. Imagine drawing an arbitrary line across Fig. 8.1b. The line intersects the ellipse E_1 at two points. Given one point, the other is uniquely determined because they are connected by the path. Thus instead of the dynamics along the path, we can examine the dynamics between the points and say the pendulum "jumps" between the two points at discrete times. This example is trivial, but the idea becomes powerful when it is generalized to higher dimensions and more complex paths.

A line is a one-dimensional surface in a two-dimensional world. Generally, a $(N - 1)$-dimensional surface in a N-dimensional state space cuts across a complicated path like a transparent knife cutting across a loose ball of tangled yarn. The cross section creates a set of points on the surface like the cut ends of the yarn sticking on the blade. The initial intersection between the surface and the path is called the state of the system at time t_1; when the path returns from the same direction to the surface for a second time, the intersection is called the state at t_2, and so on. The state at t_{n+1} is uniquely determined by the state at t_n, and the temporal relation between the successive states at the surface is represented by a map. Thus a continuous dynamic problem in N dimensions is reduced to a simpler discrete dynamic problem in $N-1$ dimensions (Fig. 8.2). Many graphics for dynamic systems are surface sections.

The surface section does not preserve the notion of temporal duration. The return time of the path to intersect the surface again may be short for one round and long for the next, but this difference is neglected. The state x_{n+1} is the successor state of x_n, regardless of the actual time lapse or the length of the path linking the two. The surface section method is widely used in dynamic problems. It indicates that the most important roles of time in dynamics are *indexing* and *ordering*, which are what the surface section preserves.

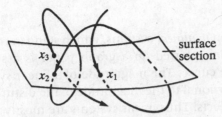

FIGURE 8.2 A Poincaré surface section cuts across a path in a three-dimensional state space. The continuous path is governed by the dynamic rule of the system. The surface section enables us to find a map that relates the states x_1, x_2, ..., at which the surface successively intersects the path from the same direction. Thus a continuous three-dimensional dynamic problem is reduced to a simpler discrete two-dimensional problem.

Invertible and Noninvertible Systems

In all the systems we consider, the dynamic rule is a single-valued function that specifies a *unique* successor to a given state and hence a unique path for ascending temporal indices. Such dynamic systems are *deterministic*. They exclude whatever influences prevent the unique determination of succeeding states.

The possible paths of a deterministic system cannot split, nor can they intersect at any particular time. For if two possible paths intersected, then the successor to the state at the intersection would not be unique. Sometimes, however, the possible paths of a system can merge. If the possible paths of a dynamic system do not merge, that is, if each state has a unique predecessor state, then the system is *invertible*. Otherwise it is *noninvertible*. In a noninvertible system, several states can evolve into a single state; hence we cannot determine from which one the system originated by using the inverse of the dynamic rule. We lose our way back, so to speak. Many maps are noninvertible, for instance, the logistic map discussed in the following section. Deterministic flows are always invertible.

Invertibility should not be confused with reversibility. A dynamic process is *reversible* if it is indistinguishable from the process obtained by substituting $-t$ for t in its dynamic rule (§ 40). Reversibility is characteristic of mechanical processes but not of statistical mechanical processes and live processes.

Regular and Chaotic Systems

A deterministic dynamic system can be regular or chaotic, depending on the relative behaviors of groups of its possible paths. A system behaves *regularly* if nearby paths stay close to each other as they evolve, diverging at most linearly with time. It behaves *chaotically* if nearby paths separate at an exponential rate. The divergence of paths can be quantified in terms of the *Lyapunov exponent*, which gives the mean exponential rate of separation as time goes to infinity for paths that are initially within an infinitesimal region of each other. Paths with positive Lyapunov exponents are chaotic. Chaotic paths are not periodic and are usually very complicated (§ 31).

Linear and Nonlinear Systems

Chaotic behaviors are found only in *nonlinear* systems, which stand in contrast to *linear* systems. A linear function plotted on a graph paper is a straight line; a nonlinear function is some curve. The magnitude of a linear system's response to a disturbance is proportional to the magnitude of the disturbance, so that small causes have small effects. Thus linear systems are mostly stable. Most small perturbations from equilibrium can be treated as linear in the first approximation; that is why linear phenomena are so prevalent. Many phenomena in optics, electromagnetism, and strain and stress of materials are linear.

Linear systems obey a *superposition principle*, in which the combination of solutions yields another solution. Thus scientists can analyze a complicated problem into many simple ones, find the solutions for them, then superpose the solutions to get the answer to the original problem. Analytical tools such as the Fourier transform, by which any periodic function can be analyzed into a superposition of simple sine waves, are the work horses in many sciences.

The superposition principle usually fails for systems operating far away from equilibrium, where nonlinearity is unavoidable. Since two solutions of a nonlinear system cannot be added together to yield another solution, the nonlinear problem must be solved as a whole. In the sense of emergence discussed in § 22, nonlinear systems exhibit emergent behaviors, whereas the behaviors of linear systems are resultant.

Nonlinear systems are far more complex than linear systems and are just beginning to yield themselves to detailed investigation. They are prone to instability; the smallest disturbance can produce huge effects on them, sometimes transforming their behaviors from regular to chaotic. Their dramatic responses to triggering perturbations underlie interesting natural phenomena ranging from avalanche to turbulence. They also undermine the stable worldview prompted by linear systems.

Conservative and Dissipative Systems

A deterministic dynamic system can be either conservative or dissipative. A pendulum without friction is a conservative system; it goes on swinging forever because its energy is conserved. The addition of friction turns the pendulum into a dissipative system; its motion is gradually damped out as its energy is dissipated as heat, until finally it comes to rest. The paths of the damped and undamped pendulums are very different. A path for an undamped pendulum forever repeats an ellipse, so that the volume of the state space it occupies remains constant. A path for a damped pendulum deviates from the ellipse and spirals in as the pendulum's displacement and momentum decrease, until it ends in the point $\theta = 0$, $p_\theta = 0$. The volume of the state space it occupies contracts.

Generally, a dynamic system is *conservative* if its dynamics preserves the volume of its state space. It is *dissipative* if its dynamics leads to a contraction of its state-space volume. Dissipative systems are usually not invertible, for paths converge in the state-space contraction, as all damped oscillators eventually come to rest. Note that here *conservative* and *dissipative* need not imply the conservation or dissipation of energy. Conservative and dissipative systems both exhibit regular and chaotic behaviors. Chaotic conservative systems are usually more difficult to treat because of the constraint on the state-space volume.

Consider a dissipative system in an arbitrary initial state. A path of the system either wanders off to infinity or approaches a bounded region of the state space. In the former case the system is unstable. The latter case is more interesting. The bounded region to which paths issuing from different initial states converge asymptotically is called an *attractor*, and the initial states for all converging paths constitute the attractor's *basin of attraction*. Attractors on which dynamic systems behave chaotically are called *strange attractors*. Systems on an attractor stay on it forever; systems initially not on it get so close to it after some time that for practical purposes they are regarded as on it. A dynamic system can have two or more attractors, whose basins of attraction are separated by boundaries called *separatrices*. If an attractor is like a sea to which all streams in its basin of drainage converge, then the separatrix is like a continental divide between two watersheds (Fig. 8.3). Highly complex systems can have arrays of attractors with different characteristics. An example of a complex system of attractors is the rugged-fitness landscape discussed in § 24. For such cases it is usually difficult to find the particular attractor into which the dynamic system settles.

The meaning of attractors is most apparent in physical systems, where dissipative systems often involve the dissipation of energy. If a system has an energy source to replenish what it dissipates, then it need not come to rest but can settle in a dynamic attractor such as the periodic motion of a metronome or a grandfather clock. The energy source usually imposes some structure, which the system tries to dissipate and then returns to its natural behavior. The attractors are the centers of attention for many theories; they represent

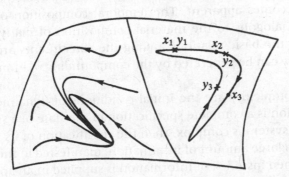

FIGURE 8.3 The state space of a dissipative system is divided into two basins of attraction by a separatrix. In the left basin, all paths converge on an elliptical attractor, resulting in a contraction of state-space volume, because the ellipse is smaller than the basin itself. A strange attractor is depicted alone in the right basin. At time t_1, two paths are close to each other at x_1 and y_1. Later their exponential divergence leads them far apart at x_3 and y_3.

the "natural" and "settled-down" behavior of the system in which transient disturbances imposed by various initial conditions have died out.

The evolution of a system onto an attractor signifies a simplification of its behavior and a reduction in its number of degrees of freedom. For instance, a fluid in motion is a complex system with many degrees of freedom, but when it comes to rest its behavior can be described simply. Thus the dimensionality of an attractor is usually much smaller than the dimensionality of the state space. The *dimension* of a set is associated with the amount of information required to specify the points on it accurately. Hence the dimensionality of a system's attractor is often used as a measure of its complexity.

Complexity in the Temporal Dimension

The dynamics of deterministic systems can be chaotic and highly complex. Furthermore, chaotic behaviors are not limited to systems governed by complicated dynamic rules. Chaos can be exhibited even by "simple" systems. What does simplicity mean here? Where does the complication come from?

We tend to call a dynamic system "simple" if it is characterized by a small number of state variables and governed by a dynamic rule of simple form. Most chaotic systems that have been studied have only a few variables. For large many-body systems, we are mainly interested in the dynamics of their macrostates. A discrete-time system with only one variable is capable of behaving chaotically. Continuous-time systems require more variables, but three are sufficient for chaos. The Lorenz model of the weather, which contains only three variables, is chaotic.

The basic dynamic rule relating the consecutive states of a system or the consecutive stages of a process need not be complicated. In many cases the rules are disarmingly innocuous, as can be seen in the examples in the following sections. In such cases, complexity arises from the repeated iteration of the basic rule, which yields the relation among states separated by many periods, during which chaos becomes apparent. The temporal composition of long dynamic processes is analogous to the material composition of many-body systems. In both cases, the basic relations among the constituents are simple, but great complexity can be generated by the compounding of numerous basic relations.

Another source of complexity are the initial conditions of dynamic processes. An initial condition is a complete specification of the state of a system at a specific time. If the system is complex, then the specification of its initial condition must require a large amount of information, manifested in the high accuracy demanded. When insufficient information is supplied in an approximate initial condition, the deficiency will be revealed as the system evolves. The structure of initial conditions is as important as the structure of dynamic rules for the comprehension of complex dynamic systems.

Dynamic theories illustrate a principle of scientific research: There is no free lunch. If the object is complicated, then cheap ways to evade the difficulty cannot succeed. We must face up to the relevant information, if not in the dynamic rule, then in the initial conditions. If we want to bypass some details, then we must be satisfied with a coarser view. On the other hand, if we are not microreductionists and are willing and able to make the tradeoff, we can discern patterns, conceptualize regularities, and define tractable problems on other levels of generality and organization. Chaos, fractals, exponents, and strange attractors are examples of concepts on a higher level of abstraction that make intelligible an otherwise intractable mass of details. Perhaps this is the best way of understanding "order out of chaos."

30. Stability, Instability, Bifurcation

In dynamic theories as in many-body theories, we ideally separate a system from the world to be our topic of investigation. In reality, systems are never truly isolated. Here we consider the residual influence of the external world on the behavior of dynamic systems, especially their stability.

The primary notion of change for a dynamic system is the change of its state according to the dynamic rule, which generates a process. This section considers a secondary sense of change, or the change of the process in response to the variation in the control parameters representing the influence of the external world. Secondary changes do not always occur, because the control parameters are often fixed. However, when the control parameters are varied, they can produce dramatic effects. A slight variation in the value of a control parameter can sometimes cause the system to alter its entire pattern of behavior by jumping from one type of process to another type. Such instabilities in behavioral pattern are called *bifurcations*. Some of the most interesting results of dynamic theories concern the stability or instability of systems and the way a cascade of bifurcations lead to chaos.

Stability and Invariant Sets

Stability in the face of perturbation is a generic concept that applies to many entities: paths, equilibrium states, invariant sets, systems. A pendulum has two equilibrium states. The equilibrium state at $\theta = 0$, $p_\theta = 0$, which corresponds to the pendulum at rest with its bob in the vertically downward position, is stationary and *stable*; the pendulum returns to it after a small perturbation. The state at $\theta = \pi$, $p_\theta = 0$, which corresponds to the pendulum at rest with its bob in the vertically upward position, is stationary but *unstable*; the slightest perturbation will send the bob swinging down.

A *path is stable* if a "tube" with a fixed small radius can be defined around it so that if the initial condition is perturbed slightly, the perturbed path stays within the tube at all times. It is unstable if no such "tube" can be defined. This criterion of stability is weaker than the asymptotic stability that requires the perturbed path to converge to the unperturbed one as time goes to infinity. Stable paths are regular, and chaotic paths are unstable, but the converse of either statement is not generally true.

To define the stability of a system, we must first grasp its forms of motion. The possible processes a system undergoes can be classified into broad types represented by the characteristics of various *invariant sets* of the system. A set of states is invariant under the dynamic rule if the path through any state in the set includes only other states in the set, even as time runs to infinity. The ellipse in Fig. 8.1b is an example of an invariant set. An attractor is an invariant set, but the reverse need not hold; conservative systems have invariant sets but no attractors.

The simplest invariant set is a *fixed point*, which is a set containing a single state. A system in the fixed state stays in it forever. The pendulum has two fixed points: the state at $\theta = 0$, $p_\theta = 0$, and the state at $\theta = \pi$, $p_\theta = 0$. *Cycles* are the next simplest invariant sets. A cycle that is an attractor is called a *limit cycle*. The behavior of systems confined to cycles is *periodic* with certain frequencies or periods, as is the swing of a pendulum. Cycles can also be discrete. For a discrete cycle of period p, iterating the dynamic map p times returns the system to the initial state. Thus a period-two cycle consists of two states, between which the system jumps back and forth. A more complicated invariant set is a *torus* formed by the product of two cycles, the ratio of whose frequencies is an irrational number. Invariant sets are generally called *tori*: Fixed points are zero-dimensional tori, cycles one-dimensional tori, doughnutlike sets two-dimensional tori, high-dimensional sets in higher dimensions higher-dimensional tori. Paths on limit cycles and tori are all regular.

The behavior of a dynamic system depends on the values of its control parameters. A *system is stable* if minor variations in the values of the control parameters do not change the structure of its invariant sets. A system can be stable for a range of values of the control parameters but become unstable at other values. At the points of instability, invariant sets can appear, disappear, or change qualitatively when one or more parameters are varied. If the system is operating on an invariant set, as it often is, then its behavior changes dramatically. The instability in which a system goes from one invariant set to another is called a *bifurcation*. If the change in control parameters is called a *cause*, then a small cause immediately produces a large effect on a system in a bifurcation.

Bifurcations occur in both conservative and dissipative systems. There are several kinds of bifurcation. Thus bifurcation is more general than its original Latin meaning of "forking into two," although the original meaning aptly describes the period-doubling bifurcation discussed in the following.

Bifurcation in Dissipative Systems: The Logistic Map[2]

The prototype system displaying a cascade of bifurcations leading to chaos is the logistic map, whose application to population biology will be discussed shortly. The *logistic equation* $x_{n+1} = rx_n(1 - x_n)$ represents one-dimensional, discrete-time, dissipative, noninvertible systems. It maps a system in state x_n at time t_n to the state x_{n+1} at t_{n+1}, and its control parameter is r. For $0 < r \leq 4$, the state space of the system is the interval $0 \leq x_n \leq 1$.

The motion of a logistic system can be regular or chaotic, depending on the value of r. A logistic system in regular motion has an attractor x^* whose basin of attraction is the entire state space. From any initial value x_0 on the interval $(0, 1)$, the system approaches x^* as time increases and remains on x^* forever.

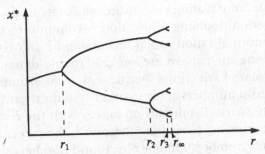

FIGURE 8.4 The curves depict the stable attractors of the logistic map as a function of the control parameter r. Only the first three bifurcations are shown. An infinite number of bifurcations occur for $r_3 < r < r_\infty$. Chaotic behavior commences beyond the accumulation point r_∞.

The attractors for various values of r can be classified into infinitely many types, each a cycle with a distinct period (Fig. 8.4). A period-n cycle means x^* has n values and the system goes from one value to another periodically. A bifurcation occurs when, on a slight variation in r, the system jumps from a cycle with a certain period to a cycle with a different period.

Regular motion occurs for $0 < r < 3.5699$ and falls into many types. For $0 < r < 3$, the attractor is a period-1 cycle or a fixed point. At $r = r_1 = 3$, a *period-doubling bifurcation* occurs. The fixed point attractor becomes unstable and a stable attractor in the form of a period-2 limit cycle appears. The system jumps to the new stable attractor and, for $3 < r \leq r_2 = 3.4495$, oscillates between the two states. At $r = r_2$, a second bifurcation occurs; the period-2 cycle becomes unstable and a stable period-4 attractor appears. As r increases further, an infinite cascade of period-doubling bifurcations occurs. The range of r in which the period-2^m cycle is stable decreases almost geometrically with m. They become so short it is difficult to depict them in the figure. Beyond $r = r_\infty = 3.5699$, the system becomes chaotic. For $r_\infty < r < 4$, chaotic bands intersperse with regular cycles with odd periods. At $r = 4$, the system is uniformly chaotic for almost all initial conditions, with values of x covering the whole range of $(0, 1)$.

Period-doubling bifurcation is observed in experiments ranging from non-linear electrical circuits and semiconductor devices to the Bénard convection of fluids. In the Bénard convection, a fluid contained between two horizontal, thermally conducting plates is heated at the bottom. The control parameter is the temperature difference between the top and bottom plates. For small temperature differences, thermal conduction occurs but the liquid is static. At a critical temperature difference, convective flow in the form of rolls set in; hot liquid in the middle rises and cool liquid descends at the side. As the temperature difference increases further, the rolls become unstable by developing waves. The instability waves exhibit period-doubling bifurcation as the temperature difference increases. In convection experiments with mercury in a magnetic field, at least four period doublings have been observed.

Despite the diversity in dynamic systems that exhibit period-doubling bifurcation, certain features of the period-doubling cascade are universal and independent of the details of the system. The way in which period-2^m cycles converge to chaos follows definite scaling rules. Mitchell Feigenbaum noted that the same pattern of period-doubling bifurcation is common to a large class of maps. Using the renormalization group technique, he showed that several numbers characterizing the pattern are *universal* to one-dimensional dissipative maps with maxima, of which the logistic map is an example. He also argued that these universal numbers apply to other systems regardless of the dimensions of their state space. His prediction agrees with the results of the mercury convection experiments to within 5 percent.[3] The agreement is remarkable, for Feigenbaum's numbers are not nice round numbers such as 2 or 2/3 but jagged numbers such as 4.6692, and the complicated mechanism of convection is nothing like what the simple logistic map describes. The applicability of Feigenbaum's numbers to such diverse phenomena justifies the epithet *universality* and reminds us of the universality of scaling parameters in phase transition (§ 23).

Examples in Population Biology[4]

Some form of the logistic equation has long been used by ecologists and population biologists to describe the wax and wane of populations under the constraint of limited resources. The newly discovered complexity in the behaviors of logistic systems introduces more possible explanations for population dynamics.

For brevity let us consider only populations with nonoverlapping generations, which are found in many insects that live only one season. Such populations grow in discrete steps and can be represented by a difference equation. The logistic equation in this application is usually written as $N_{n+1} = N_n[1 + r(1 - N_n/K)]$, where the state variable N_n is the density of a population at time t_n. The model contains two exogenous parameters, the per capita growth rate r and the carrying capacity of the environment K. According to traditional reasoning, when the population density is low, the nonlinear term rN_n^2/K is small compared to the linear term $N_n(1 + r)$, and the population grows exponentially with the rate r. As the population density increases, the nonlinear term becomes important and puts the brake on further growth; consequently the population levels off at a constant level determined by K.[5]

In the mid-1970s, Robert May pointed out that the simple interpretation is misleading, for the logistic equation can lead to complicated dynamics. Depending on the values of r and K, the density of a population can remain stationary, oscillate in stable cycles, or behave chaotically. At least that is what the theoretical analysis of the logistic map shows. Is the theoretical complexity reflected in reality?

Many populations in the wild are stable. Others fluctuate widely, but the fluctuation is often caused by weather and not by intrinsic population dynamics. Some populations undergo dramatic periodic increase and decline; the four-year cycle in the abundance of mice and lemmings in many northerly regions is legendary. Cyclical variation in population density is more readily observed in laboratory experiments, where the ideal of an isolated population under controlled resource constraints is satisfied.

Empirically, the values of the exogenous parameters r and K can be extracted by fitting the field or laboratory data to the logistic equation. Theoretically, numerical analysis of the logistic equation yields the ranges of r and K for which the population dynamics is stable, cyclical, or chaotic. The empirical and theoretical results can be combined in a diagram of the exogenous-parameter space. A compilation of data for seasonally breeding insects finds all field populations stable. Over 90 percent of the populations have fixed point attractors. The only species exhibiting a period-2 limit cycle is the Colorado potato beetle, whose contemporary role in agroecosystems lacks an evolutionary pedigree. Laboratory populations are more volatile, but only one falls in the chaotic range.

To explain the boom and bust cycles of lemmings, the logistic equation must be modified to take account of the delay with which many population-regulatory mechanisms take effect, for instance, the time lag between conception and maturation. Such delays can cause oscillations in the same way as many feedback mechanisms do; the system tends to overshoot in its corrective action and to overcompensate. Even in the refined model, field data show regular limit cycles but no chaos. Despite the enthusiasm for chaos, population biological data provide little evidence of it.

May argued that the possibility of high-period limit cycles and chaotic behaviors of logistic systems has great consequences for ecological models, which depend heavily on the logistic equation. It implies the practical impossibility of distinguishing the intrinsic randomness of population dynamics from the stochastic variation in the environment or the error of measurement. Mathematically, logistic models do yield many interesting results. However, as Eric Renshaw argued in his review of population dynamics: "The trap for the unwary lies in being carried away on a wave of *mathematical* enthusiasm, since there is absolutely no reason why model (1.2) [the logistic equation] should give rise to an equally diverse range of *biological* behavior. Indeed, as nature is inherently stochastic we have to investigate what mathematical features remain when a random component is added to completely deterministic equations such as (1.2)."[6]

Concepts in Theories of Dissipative and Equilibrium Systems

I end the section by briefly examining some philosophical interpretations of bifurcation and the behavior of dissipative systems. Ilya Prigogine, who contributed much to the research on dissipative systems, argued that their bifurcation introduces a new sense of time associated with progress, the accumulation of change, and the increase in complexity.[7] I find his thesis misleading.

The independent variable in bifurcation is not time but the control parameter. The new features resulting from bifurcations are not the fruit of "time's creative hand" but the response to the control parameters, which can be set in any temporal order. Suppose a logistic system executes a period-4 oscillation at $r = 3.5$. We cannot assume that it has passed through two bifurcations and accumulated features on the way. The exogenous parameter r can happen to be fixed at 3.5. Suppose r varies with a rate that is slow compared to the characteristic dynamic time of the system. Still there is no sense of progress, for *r* can decrease so that the bifurcations make the system behavior simpler, or it can oscillate and produce no trend at all. What is mistaken for a new sense of time is merely a classification of processes coupled with the temporal scale of the dynamics of the control parameter, which is not considered in the theory of dissipative systems.

The suggestion that dissipative systems introduce a notion of history into physics is equally spurious. The motion of dissipative systems toward their attractors does give them a temporal directionality, which is absent in many basic physical processes. However, the directionality is common to all thermodynamic systems and has nothing to do with bifurcation or progress. If history means no more than a path, then it is found in all of physics. If it means the influence of contingent factors, then it is found in many physical systems. If it means retaining some memory of the past, then dissipative systems are even less historical than some equilibrium systems.

Instabilities analogous to bifurcations also occur in equilibrium systems, as in phase transitions. The bifurcation of dissipative systems is different from the symmetry breaking of equilibrium systems. In symmetry breaking, the system realizes a particular state out of a plethora of possibilities, and the accidental factors that determine the particular state are frozen into the structure of the system. Equilibrium systems remember, so to speak, and their memories determine their future responses, as exhibited in the phenomenon of hysteresis. In contrast, dissipative systems forget. Dissipative systems that originated in different states in the basin of attraction all tend toward the same attractor. In so doing they forget their diverse origins. Thus dissipative systems are unhistorical because they have no memory.

The spontaneous formation of new structures in the symmetry breaking of equilibrium systems is often called *self-organization*. The equilibrium structures are autonomous in the sense of being intrinsic to the nature of the system and independent of boundary conditions. Ice has the same structure and the same lattice constant whether it is formed in a lake or in a cup. Spontaneous formation of spatial structures also occurs in dissipative systems; an example is the convective cell in Bénard convection. However, structures in dissipative systems are very sensitive to their boundary conditions. For instance, the geometrical characteristics of the Bénard cells are determined by the geometrical characteristics of the container that holds the liquid. Occasionally, as when the height-to-width ratio of the container is too great, cells may even fail to form. There is a spontaneous element in the formation of dissipative structures, but the lack of autonomy in these structures warns us to be careful about the name *self*-organization.

31.   Chaos and Predictability

Chaos has acquired a mystique in the popular literature, which is partly due to a confusion of the technical and ordinary meanings of the word.[8] The technical meaning of *chaos* means neither utter confusion nor lack of fixed principles, as "chaos" or "chaotic" connotes in ordinary parlance or much of philosophy. A deterministic dynamic system is *regular* if initially nearby paths stay close together, *chaotic* if they separate at an exponential rate. Exponential divergence makes chaotic processes unpredictable in the long run because it amplifies tiny errors in the initial conditions. This is the full technical meaning of chaos.

The exponential divergence of paths by itself is not surprising; it occurs in many cases where the state space is unbounded. Chaotic dynamics are remarkable because exponential divergence occurs even when the paths are confined to a bounded region of the state space, for instance, on the strange attractor of a dissipative system (Fig. 8.3). Within the confinement, the diverging paths are mixed up and interspersed, generating complicated and fantastic geometrical patterns. The mixing also lays the foundation for the applicability of statistical concepts.

The Binary-shift Map[9]

Perhaps the most transparent examples of chaotic systems are the shift map and the baker transformation. Simple as they are, they display many features of chaotic dynamics and will be referred to again in subsequent sections. They themselves are abstract, but they hide in some maps that represent realistic systems. For instance, the chaotic range of the logistic map, which is often used to represent population growth, is a kind of shift map.

The *binary-shift map* $x_{n+1} = 2x_n$ modulo 1 represents a one-dimensional deterministic dynamic system, a state of which is a number and the state space of which is the line interval $(0, 1)$. The subscript n of the state x_n represents discrete time t_n. The dynamic rule includes two steps: First a number x_n in the interval $(0, 1)$ is multiplied by 2. Then the modulo operation drops the integer of the result and keeps the decimal part as x_{n+1}, which is again within the interval $(0, 1)$. For example, 0.4 is mapped into 0.8, and 0.8 into 0.6. The binary-shift map is nonlinear because the modulo operator introduces a discontinuity at 0.5.

The meaning of "binary shift" is clearer when the numbers are written in binary decimals such as $1/2 = 0.10000\ldots$, $1/4 = 0.01000\ldots$, $1/8 = 0.00100\ldots$. In binary decimals, multiplication by 2 shifts the decimal point one place to the right, just as multiplication by 10 shifts the decimal point in ordinary numbers. If we write the initial number generally as $x_0 = 0.a_1a_2a_3\ldots$, where each a_i can be either 0 or 1, then $x_1 = 0.a_2a_3a_4\ldots$, $x_2 = 0.a_3a_4a_5\ldots$, and so on. Each iteration of the map shifts the decimal point one place to the right and drops the integer before the decimal.

The binary-shift map embodies the stretching and folding operations found in many chaotic dynamics. The multiplication stretches the line interval $(0, 1)$ out to twice its original length; the modulo operation cuts the stretched line into two equal parts, stacks them on top of each other, and identifies them to recover the interval $(0, 1)$. As the line is stretched repeatedly, the difference between two nearby points is amplified exponentially. On the other hand, the stacking and identification keep the state space bounded, generate complications within it, and eliminate certain information. Consequently the shift map is not invertible. Given x_n, we cannot go back and recover x_0.

The Baker Transformation and Bernoulli Systems[10]

Noninvertibility is not essential to chaos. An invertible system can be obtained by putting two numbers back to back to produce a doubly infinite series, $\ldots a_{-2}a_{-1}.a_0a_1a_2 \ldots$, where the subscripts represent time from the infinite past to the infinite future. The dynamic rule shifts the decimal point one place to the right for each forward temporal step, one place to the left for each backward step. This invertible shift map is a *Bernoulli system*.

A representation of the Bernoulli system is the *baker transformation*, so called because it is like a baker kneading the dough by rolling it into a sheet, folding it, rolling it out, and folding it again and again. A state of the system is an ordered pair of numbers, and the state space is the unit square, the product of the interval (0, 1) with itself. The baker transformation stretches the square to a rectangle, cuts it in half, then stacks up the two halves to recover a square. It stretches and stacks like the shift map, but because its state space is two-dimensional, it need not identify the stacked-up halves with each other, thus preventing the information loss in the identification. This accounts for the doubly infinite series, the preservation of information, and the invertibility of the map.

Suppose originally there is a pattern on the square, say a smiling face. The face shrinks vertically, expands horizontally, and fragments in the baker transformation. After many transformations, it almost becomes a series of horizontal lines, which are finely meshed in all parts of the square. Now imagine that similar operations are performed on the state space of a system. The stretching that pulls the paths apart generates chaos; the folding that gathers the paths keeps them in a bounded region and mixes them up.

Similar stretching and folding operations are responsible for the intricate geometric structure of strange attractors, pictures of which can be found in many books, technical and popular. Since paths for deterministic systems cannot intersect, they keep to certain rules even in the mixed-up condition. Thus strange attractors exhibit the symmetry of *self-similarity* or invariance to scale transformation. When a fragment of a strange attractor is enlarged, it exhibits the same pattern as the whole attractor; when a fragment of the enlargement is magnified again, the same pattern shows up again; and so on, down to arbitrarily small scales. For instance, the Hénon attractor looks like three sets of lines bent into a crescent; when one of the sets is magnified, it reveals three more sets, and when one of the daughter sets is magnified, it reveals more triplets, and so on. Objects with such self-similar characters with respect to scale transformation have dimensions that are not integers. They are called *fractals* (see note 12 in Ch. 6).

Initial Conditions and Prediction

Consider two paths of the binary-shift map with initial conditions that differ by one part in a billion. Writing the state in binary digits, $x_0 = 0.a_1a_2a_3 \ldots$, the two initial conditions have the same values for a_1 through a_{29} but not for a_{30} and beyond. Under the dynamics of the shift map, the paths diverge exponentially, as the difference between them doubles for each succeeding period. Finally, the original minute difference becomes all-important when a_{30} comes to the fore at the twenty-ninth iteration, when the magnitude of their difference becomes the same order of magnitude as the scale of the state space. From then on the two paths bear no relation to each other at all.

We rely on the results of dynamic rules for prediction. The sensitivity to initial conditions puts a limit on the predictive power of dynamic rules. For each iteration of the binary-shift map, the error in the initial condition is doubled. Soon the error will exceed whatever the preset tolerance is. If we know the initial condition to one part in a billion, then the thirtieth shift yields $x_{30} = 0.0000\ldots$, which contains no information on the further development of the system.

The path of a system is determined by the dynamic rule only if an initial condition is specified. The initial condition is seldom exact in the sense that it pinpoints a unique state out of the possible states of the system. In the mathematical formulation of most physical systems, the state spaces are continua with uncountably many states, so that the exact and exhaustive description of a state requires an infinite amount of information, which is impossible for us finite beings to supply. In the continuous case definite descriptions are *always* inexact. An inexact initial condition covers a group of states within its margin of error. With increasing accuracy, the group can be reduced in size, but it can never shrink to a singleton if the state variables are continuous. Therefore the solution of a dynamic equation with an inexact initial condition yields not a single path but a group of paths originating from proximate states. If the system is regular, the paths stay close together, so that the dynamic rule predicts the succeeding states with roughly the same accuracy as that of the initial condition. If the system is chaotic, the paths diverge exponentially, so that the accuracy of prediction based on the dynamic rule deteriorates. An infinitesimal inaccuracy in the initial condition, amplified exponentially through time, will eventually become so gross the dynamic rule ceases to provide acceptable prediction.

Chaotic dynamic systems are deterministic; the same initial condition always yields the same path. However, they demand that "same" be taken literally and "extremely similar" cannot be substituted for it. Similar initial conditions do not yield similar paths. The finite accuracy in the specification of initial conditions translates into a finite temporal horizon beyond which the results of the dynamic rule are swamped by error and cannot serve as the basis of prediction. The horizon of prediction can be extended by improving the accuracy of the initial conditions, but the extension is difficult because the loss of accuracy is exponential. In the binary-shift map, if we improve our accuracy by a factor of 2, we extend the horizon of prediction by only one period.

Example: The Long-term Unpredictability of Weather and Planetary Motion[11]

The *Lorenz model of hydrodynamic flow*, motivated by long-term weather forecasting illustrates the chaotic behavior of deterministic systems. When air near the ground is heated, it rises, cools, and eventually sinks back. The air mass undergoes a convective flow, driven by heating from the ground and dampened by dissipation due to viscosity. Its motion is governed by the hydrodynamic equations, which are partial differential equations accounting for the air mass's temporal evolution and the spatial patterns of its density distribution. Drastically simplifying and truncating the hydrodynamic equations, Edward Lorenz obtained three coupled ordinary differential equations that nevertheless capture the bare essence of driven and dissipative convection. The Lorenz equations represent the weather as a deterministic dynamic system characterized by three state variables: the circulatory flow velocity and the horizontal and vertical temperature differences. There are three exogenous parameters: the strength of thermal driving, the ratio of the viscous force to the thermal force, and the difference between the horizontal and vertical damping rates. The Lorentz equations have a strange attractor that looks like a butterfly when plotted on paper.

The Lorenz equations are solved numerically. In one case, a computer solution tracks the weather from 10,000 initial conditions and takes snapshots at regular intervals. Each snapshot shows 10,000 points in the appropriate state space representing the possible states of a meteorological system at the specific time. Suppose we know the initial condition of a meteorological system only to the accuracy the eye can distinguish in the snapshot. We want to make a weather forecast based on it and the Lorenz equations. Initially the 10,000 possible states are so close together they appear as a single point that the eye cannot resolve into parts. In this early stage, prediction is excellent. After a couple of hundred units of time lapse, some distinct points appear but they form a tight cluster, so that prediction with a tolerable margin of error is possible. After a few thousand units of time lapse, the snapshot shows 10,000 points scattering all over the allowed region of the state space. Now the dynamic rule has lost its predictive power; it has answers covering all outcomes to suit all tastes and is not superior to a crystal ball.

To gain a more realistic feel for the horizon of prediction, we distinguish between the gross and fine structures of weather patterns. Gross structures, which meteorologists try to predict, can be resolved by our network of weather-surveillance detectors. Fine structures, such as the positions of particular clouds, are usually not the subject of forecasting, but they are the real killers in long-range forecasting.

Using deterministic models, it is estimated that small errors in the initial conditions for gross structures double in about three days, after which the growth rate of errors slows. Thus doubling the accuracy in measuring the gross structures would extend the horizon of prediction by three days. On the basis of this, it is reasonable to expect that advances in surveillance technology will facilitate weather forecasting several weeks ahead with tolerable accuracy.

Small errors in the initial conditions of fine structures grow much faster, doubling in an hour or so. Although fine structures are insignificant in themselves, their errors are quickly amplified to a scale that affects the gross structures. Furthermore, doubling the precision of specifying fine structures, which is much more difficult than doubling the precision of measuring gross structures, extends the horizon of prediction by only a matter of hours. The chaotic behaviors of fine structures foil the hope of accurately predicting the weather for more than a couple of weeks. Thus even if the weather is governed by deterministic dynamic rules, its chaotic behavior sets a limit on long-range forecasting.

The long-term unpredictability of chaotic systems is perfectly compatible with their short-term predictability. Recent numerical results on the solar system, which cover hundreds of millions of years, show that planetary motions are chaotic; their paths diverge with a positive Lyapunov exponent, but fortunately the divergence rate is very slow on human scales. Chaotic behaviors are more pronounced for the inner planets Mercury, Venus, Earth, and Mars, whose orbits we cannot predict for more than a few tens of millions of years. The finite predictive horizon does not prevent physicists from predicting planetary alignments that occur within our lifetime to within seconds, or to predict their occurrence a few thousands of years ahead with full confidence.

The Constructive Side of Chaotic Dynamics

Regarding chaotic dynamics, pessimists lament the loss of power to make long-term predictions; optimists celebrate the possibility of understanding complicated processes and making short-term predictions. When we realize that systems representable by a few variables and governed by simple dynamic rules can exhibit chaotic behaviors, apparent complexity no longer drives us to dispair. We can hope to see through the complexity and find out the underlying dynamic rule with appropriately designed experiments. Then we can use the rule to predict and control the behaviors of the system within the horizon of prediction. The horizon is limited, but it is better than no prediction at all.

An important condition for the feasibility of experimentally uncovering the underlying dynamics is that the number of variables characterizing the system is small. In other words, the system has a small dimension. This condition is most readily satisfied for the long-term behaviors of dissipative systems. The strange attractors of dissipative systems usually have smaller dimensions than the original state spaces because of state-space contraction. They are interesting because they exclude transient effects and represent the "natural" behaviors of the systems. In many experiments, strange attractors are successfully reconstructed from time-series data.[12]

Chaotic dynamics also opens the possibility for time and patience to compensate for the crudeness of observation. Repeated measurements accumulate information on chaotic systems but not on regular systems. For regular systems, paths with proximate initial conditions stay close to each other. If the resolution of our equipment fails to detect the initial differences, then it cannot distinguish them even after patient observation. In contrast, tiny differences in chaotic systems that elude the resolution of our measuring device will be amplified and eventually become observable. Suppose we observe binary-shift system by an instrument that can only resolve differences to one part in a thousand; then on the twentieth period we can discern states that initially differ by one part in a billion.

Although we cannot make long-term predictions on chaotic systems, we can increase our knowledge of them by long-term observation. Thus another way to distinguish regular and chaotic systems is that successive measurements on chaotic systems generate information. The average time rate of information generation is represented by the *metric entropy*. Like other notions of entropy, the metric entropy is a statistical concept. It is closer to information entropy that measures the capacity of communication channels than to thermodynamic entropy. The metric entropy is closely associated with the exponential rate of divergence and is positive only for chaotic systems with exponentially diverging paths. For the binary-shift map, the metric entropy is log 2, which is 1 bit in the standard unit of information.

32.   Uniting Deterministic and Stochastic Concepts

Deterministic dynamics accounts in detail for the moment-to-moment evolution of a single system or the stage-to-stage unfolding of a single process represented by a single path in the state space. Without relinquishing this fine-grained view of processes, the geometric formulation also grasps all possible paths as wholes and their general classifications. The synthetic view of dynamic processes makes possible the definition of chaos in terms of the divergence rate between paths and of dissipative systems in terms of state-space volume contraction. It also paves the way for statistical concepts such as distribution and proportion, which enable us to disengage from the details of the complicated paths and capture their salient features.

Statistical concepts forsake the fine-grained description of the temporal unfolding of single paths and provide coarse-grained descriptions of the patterns of ensembles of paths. They can and usually do stand on their own without appealing to dynamic rules, as they do in the theory of stochastic processes discussed in the following chapter. Thus people often think that stochastic processes are diametrically opposite to deterministic processes and probably governed by chance instead of law. Such thinking incites debates on determinism and tychism. I argue that both metaphysical doctrines are quite irrelevant to our understanding: Stochastic and deterministic theories are two ways to view a complicated world. Deterministic theories, which are logically simpler, are preferred whenever practical. However, the detailed specification they demand is not practical or possible for many complex processes, where stochastic theories come to the fore. The separate applications do not imply that stochastic and deterministic concepts are mutually exclusive. Their logic is different, but they can be applied to the same process. The dual applications that unify deterministic and stochastic concepts are found in the ergodic theory.

Statistical concepts were first introduced into deterministic theories by Maxwell and Boltzmann, who tried to justify the conceptual framework of statistical mechanics (§ 11). In the early 1930s, some of their seminal ideas were taken up by George Birkhoff and John Von Neumann. Since then, the *ergodic theory* has developed into a thriving branch of mathematics that encompasses both deterministic and stochastic notions and hence offers the appropriate context to analyze their meanings and functions. We will examine how and where the ideas of randomness and uncertainty arise in the study of deterministic systems. As it develops, ergodic theory becomes detached from statistical mechanics. Yet it still helps to clarify foundational problems in statistical mechanics. Some of the problems will be discussed in the following section.

Paths and Distributions of the Logistic Map[13]

Let us start with deterministic processes. Consider the logistic map $x_{n+1} = rx_n(1 - x_n)$ with the parameter value $r = 4$. For this value the logistic system behaves chaotically, but it nevertheless is deterministic; given x_n, a child can calculate x_{n+1}.

The state space of the logistic system comprises real numbers in the interval between 0 and 1. We pick an arbitrary number in the interval as the initial state x_0 and iterate the map many times to generate a path for the logistic system. If we are lucky enough to pick $x_0 = 3/4$, we get a constant path. We may hit other jackpots for which the paths oscillate regularly: a period-2 cycle for $x_0 = 0.346$ or 0.906; a period-3 cycle for $x_0 = 0.188$, 0.611, or 0.950; and so on. However, these special initial states are atypical and occur much less frequently than jackpots in casinos. They are also unstable; the cycles are destroyed by the slightest variation in the values of x_0. For the overwhelming majority of initial states, the paths typically show points jumping wildly in the interval $[0, 1)$ with no discernible pattern at all (Fig. 8.5a). Worse, if we change the value of the initial state by a tiny bit, we get a new path, equally erratic but completely different from the old one. Because of such behavior, the notion of an individual path becomes ineffective in the study of chaotic dynamics. Statistical concepts prove to be more powerful.

We introduce statistical concepts by two *coarse-graining* steps. First, we coarse grain the state space by partitioning it into cells, each of which contains many states. Second, we give up asking about the states of a path at specific times. Instead, we count the number of states falling within each cell regardless of when they occur, thereby generating a *distribution of states* for the path. We find that the distribution of states over the cells is quite smooth if the path is long enough, despite the erratic succession of the states. As illustrated in Fig. 8.5b, the density distribution of the erratic path is quite regular, showing a minimum for cells around $x = 0.5$ and maxima for cells near $x = 0$ and $x = 1$. Furthermore, for infinitely long paths, the distribution is shared by all paths, except those few issuing from the atypical initial states. The sensitivity to initial conditions characteristic of chaotic dynamics is typically not reflected in the distribution of states in long paths.

In the geometric view of deterministic dynamics, we consider groups of paths. *Ensembles of paths* become even more important in the statistical view. Instead of counting states at various times in a single path, we take an ensemble of paths with various initial states, count the number of paths that fall within each cell at time t_n, and define a *density distribution* of the ensemble. The

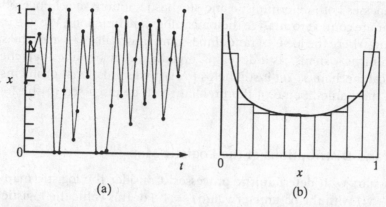

FIGURE 8.5 **(a)** The deterministic description of a logistic system keeps track of its "path," which is its state x at successive times t. (The states are joined to aid the eye.) For certain values of the control parameter, the system is chaotic; the values of x jump erratically over the state space $[0, 1)$. To transform the deterministic description to a statistical description, we neglect the time element, and partition the state space, which is the dependent variable in the deterministic description, into cells. **(b)** In the statistical description, we use the cells as an independent variable, count the states in the path that fall within each cell, and define a distribution of states, which is a histogram. The distributions for all long typical paths share the same form, which is shown by the continuous curve when the cells are shrunk into infinitesimal sizes.

density distribution generally changes with time. We study its temporal evolution and are especially interested in cases where it maintains a steady state.

For instance, consider an ensemble of a billion paths with initial states uniformly distributed among the cells. The logistic map sends each of the billion states into a new state, producing a new density distribution for t_1 that differs from the initial uniform distribution by peaking sharply for cells near $x = 1$. The distribution at t_2 is again different, peaking sharply at cells near both $x = 0$ and $x = 1$. From t_2 onward, the density distribution remains more or less the same and rigorously converges to an invariant form as time goes to infinity. Furthermore, the invariant density distribution for the ensemble of paths is equal to the distribution of states obtained by counting the states in a *typical* path that is infinitely long. The *invariant density distribution*, which contains the statistical information on the paths, becomes the focus of the ergodic theory.

The example of the logistic map illustrates several features of statistical theories. First, we have partitioned the state space into cells. The cells can be as small as the infinitesimal regions dx around x, yet they present a coarser view of the state space than the exact specifications of the states. Second, we have changed the independent variable. In deterministic theories, the independent variable is time t, whose values pick out individual states in a path. In statistical theories, the independent variable is the cell, whose values designate coarse-grained characters of states, for instance, the character of having a value between 0.1 and 0.2. Since the independent variable designates the subject of discourse, the change makes the logical structures of deterministic and statistical theories different. The difference is akin to that between thinking about individual persons and thinking about those who have blue eyes. The statistical account is coarser and contains less information. Third, we are no longer concerned with the temporal order of the states; instead, we want to find out their statistical distribution. Fourth, we concentrate on long paths or large ensembles, for which the statistical method is appropriate. Fifth, we consider only the typical cases and ignore the small group of atypical paths that do not share the statistical regularity.

The Ergodic Theory[14]

The *ergodic theory* studies the long-term statistical behaviors of abstract deterministic dynamic systems. It retains the structure of deterministic dynamics – time, state, dynamic rule, exogenous parameter – and adds the statistical structure of measure theory. It concentrates on those dynamic rules that leave the density distributions invariant and classifies the systems obeying these rules according to their statistical characteristics.

In the ergodic theory, the state space of a deterministic system is enriched with the structure of a measure space. A measure space contains two main ideas. The first introduces the idea of *parts*, which is not contained in the usual definition of state spaces; measure theory partitions the state space in the most general and systematic way. The second idea introduces the concept of *proportion, relative magnitude, or statistical weight* (§ 11). The *measure* of a part of the state space is the statistical weight of the states in the part relative to the statistical weight of all the states in the state space. If all states have equal weight, then the measure of a part is just the proportion of the number of states it contains. The measure also means the systematic assignment of a measure to each part of the state space that results from the general partitioning. Thus a measure gives a density distribution among various parts of the state space and represents the statistical character of the system. The measure is *invariant* if it is not changed by the dynamics of the system. The ergodic theory studies those systems whose dynamic rules preserve their respective measures.

Measure theory is the bedrock of the probability calculus, where the measure is by definition the *probability*. The technical definition of probability is a measure and always indicates the statistical weight of a part of a whole. It differs from the natural meaning of probability, which is about a single state.

The ergodic theory is concerned only with *typical* states; that is, it ignores a set of *atypical* states with measure zero or vanishing statistical weight. For example, the atypical states in the logistic map are those that initiate regular paths. To remind us that atypical states and paths are excluded from consideration, the results of the ergodic theory are qualified with "almost always," "almost everywhere," or "with probability 1." The justification for these results is that almost all states are typical and that the atypical states are *statistically* insignificant or have zero probability.

In the technical sense, probability 1 is not certainty, probability 0 does not mean impossibility, and statistical insignificance does not imply physical insignificance. Suppose the state space is a segment of the real line, say the real numbers within the interval [0, 1]. Then all the rational numbers, which are representable as fractions, constitute a set with measure zero. There are infinitely many rational numbers, but embedded in the real number system, they are still atypical and statistically insignificant. Despite the statistics, however, it can happen that the quantities represented by rational numbers are physically important for some individual cases. The emphasis on typicality and the qualification "with probability 1" remind us that our topics are not individuals but statistical systems.

The measure–space structure of the state space enables us to define various distributions over states and times. The characters of the distributions distinguish the degrees of randomness of various processes: *independent* (random), *Markovian* (having short memory), *statistically deterministic* (predictable with probability 1 given the entire past record). These distinctions are made without mentioning dynamic rules and hence are more frequently associated with stochastic processes.

The ergodic theory is peculiar for its effort to relate the stochastic concepts to the behaviors of deterministic systems. Among its results is a hierarchy of deterministic systems with varying statistical characteristics: the ergodic system, mixing system, K system, C system, and Bernoulli system, each more random than the preceding one and implying it. K, C, and Bernoulli systems are chaotic; their paths diverge exponentially. Although many of these systems generate independent or Markov processes, they all retain their deterministic nature as governed by dynamic rules, so we can still examine individual paths to seek the sources of randomness.

Bernoulli Systems and Their Physical Equivalence[15]

The binary-shift map $x_{n+1} = 2x_n$ modulo 1 discussed in § 31 is a noninvertible Bernoulli system. To see it as a stochastic process, we introduce a coarse-grained partition of its state space. There are infinitely many ways to partition the interval [0, 1); let us choose to divide it into two equal cells separated at 0.5. We label these cells L and R. The number x_n falls in the cell L if $0 \le x_n < 0.5$ and the cell R if $0.5 \le x_n < 1.0$. Suppose we have a measuring device that prints out the labels of the cells in which x_n falls for successive n. If the initial number is 0.100110..., the device prints out the string RLLRRL ....

In the equipartition of the state space, the printout mirrors the initial number, so we can reconstruct the initial number exactly from the infinite string of data. Not all partitions can achieve this. If we had divided the state space into two unequal cells, we would get a string of R's and L's with unequal frequencies, from which we could not reconstruct the initial number. In this case the resolution of our measurement would not be good enough.

Suppose we are ignorant of the binary-shift map and the initial state but have access to the complete printout for the successive cells in which the system has been in its entire past. Can the record help us to predict in which cell the system falls next? It all depends on the initial number. If x_0 happens to be a rational number, which in decimal expansion either ends in a string of zeros or repeats itself in some way, the record is invaluable for prediction. However, rational numbers are atypical in the real number system. The typical real number is irrational, and its decimal expansion goes on infinitely with no recurring pattern, so that the past record is useless in prediction. Except the statistically insignificant cases where x_0 are rational, the measurement results are statistically indistinguishable from the result generated by the independent tossing of a fair coin.

Generally, a state of a *Bernoulli shift* is a doubly infinite sequence $\{a_i\} \equiv \ldots a_{-2} a_{-1} a_0 a_1 a_2 \ldots$. Each element a_i of the sequence can take any one of K values with a specific probability, and the value it takes is independent of what the other elements take. The state space that encompasses all possible sequences is a measure space. The dynamic rule maps each sequence to another that is shifted one place to the right, $\{a_i\} \rightarrow \{a_i'\}$ such that $a_i' = a_{i+1}$ for all i. The introduction of a coarse-grain partition turns a Bernoulli system into a stochastic process.

Although the dynamic systems studied in the ergodic theory are abstract, they can be realized in physical systems. The simplest Bernoulli-shift system has $K = 2$ and equal statistical weights for the two values. A point in its state space is a doubly infinite sequence of 0s and 1s with equal frequency. One realization of this Bernoulli system is the consecutive tossing of a coin. The repeated throwing of a loaded die realizes a Bernoulli system with $K = 6$, where each of the six values has a different statistical weight. The independent spinning of a roulette wheel with 300 even compartments realizes a Bernoulli system with $K = 300$ and equal statistical weight.

Bernoulli systems can also be realized in ideal mechanical systems. An example is the *hard-sphere gas*, a box of hard spheres that do not interact with each other unless they collide. This is a nonlinear system exhibiting chaotic behaviors. A tiny displacement in the point of impact causes the spheres to go off in a different direction. The point of impact for the next collision is displaced even more. Consequently the slightest alteration in the initial conditions of the spheres produces entirely different paths after a few collisions.

The hard-sphere gas is one of several ideal mechanical systems that are *isomorphic* or *statistically equivalent* to Bernoulli systems – a system is isomorphic to a Bernoulli system if its typical states can be put into a one-to-one correspondence with the states of the Bernoulli system such that both the statistical weights and the dynamic paths are preserved. Another such system is a billiard ball moving on a square table with a finite number of convex obstacles. A third is a particle moving on a frictionless surface of negative curvature without external force.

Consider the case of the *billiard*. A billiard ball moves on a square table with a convex obstacle. The state of the ball is specified by its position and velocity, and its state space encompasses all positions and velocities within the constraints of the table and its energy. Given an initial position and velocity, the ball's path is determined by the laws of classical mechanics. Now suppose we measure the path at fixed intervals of time. To represent a measurement device with finite accuracy, we partition the state space into a finite number of disjoint cells, each designated by a single value of position and velocity.

Seen through the device, the path of the ball becomes a sequence of discrete values of position and velocity, which I summarily call p_n for the time t_n. The resolution of the partition is refined enough that the path can be reconstructed with probability 1 from the data taken from the infinite past to the infinite future.

The billiard ball seen through the measuring device undergoes a Markov process. The ball stays at a value p_n for a period δt_n, and then randomly jumps to another value, randomly meaning we cannot tell whether the destination is decided upon by spinning a roulette wheel with unequal compartments corresponding to our partition of the billiard's state space. Thus statistically, we cannot distinguish whether the observed data are generated by a biased roulette wheel or a deterministic system governed by Newton's laws.

Randomness and Complexity[16]

The binary-shift map is a trivial rulelike operation, yet its results in the coarse-grained description are indistinguishable from the tossing of a coin. Where does the randomness come from? What does randomness mean?

The preceding discussion suggests that something is random if it is or may have been generated by a random mechanism. As to random mechanisms, it offers paradigm examples instead of definitions. This line of reasoning is common in philosophical papers. However, many familiar random mechanisms are not so random on closer examination. An example is card shuffling; hand-shuffled bridge hands fail the chi-square test of randomness; that is why tournament bridge switched to computer shuffling in the late 1960s. Thus the reasoning is turned around. A mechanism is deemed random because the results it generates are random. There are many statistical criteria to test the randomness of the results. The criteria are not perfect, but they are more clear and precise than those for the randomness of physical mechanisms. In the following, randomness is taken to be an attribute of the results, not the mechanism that produces them.

An influential mathematical definition associates randomness with complexity. Randomness usually means the lack of salient patterns. Intuitively, when patterns are discerned, concepts can be introduced to describe them economically. Suppose we want to write a program instructing a computer to print out a long sequence of binary digits. If the sequence exhibits certain patterns, then the shortest program can be quite simple, such as "Print 011 a billion times." If the sequence is random, then the shortest program would have to encode the sequence itself, as "Print 01001011 . . . ." Thus the shortest program, which is used as a measure of complexity, can similarly be used as a measure of randomness. A *random* sequence is one with maximal complexity, for its information content cannot be simplified and put into more compact form, and the smallest program required to specify it has about the same number of bits as it has. By using the notion of complexity, various degrees of randomness can be defined.

Shift maps are simple dynamic rules. The complexity of the systems they represent hides in the initial conditions and is unfolded in time by the dynamic rule. An initial condition for a system is just the complete specification of it at an instant of time. The specification of a complex system requires a large amount of information, which is reflected in the demand for high accuracy. When the amount of information taxes our ability to conceptualize, we deem the system random.

33. The Foundation of Statistical Mechanics

The chaotic and unstable behaviors of deterministic systems are propitious for the foundation of statistical mechanics. Statistical mechanics studies mechanical systems with the additional help of a statistical structure that screens for significant macrophenomena and connects them with the deterministic laws of mechanics. The statistical procedure has several basic assumptions, the justification of which becomes the problem for foundational research (§ 11). The credibility of the assumptions receives a boost if the underlying deterministic dynamics is unstable or chaotic. Instability is invoked in the arguments of the founders of statistical mechanics and later clearly formulated in the ergodic theory. It does provide theoretical explanations for crucial concepts such as the increase in entropy and the approach to equilibrium. However, it is doubtful whether the dynamics of most realistic physical systems are unstable. Thus the foundational problem of statistical mechanics remains.

Ergodicity and Mixing[17]

A basic assumption of equilibrium statistical mechanics is that macroscopic quantities can be computed by averaging over the ensemble of microstates compatible with certain appropriate macroscopic constraints. Another assumption is that in coarse-grained descriptions of the microstate space, a system in equilibrium occupies the maximum microstate-space volume compatible with the constraints. A system out of equilibrium occupies a much smaller volume of the microstate space, and the volume expands as it approaches equilibrium. The expansion of the microstate-space volume occupied corresponds to the increase of entropy.

In terms of deterministic dynamics, the macroscopic constraint for a system out of equilibrium imposes an inexact initial condition covering a group of roughly similar microstates clustering in one corner of the microstate space. Statistical mechanics assumes that the paths originating from these initial states spread out and find their way into every corner of the microstate space when the system settles in equilibrium. Chaotic dynamics, in which adjacent paths diverge exponentially, is ideal to explain the spreading – if it applies. It frequently does not. So physicists look to *ergodicity* and to *mixing*, which, at the bottom of the hierarchy of random notions, poses less stringent conditions.

Ergodicity helps to justify the ensemble average of equilibrium statistical mechanics. If a dynamic system is ergodic, then the infinite time average of a macrovariable over a typical path of the system is equal to the statistical average of the variable with respect to a suitable ensemble of microstates with equal statistical weight. The justification is not perfect, as discussed in § 11, but it gives a clear theoretical connection between deterministic paths and their statistical manifestations that strengthens the adequacy of the ensemble average.

Ergodicity is insufficient to explain how the equilibrium is attained. The paths of an ergodic system need not be unstable. Nearby paths can stay close to each other as they cover the state space in infinite time; it is like a machine winding threads on a spool, in that the threads cover the spool in an orderly and coherent manner. Since the paths stay together, the system does not approach equilibrium, for to settle in equilibrium the microstates must disperse throughout the microstate space.

To explain the approach to equilibrium, the stronger stochastic notion of *mixing* is required. A mixing system is ergodic, but not vice versa. The idea of mixing is intuitively explained by Gibbs's analogy. Imagine that a little cream is poured into a cup of coffee so that the cream makes up 5 percent of the total volume. The cream initially concentrates in a small region but becomes thoroughly mixed in the coffee after a little stirring. A fine-grained accounting shows that the volume of the cream remains the same. If we examine a drop of the mixture under the microscope, we find that it is 5 percent cream and 95 percent coffee, with regions occupied by cream and coffee finely intermingled but clearly distinguishable; each molecule is either cream or coffee. However, in the coarse-grained account of the mixture, the cream "occupies" the entire cup because it is found everywhere.

In statistical mechanics, the initial concentration of the cream is analogous to the distribution of microstates in the microstate space conforming to the initial nonequilibrium macroscopic condition, the stirring to the dynamic evolution of the microstates after the removal of the macroscopic constraint, and the mixture to the equilibrium distribution of the microstates. Suppose at $t = 0$, the initial microstates of a system cluster in a region A of the microstate space (the Γ-space). The microstates evolve according to the laws of mechanics f, so that a measurement at a subsequent time t finds them occupying the region $f_t(A)$ of the microstate space. For a mixing system, $f_t(A)$ is highly fibrillated, with fine filaments sprawling all over the microstate space. The microstate in $f_t(A)$ spreads out uniformly in the microstate space as t goes to infinity. More precisely, if a system is *mixing*, the portion of microstates found within an arbitrary region B of the microstate space is equal to the volume of B relative to the total volume of the microstate space (Fig. 8.6b). Mixing systems are not chaotic; their paths diverge and spread throughout the state space as time goes to infinity, but the divergence is less than exponential.

Mixing alone is insufficient to account for the approach to equilibrium, for the laws of mechanics dictate that $f_t(A)$ can change only in shape, not in volume. The volume of the microstate space occupied by $f_t(A)$ must be equal to the volume of A; the cream does not expand. Thus the "expansion of microstate-space volume" corresponding to the increase in entropy makes sense only in a coarse-grained view of the microstate space. We partition the microstate space into cells, call a cell occupied if it contains some microstates of $f_t(A)$, define the volume of occupied microstate space in terms of

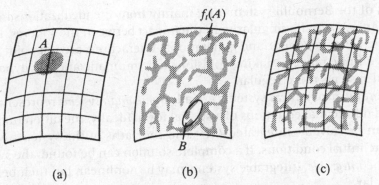

FIGURE 8.6 (a) An ensemble of microstates compatible with a macroscopic constraint lie in region A of the dynamic system's microstate space. The constraint is removed at time $t = 0$. (b) At time t, the microstates in A evolve under the dynamic rule f into the states in the region $f_t(A)$. The laws of mechanics dictate that A and $f_t(A)$ must be equal in volume but can differ in shape. If the system is *mixing*, $f_t(A)$ is sprawling and, as t goes to infinity, the portion occupied by $f_t(A)$ in any region B is equal to the portion occupied by A in the entire microstate space. (c) The microstate space is partitioned into coarse-grained cells. The *entropy* of the system is defined in terms of the number of occupied cells, where an occupied cell is one that contains some microstates in $f_t(A)$. For typical initial conditions, the entropy is larger at t than initially, although the volumes of A and $f_t(A)$ are equal.

the number of occupied cells, and associate the occupied volume in the coarse-grained view with the *entropy* of the system. Since the microstates of the initial nonequilibrium system cluster together, only a small coarse-grained volume is occupied and the entropy is low. As the microstates disperse into more and more cells in mixing, the entropy increases. Finally, all cells are occupied as the microstates distribute themselves uniformly throughout the microstate space, and the entropy reaches maximum in equilibrium (Fig. 8.6c).

The Stability of Hamiltonian Systems: The KAM Theorem[18]

Ergodicity and mixing as defined are abstract. They can contribute to the justification of statistical mechanics only if most realistic systems are indeed ergodic or mixing. We have seen in the previous section that even simple mechanical systems such as the hard-sphere gas or the billiard ball on a table with a convex obstacle are Bernoulli. This prompts us to expect larger systems to be at least ergodic or mixing. Wait.

Perhaps some Americans still remember the 1970 draft lottery in which capsules for the 365 birthdays were put into an urn, shaken for several hours, then drawn out one by one. The result made many people cry foul; the low numbers contain disproportionally more birthdays in December and November and the high numbers more in January and February. The reason is that the capsules were put into the urn in a pattern, and the hours of turning failed to randomize them thoroughly. It is not always easy to mix things up.

The difference between ideal and realistic systems again becomes important. The chaotic behaviors of the hard-sphere gas and other ideal mechanical models of the Bernoulli system arise mainly from the idealizations involved, for instance, from the singularity in the impact between the spheres. Realistically, the electromagnetic and gravitational interactions are not point impacts but have finite ranges, making collisions less traumatic and the dynamics of physical systems more regular.

Generally, mechanical systems are all *Hamiltonian systems* representable by coupled differential equations of a special form. Ideally, the differential equations can be solved – integrated – to yield the paths of the systems with appropriate initial conditions. If a complete solution can be found, the system is said to be *integrable*. Integrable systems may be nonlinear, but their behaviors are regular.

Most systems with n degrees of freedom are not integrable if n is greater than 2. The major question for us is whether nonintegrable systems satisfy some of the instability and randomness criteria needed for the justification of statistical mechanics.

The traditional way to treat nonintegrable systems is perturbation expansion. The solution to a corresponding integrable system is found, and the factors that make the system nonintegrable are added as perturbations. The perturbation technique is powerful and is still widely used. However, it has serious mathematical difficulties; under certain conditions of resonance, the perturbative terms go to infinity.

Nonintegrable systems were intensively studied in the late nineteenth century, especially in the context of the long-term stability of the solar system. A big breakthrough in the problem of stability is achieved by the KAM theorem, initiated in 1954 and proved eight years later. Andrei Kolmogorov, Vladimir Arnold, and Jürgen Moser deviated from the traditional method of trying to obtain solutions to the perturbed system based on the unperturbed solutions. Instead, they aimed for a general answer by directly proving for weakly perturbed systems the existence of stable tori, motions on which are regular. The *KAM theorem* states that under small perturbations, the majority of invariant tori that satisfy certain conditions are preserved, although they are slightly deformed. Consequently the majority of paths for Hamiltonian systems are regular and not chaotic.

The small number of paths that do not satisfy the KAM conditions become chaotic. In a Poincaré surface section of the state space, regular paths appear as smooth and closed orbits and chaotic paths appear as a splatter of points. The splatter occurs mostly near unstable fixed points and separatrices between various modes of motion. They are local, constituting a thin layer between stable KAM surfaces. Thus although there is local chaos within the splatter region, the system is stable overall, at least under perturbations of moderate strength.

The stability and regularity of Hamiltonian systems frustrate the hope for a secure mechanical foundation of statistical mechanics, which needs instability to account for the spread of microstates. But one cannot argue with nature. In a review on regular and chaotic motions of dynamic systems, Arthur Wightman said: "The flows defined by most of the models of classical statistical mechanics, for example, N particles in a box interacting via a Lenard–Jones potential, are almost certainly not ergodic, not to speak of Bernoulli. Almost certainly means that there is no published proof but most of the experts regard the statement as likely to be correct."[19] Workers on chaos like to say that non-chaotic systems are as rare as hen's teeth. As with many things in life, chaos is everywhere, except where it is needed.

34. Causality but Not Determinism

Deterministic theories are widespread in the sciences, but they are far from holding a monopoly. We have seen that deterministic processes can be coarsely represented in terms of statistics, and some become random in the coarse-grained description. We will see in the following chapter that many processes are more appropriately described in statistical and probabilistic terms without adverting to dynamic rules. The success and limitation of deterministic and probabilistic theories have fueled two contending metaphysical doctrines: determinism and tychism.

Determinism and tychism are both global doctrines about the conditions of the world at large. In contrast, determinacy and stochasticity are local characteristics attributed to individual systems or processes. Many assumptions are required to go from the existence of some deterministic systems to the doctrine that the world consists only of deterministic systems, or from the existence of some stochastic systems to the doctrine that chance reigns. I argue that most extra assumptions, which are the crux of determinism and tychism, are unwarranted. The compatibility of deterministic dynamics with instability (§ 30), unpredictability (§ 31), randomness and statistical indeterminacy (§ 32) can be articulated with mathematical precision. Therefore deterministic dynamics does not imply the negation of these notions, as determinism does. Conversely, instability, unpredictability, and randomness do not imply the nonexistence of dynamic rules, as tychism insists. Deterministic dynamic rules are often interpreted as causal laws; thus the rejection of determinism does not imply the rejection of causality. This section examines the philosophical significance of deterministic dynamics and disentangles the notions of determinacy, causality, and determinism.

Causal Relation, Causality, Causation

Bertrand Russell once argued, referring to physical theories, that the idea of causation was a relic of a bygone age and had become otiose in science. Russell later changed his mind for the better.[20] Causation is not a relic but thrives in scientific reason. When formulating theories, scientists face open situations with a multitude of entangled events and processes. They must weigh various factors and judge what to include in their models, thus engaging in cause–effect analysis.

Nevertheless Russell did make a keen observation. Although dynamic rules are often called causal laws, the idea of causation is mostly absent in theories and models of deterministic dynamics. In the preceding sections, causes are noticeable only as the changes in the control or exogenous parameters representing the interaction of the dynamic system with the external world. Control parameters carry the commonsense ideas of manipulation and intervention.

When scientists say that a small cause produces a great effect on an unstable system, the effect referred to is the system's response to a change in the control parameter (§ 30). However, changes of control parameters are not essential to deterministic dynamics; most processes have fixed control parameters or no parameter at all.

To see how deterministic dynamics can be so often identified with causality and yet be cited as spurning causation, let us examine some causal ideas and see how they are embodied in deterministic systems. *Causal relations* are not something extra added to predefined noncausal objects. They appear simultaneously with objects in a world that becomes, as a result of systematic individuation, a complex *causal network* of things and events (§ 10). Causal relations obtain among states of things in static conditions and events in dynamic conditions. An example of a static causal relation is the suspension of the Golden Gate Bridge by steel cables. Two states or events are causally relatable if they are connectible by a process, which can be stationary, related if they are so connected. If the connecting process is the change of a thing, then the thing is the agent of interaction.

The general concept of a causal network is implicit in our everyday understanding and is made more explicit in the sciences, which also fill in many substantive features of the network. In fundamental physics, all processes that possibly connect two field points contribute to the causal relation between them, generating the pervasive dynamic of fields. The delimitation of field points that are causally relatable from field points that are not is called *microcausality*. *Causality* is generally a causal relation with the additional connotation of temporal ordering.

Causation I reserve for the relation between causes and effects. A network of causal relations is complex; an event in it is causally connected to many others, directly or indirectly. Causation introduces a cruder view of the causal network by singling out some salient factors as the causes of an event and lumping the remaining contributing factors under the label of standing conditions. For instance, a faulty wire is singled out as the cause of a fire, whereas the electric current and the presence of oxygen and flammable material, without which the fire cannot roar, are relegated to standing conditions. Often we count only one aspect of a complex event as the cause. If an unlicensed driver speeds and crashes in a tight turn, we may cite the speeding or the unlicensed driving or the sharp corner as the cause of the accident, although they are all aspects of the same concrete event of driving. The distinction is not frivolous; imagine the arguments in a lawsuit. A major part of the philosophical analysis of causation focuses on the general criteria for a causal factor's being a cause instead of merely a standing condition. Causes are further classified as proximate and remote.

Causation contains two ideas of asymmetry that are not explicit in the notion of a causal network. The first is temporal; a cause precedes its effect. Thus some people argue that the idea of simultaneous causation is spurious; there is no reason to prefer saying the suspending cables cause the roadway of a suspension bridge to stand in midair to saying the roadway causes the cables to be taut. The second asymmetry arises from the ideas of physical manipulation and intervention. We can manipulate a cause to bring about an effect, but not the other way around. The idea is tacitly generalized to mean the direction of action. The asymmetry is reflected in the asymmetry of explanation, which implicitly assumes causation. The position of the sun, the height of a pillar, and the length of its shadow form a causal network. We readily say the shadow has a specific length because of the sun's position and the pillar's height, but we buck at saying the pillar has its height because of the shadow's length and the sun's position. That is the way *because* is used: The pillar produces the shadow; the shadow does not produce the pillar.

Since Hume's critical analysis of causation, philosophers generally agree that particular causes and effects are acknowledged as instances of certain generalizations. What distinguishes "*A* causes *B*" and "*A* precedes *B* and is spatially contiguous to *B*" is a universal law of which the relation between *A* and *B* is an instance. Thus an event *B* is the effect of an event *A* if and only if there is a set of causal laws from which the occurrence of *B* can be derived, given the occurrence of *A* and certain auxiliary conditions. Dynamic rules in deterministic dynamics are often regarded as causal laws.

Closely associated with the thesis of lawful causation are various conditional analyses. Causes are variously said to be the necessary and/or sufficient conditions of their effects. John Mackie proposed that a cause is an insufficient but necessary part of an unnecessary but sufficient condition (INUS condition) for the effect; an example is a burning cigarette identified as the cause of a fire. Although the cause itself is insufficient to determine the effect, the criterion is deterministic because it implies there is a sufficient condition for the effect.[21]

Dynamic Systems as Enduring and Changing Things

Causation – not causality – eclipses within models of deterministic dynamics for several reasons. Causes and effects are crude circumscriptions of events that are distinctly recognizable, as the metaphor "a chain of causes and effects" suggests. In most dynamic theories, the successive states are so finely interconnected they appear as a continuum, and in a "causal rope" the links of causes and effects are hardly discernible. More important, the topics of scientific models are mainly closed or almost-closed systems, and the idea of causation is overshadowed by the idea of *a changing thing*. The ideally closed system featured in a scientific model may be complex, but it is a unitary system whose change is described by a dynamic rule. Within the model, all causal factors included are exhaustively represented by the dynamic rule, which answers Mill's idea of a total cause: "The sum total of the conditions, positive and negative taken together, the whole of the contingencies of every description, which being realized, the consequent invariably follows."[22] The rope of total causes and total effects is assimilated by the idea of a system changing by itself.

Instead of regarding a thing's state at a time as the cause of its later states and the effect of its earlier states, we more customarily think of the thing as undergoing a causal process without breaking it into distinctive causes and effects. The causality of the process, an objective relation among its successive stages corresponding to the successive states of the thing, is an integral element in the notion of an enduring and changing thing and not some logical relation we introduce to cement disjoint stages. The concepts of causality and enduring thing are inseparable, but the thing dominates when it is prominent. Thus deterministic dynamics realizes a most entrenched idea in our everyday thinking: A thing carefully isolated and left to itself evolves in its own determinate way.

The hallmark of a deterministic system is the uniqueness of the succeeding states given its state at an earlier time. The uniqueness is possible only because total causes are specified within deterministic models. All active external influences that may cause ambiguity are either ideally excluded or precisely specified in the control parameters. The importance of isolation highlights the *local* nature of deterministic dynamics, which applies to individual things and only to those things that are delineated without ambiguity. Whenever there is ambiguity in the specification of the system, the uniqueness characteristic of determinacy is threatened. Ambiguity in the initial condition leads to chaos. Ambiguity of coarse graining introduces statistical indeterminacy. When some ambiguity is left in dissipative systems, which are isolated only from active disturbances but not from a passively receptive environment, uniqueness is unattainable for the preceding states; the dynamics is deterministic but not invertible. The intolerance to the ambiguity of specification makes it highly unlikely that deterministic dynamics can be extrapolated to cover the world with all its bustling activities and interaction, as determinism demands.

Determinism[23]

Determinism is sometimes used as the noun for *deterministic*, as in "Chaos implies the compatibility of determinism and randomness." The usage is confusing, for determinism is a definite metaphysical doctrine with much wider implication than the determinacy of some systems.

The word *determinism* was coined in the late eighteenth century. The historian Theodore Porter: "'Determinism' was until the mid-nineteenth century a theory of the will – a denial of human freedom – and some dictionaries still give this as its first meaning. Partly as a result of the statistical discussion, it assumed during the 1850s and 1860s its modern and more general meaning, by which the future of the world is held to be wholly determined by its present configuration." A different view was offered by the philosopher Ian Hacking, who argued that the doctrine of necessity was entrenched long ago: "Determinism was eroded during the nineteenth century and a space was cleared for autonomous laws of chance."[24]

I subscribe to Porter's version. Undoubtedly Hacking is right that the idea of necessity is ancient, but it is too vague to be a philosophical doctrine. The vague idea of chance is just as old. The inexorability of fate and the fickleness of fortune have been expressed in many cultures since antiquity. The Homeric *moira* (fate) was succeeded by the deep sense of insecurity in the Archaic literature, and the cult of *tyche* (luck) was widespread in the early Hellenic age. In medieval times, the main source of necessity was the predetermination of divine creation, but freedom of the will within the predetermination was argued for by philosophers such as Augustine of Hippo. Since the Scientific Revolution, the authority of God gradually has given way to that of natural law. Yet shortly after Descartes and Hobbes advanced their mechanistic and materialistic worldviews, Pascal and Fermat wrote to each other on games of chance. Newtonian mechanics demonstrated clearly how some systems are deterministic, but it did not prevent scientists from working on the theory of probability. Laplace was not the only one who contributed to both mechanics and probability. In the time when mechanics was gaining prestige, a lottery craze swept Western Europe and speculation was rampant. Newton, who was reputed to have said, "I can calculate the motions of heavenly bodies but not the madness of people," lost heavily in the South Sea Bubble. During much of the eighteenth century, underwriting insurance was regarded as betting on other people's life and fortune. Buffered by the vagueness of thought, the ideas of necessity and chance coexisted peacefully for a long time, like two cultures separated by large expanses of wilderness. When the ideas were increasingly being clarified and substantiated, they clashed like two expanding empires, and philosophical doctrines were precipitated in the conflict.

The probability calculus, the classical form of which was fully developed by the beginning of the nineteenth century, enables people to reckon risk in quantitative terms. Many of its important results, including the strong law of large numbers, assert conditions that hold with probability 1, which is misleadingly interpreted as certainty. These results are instrumental in bolstering the idea of statistical determinacy. The spread of social statistics in the nineteenth century and its interpretation based on the probability calculus revealed the regularities in social phenomena. With the idea that social and perhaps human affairs are as subject to laws as natural events are, the freedom of will or action is again threatened, this time not by God. A fierce debate ensued; it ended in a stalemate but still flares up occasionally. For instance, Maxwell's objection to determinism was motivated not by his statistical mechanics but by his belief in free will. During the debates on free will and other issues accompanying the rise of the sciences of mass phenomena, the doctrine of determinism and its opposition were clearly articulated.

William James said in 1884: "What does determinism profess? It professes that those parts of the universe already laid down absolutely appoint and decree what the other parts shall be. The future has no ambiguous possibilities hidden in its womb: The part we call the present is compatible with only one totality."[25] Determinism is different from predetermination; it does not hold if all events in the course of the universe were predestined by God's caprice. The arbitrarily determined course is like the record of a sequence of coin tosses; all outcomes in the record are fixed, but the first few outcomes do not decree what follows. Determinism's emphasis on appointment and decree implies some lawlike generality running through the course of events. The lawfulness is obvious in the dynamic rules of deterministic systems. Thus the philosopher Patrick Suppes remarked, "One formulation of determinism grows naturally out of the theory of differential equations."[26]

The context in which the debate on determinism rages shows that determinism demands the absence of ambiguity in all details of the world as it is understood in our general conceptual framework. Only thus does it conflict with the freedom of action of such tiny entities as human beings. Therefore determinism asserts much more than the existence of some deterministic systems. *Determinism* implies at least that the world consists *only* of invertible deterministic systems, the deterministic nature of which is not disturbed by the interaction among systems, so that the universe itself – in all its detail and organization levels – is nothing but a huge and complex invertible deterministic system. Some versions of determinism would add the requirements of stability and regularity.

Indeterminism is not as clearly defined. Its strong forms maintain there is neither deterministic system nor causal law. It can also include probabilism and tychism. *Probabilism* asserts that the fundamental laws of nature are essentially probabilistic. *Tychism* upholds the probabilistic nature of causation and the existence of objective chance as some form of propensity. Indeterminism is distinct from the affirmation of free will, for the freedom of action is not merely randomness or arbitrariness of action.

Determinism and strong indeterminism are not the only possible metaphysical positions. In the free-will debate, there is a third position called *compatibilism*, which argues that freedom of action is compatible with the determinacy of the physical world. Compatibilism also suggests that deterministic and probabilistic concepts are neither incompatible nor mutually exclusive.

Predictability

Laplace wrote in 1814: "We ought then to regard the present state of the universe as the effect of its anterior state and as the cause of the one which is to follow. Given for one instant an intelligence which could comprehend all the forces by which nature is animated and the respective situation of the beings who compose it – an intelligence sufficiently vast to submit these data to analysis – it would embrace in the same formula the movements of the greatest bodies of the universe and those of the lightest atom; for it, nothing would be uncertain and the future, as the past, would be present to its eyes."[27]

This passage is often quoted as the "manifesto of determinism," although it does not use the word *determinism* and appears not in Laplace's treatise on celestial mechanics but in the preface to his treatise on probability, followed by the argument that we need the probability calculus because we are not the demon.

Laplace's statement has two parts; the first asserts the universe's *causality*, the second its *predictability*. Causality and predictability have been separately identified with determinism by some authors or others. Both identifications muddle the issue.

By itself, the first sentence of Laplace's statement merely asserts that every state of the universe is caused by its anterior state. It permits the thought that the chain of causes and effects has several possible courses. Thus it is not stringent enough to characterize deterministic dynamics, even less determinism. The uniqueness of the causal chain is suggested by the second sentence, but there it is entwined with the idea of predictability. Laplace was careful to impute the predictive power to the demon. The move shifts the burden of determinism to the demon's cognitive ability. Since we can attribute any power to the demon and any character to the formula it can handle, the story holds today, chaos and quantum mechanics notwithstanding. Chaos is no problem if the demon is an infinite being capable of specifying the initial condition to any degree of accuracy required. Quantum mechanics poses no difficulty if the demon is not restricted to the kind of experiment we humans can perform and the kind of classical predicate we must use to describe what we can measure. It can simply observe and comprehend the universe as a quantum state evolving deterministically according to the Schrödinger equation. However, such a fancy of omniscience is neither philosophically nor scientifically interesting.

Philosophers such as Karl Popper adopted the second part of Laplace's statement, replaced the demon by human scientists with limited capabilities to specify initial conditions accurately, and turned it into a definition of determinism. The definition is confusing. Dynamic theories have shown clearly that a deterministic system governed by a dynamic rule can be unpredictable in the long run or even in the short run if its characterization is coarse-grained. Determinacy and predictability are two distinct concepts for individual dynamic systems and should not be conflated when applied to the world as a whole.

The definition of determinism in terms of scientific predictability is unclear because the predictive power of science changes. Determinism is trivially false if it means predictable by current science; everyone agrees that current or forseeable science cannot predict everything. For determinism to be meaningful, the definition must appeal to some ideal limit of science or some kind of in-principle predictability. Then we are faced with all the difficulty of speculating on the ideal or perfect science.

Determinism is a metaphysical concept about what the world is like; predictability involves substantive epistemological criteria on what gods or demons or humans can know about the world. The two are categorically different. Predictability has its own clear meaning; there is no reason to muddle it with the big and fuzzy name of determinism.

Causation and Determinism

The identification of determinism with causality is no less confused and unwarranted. "All events are caused" is a far weaker and more general doctrine than determinism, although the two are often confused. The old philosophical doctrine that every event has a cause merely means that the world is a causal network, so that something popping out of thin air without any prior causal relation with anything else is incomprehensible. It does not forbid an effect to have many possible causes or a cause to have many possible effects, so that it does not demand any cause or effect to be necessary. The only necessity it implies pertains to the general concept of causal relation in our thinking. It is so general even tychists tacitly subscribe to it; if random events are intelligible without any prior causal relation, then there is no need to posit propensity or objective chance, which seems to have no other function than to bring about the effects. We stick to the general idea of causal precedence, even as we speculate on chance, magic, or miracle as substantive causes. Similarly, in the interpretation of quantum mechanics, physicists talk about observations, uncontrollable jumps, and the like, as the causes of the random outcomes of measurements.

"The same causes always have the same effects" is no closer to determinism. If *same* means exactly the same, the definition is weak enough to be applicable to the most random world. As we have seen in Part I, the numbers of possible configurations for systems of any considerable size and complexity are enormous, so enormous they far outstrip the numbers of actual instances. The diversity of events exceeds that of things, because events are changes of things, and the temporal dimension adds extra variation. Consequently actual macroscopic things and events are almost all unique; only extremely rarely are two things or events exactly the same. This fact has long been noticed. Leibniz made its extreme form – no two substances are exactly alike – into a cornerstone of his philosophical system. "Same causes, same effects" imposes little constraint if causes are never the same. It cannot be a formulation for determinism, for there would be no pattern or lawlike generality if everything were unique. This point is used by many people, including Maxwell, to argue for the compatibility of freedom of action within a physicalist worldview. It also plays a crucial role in the argument against historical determinism and the utility of covering laws in historical explanations (§ 41).

The apparent force of "same causes, same effects" arises mainly from the surreptitious slackening of *same* to similar or "roughly alike." However, weakened slackened doctrine fails even for deterministic dynamics if the systems are unstable or irregular. The effect of a minor cause may be minor when a system is operating in the stable region but catastrophic when it is operating in the unstable region. As an example of similar causes with dramatically different effects, the change of the control parameter of a logistic system from 3.448 to 3.449 produces little effect, but the change from 3.449 to 3.450 causes the system to jump from a period-2 to a period-4 limit cycle (§ 30). Cases of "similar causes, dissimilar effects" become even more numerous if we interpret the earlier states of a dynamical system as the causes of its later states. The exponential divergence of trajectories of chaotic systems implies similar causes can lead to disparate effects in the long run (§ 31). Even if the system is not chaotic, it may be nonlinear and has more than one attractor. Initial states near the boundary between the basins of attraction may end in different states, as two drops of rain falling side by side at the continental divide end in different oceans (§ 29). In short, if determinism means "Similar causes always have similar effects," then it is false. Once ambiguity is allowed in, it is not always controllable.

"The evolution of all systems is governed by causal laws such as the dynamic rules of deterministic dynamics" is inadequate as a formulation of determinism. Dynamic rules generally govern isolated systems; it is doubtful whether deterministic laws exist to account for all the coupling of complicated systems. Interactions invariably perturb the dynamic systems. Perturbation, however slight, will make many nonlinear systems unstable and chaotic systems switch paths, thus producing many ambiguous possibilities for the evolution of the systems. Anyhow, the definition is false. Our sciences feature many important systems that cannot be totally represented by deterministic dynamics. Quantum systems, which populate the microscopic world studied by fundamental physics, are notable examples.

In sum, the general relation between causality and determinism is feeble. The crux of determinism is the *exactness* of causality, as Hacking defined the fall of determinism as "the past does not determine exactly what happens next."[28] We can reject the exactitude without rejecting causality. If exactitude is understood in an absolute sense that is totally detached from the general nature of our objective thought and scientific theory, then determinism becomes an empty speculation that is not worth wasting time on. Whether the world is an "iron block" regardless of the theories we can possibly manage is irrelevant to our action or understanding. If the exactitude depends partly on our way of discerning and conceptualizing causal regularities, then determinism is false. Approximation, idealization, and coarse graining suffuse our theories and models. We have seen in Part I that they are indispensable in the characterization of momentary states of complex systems. How can the evolution of states so roughly characterized be expected to be free of ambiguity? More generally, most of our best scientific theories are framed in terms of real numbers. A real number can be specified by x_0 in the abstract, but not when it represents an objective quantity. The exact specification of an objective quantity requires an uncountably infinite amount of information, which is infeasible even in principle.

Causality Without Determinism

The rejection of determinism does not imply the nonexistence of stable deterministic systems, nor does it diminish the significance of the general concept of causality. It only militates against making certain ideas absolute. As Kant argued, legitimate metaphysics is concerned with the general conditions of the world as it is intelligible to us. One lands in metaphysical illusion if he goes after absolute conditions beyond the bounds of our possible cognitive capacity.

The world intelligible to us is conceptualized, and concepts invariably involve a certain generality. Conceptual generalization is possible only if the world does exhibit some patterns and regularity. If the world is absolutely magical, with things popping in and out of existence arbitrarily, or if each event is so peculiar that no pattern exists, we cannot have the kind of intelligible experience that we do. Although "similar causes, similar effects" fails as an absolute principle as determinism demands, we are fortunate that the world contains many systems and processes to which it applies. These systems, recognized as enduring things and causal connections, constitute the backbone of the intelligible world. Thus we have the general concept of the world as a causal network. The substantive details of the network are sparsely known and we do not demand that they be completely filled in, but we know enough to talk intelligibly about objects. We admit mysteries and brace for catastrophes, but they are recognizable as such only against a more or less stable background of causal relations, without which the world would signify less than sound and fury. The tale told by the idiot Benji in Faulkner's novel makes some sense because even Benji has some grasp of causal relations.

The commonsense ideas of things and causality are more rigorously represented by the deterministic concepts in dynamic theories. Whenever possible, scientists try to represent their subject matter by deterministic models. Classical mechanics and relativistic mechanics are deterministic. So are most models for economic growth. Many statistical distributions representing composite systems evolve deterministically; for instance, the evolution of velocity distributions is governed by the kinetic equation. A great part of quantum mechanics is deterministic; quantum equations of motion are differential equations like other dynamic rules. The domain of causality is enlarged and not curtailed by the development of nonlinear and chaotic theories, which show that even apparently random phenomena can have underlying dynamic rules.

On the other hand, the limitation of "similar causes, similar effects" warns us to note the range of validity for our causal and deterministic assertions. We have to qualify our definite descriptions with the admissible degree of similarity, not unlike experimentalists inserting error bars to indicate the accuracy of their data. Our general conceptual framework must be expanded to include ideas about the approximate nature of our theories and models, for example, robustness, degree of accuracy, and characteristic magnitudes of various quantities. It must also accommodate statistical and probabilistic notions.

Stochastic Processes and Their Explanations

35. The Calculus of Probability and Stochastic Processes

The probability calculus finds application in a large and expanding class of empirical theory in the natural, social, and human sciences. What general features do the diverse topics share that make them susceptible to representation by the same mathematics?

Chance is an easy answer; probability is intuitively associated with chance. Easy answers are often wrong. Chance is not among the primary or secondary concepts of the probability calculus; it does not even have a clear definition there. The clue to the common features of the sciences that employ the probability calculus lies in the structure of the calculus, not in its name; thus it is misleading to call its application the "laws of chance" or the "empire of chance."

The first ideas introduced in the axioms of the probability calculus are that of *part and whole* and that of the *relative magnitudes* of the parts. *Probability* is technically defined as a relative magnitude. The most general feature of probabilistic systems is that they are *composite*. Various limit theorems and laws of large numbers show that the calculus is most powerful for *large or infinite* composite systems. Large composite systems are usually very complex. The probability calculus is adept in treating a special class of relatively simple system, the constituents of which are *independent* of each other. Independence is the chief meaning of randomness in the calculus, and it is often posited as an approximation that simplifies the theories of realistic composite systems.

The composite systems studied in the mathematical theory of probability are completely general and can be realized in any system, concrete or abstract. In many empirical theories, the composite entity to which the calculus directly applies is the *state space* made up of all possible states of a system. The relative magnitude of a group of possible states is technically the group's probability. The state-space application links the technical and intuitive meanings of probability via the idea of *possibility*. It also provides the conceptual connection between stochastic and deterministic processes. The evolution of a deterministic system traces out a path in the state space. In the fine-grained description of deterministic dynamics, we account for the states in the path individually. In stochastic dynamics, the state space is partitioned according to various coarse-grained characters, and we reckon with the proportion of states in the path that fall within each part.

This chapter begins with the basic concepts of the probability calculus, especially those connected to stochastic processes. It goes on to examine how they are applied in economics, evolutionary biology, and statistical physics and ends with a reflection on the kind of question framed in statistical terms. Sometimes the same phenomenon has both deterministic and stochastic models. Stochastic theories generally do not postulate underlying dynamical rules for stochastic processes. However, many include both stochastic and deterministic elements, which refer to different aspects of a complex process.

The General Concept of Magnitude: Measure[1]

The words *probable* and *probably* are commonly used to express the uncertainty in our judgments. Jakob Bernoulli said in his treatise on the probability calculus that the degree of certainty "differs from certainty as the part differs from the whole."[2] The reference to the part and the whole is not an analogy; degrees of belief are mathematically represented as the magnitudes of the parts relative to the magnitude of the whole, which represents certainty. Probabilistic systems are mathematically represented as abstract composite systems; that explains why the probabilistic calculus is readily applied to tangible composite systems in the statistical sciences.

Quantities, in contrast to qualities, can be added to and subtracted from each other. We ordinarily do not or cannot ascribe quantity to such abstract stuff as belief or certainty; the concept of quantity seems to apply only to spatial matter and physical things. We rate beauty by a scale of 1 to 10, but the rating is qualitative; it tells "how beautiful" but not "how much beauty." The quantitative question sounds absurd; can one buy beauty by the pound? Aristophanes adumbrated the possibility of abstract quantities when he described a divine scale that assayed the weight of verses, but he was skeptical about the "novel portent beyond the ordinary man's capacity."[3] The probability calculus enables the ordinary man to quantify abstract concepts and calculate the weight of belief, the amount of risk, the volume of evidence, the mass of knowledge, the magnitude of possibility, not poetically but precisely. Not surprisingly it finds wide application in many sciences.

The basic topic of the probability calculus is an abstract set of elements endowed with the structure of a measure space. As discussed in § 32, the measure space contains two main ideas: *parts* and *magnitudes*. The set, which is the whole, is systematically divided into parts or subsets such that the union and intersection of parts are themselves parts. Then magnitudes are systematically defined for the parts such that when two parts combine, their magnitudes add. The axioms for the definition of the parts and their magnitudes capture the intuitive idea of combination of parts and the additivity of quantities. The system of magnitudes is called the *measure* of the set. Measure theory is the basis of the probability calculus.

The magnitude of a part of the set is also called the *measure* of the part. Thus the measure generalizes the notion of volume or more appropriately of weight. The geometric notion of volume is apparent and can be confused with the simple reckoning of elements in the part; weight introduces another dimension, which can be used to represent the advantage of certain elements, such as those favored by a loaded die. Mathematically, *probability* is a normalized measure or a relative magnitude; the probability of a part is its "weight" relative to the "weight" of the whole set. In many textbooks, physical mass or weight is used as a heuristic analogy for probability. Since the definitions of parts and relative magnitudes are abstract, they are applicable to anything that satisfies the general axioms of their definitions. For instance, they can be applied to physical mass, although they seldom are because they are clumsier than other available representations.

Familiar as the notion of magnitude or quantity is, it is not necessary in deterministic dynamics. The path of a planet giving its successive spatial positions does not involve the idea of quantity. Magnitude comes in when the path is integrated to yield the distance the planet has traversed.

In the probability calculus, probability has no meaning other than a relative magnitude, also called relative frequency, statistical weight, proportion, or fraction. Consider a simple example. From a deck of 52 cards, one can draw 2,598,960 different poker hands of 5 cards, of which 5,148 are flushes. The proportion of flushes among all possible poker hands is $5,148/2,598,960 = 0.00198$, or the relative magnitude of the group comprising all flushes is 0.00198. If the cards are not stacked so that all possible hands have equal weight, the proportion 0.00198 is the technical definition of the "probability of flushes." If the cards are stacked, we must calculate the weighted proportion, but the general meaning of probability is the same: The technical definition of probability is the relative magnitude of a *group* of poker hands circumscribed by a certain character. It differs from the ordinary meaning of probability, which refer to individuals, not groups (§ 39). To prevent confusion, I will reserve *probability* for the ordinary usage. The technical definition of *probability* attributed to groups I will call *relative magnitude* or its synonyms. Concepts associated with relative magnitudes I call *statistical*, not probabilistic.

From Individuals to Systems of Individuals: The Random Variable[4]

Next to the idea of measure, perhaps the most important concept in the probability calculus is the random variable. A *random variable* is a function that assigns a value to each element in a set on which a measure is defined such that a corresponding measure is defined on its range of values. Despite their name, random variables carry no intrinsic notion of randomness. Functions can represent the most regular and deterministic qualities, including all variables in classical mechanics. The requirement that the domain and range of the function support the concept of magnitude does not make the function random. However, it makes possible the shift of emphasis from the function itself, which is descriptive of individual elements, to the features of its whole range of possible values, which are descriptive of groups of elements. The shift, akin to the shift from deterministic to stochastic descriptions depicted in Fig. 8.5, is the crux of random variables.

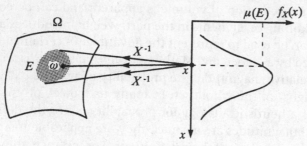

FIGURE 9.1 A measure space Ω is a set of elements ω endowed with a structure such that a relative magnitude is defined for each of its systematically defined parts. A random variable X maps an element ω into a value x in the sample space of X. Generally, many elements have the same value x. The inverse of the random variable X^{-1} picks out for x the group $E = X^{-1}(x)$ of elements having it as their value. The relative magnitude of E is $\mu(E)$. By varying x, $\mu(E)$ generates the distribution $f_X(x)$ of the random variable X over its sample space.

For instance, let the measure space Ω be the state space of a particle, the elements ω of Ω the possible states, and the function X the momentum. As an ordinary momentum function, X assigns a momentum x to each state ω, giving a fine-grained deterministic description. As a momentum random variable, its independent variable that picks out the subject of discourse changes from ω to x (Fig. 9.1, compare the change in independent variable in Fig. 8.5 of § 32). The collection of all possible values of a random variable is its *sample space*. For each momentum x in the sample space, the inverse of X picks out a group E of states whose momenta fall within the infinitesimal range dx around x. The proportion or the magnitude of the group E relative to the magnitude of Ω is technically called the probability of E. As x varies, X^{-1} picks out various E, and the various relative magnitudes of E form the momentum distribution $f_X(x)$, which gives the proportion of states with momentum falling within the range dx around each x. From the distribution, various averages and correlation relations can be calculated.

The distribution is at the center of attention of the probability calculus. Most applications do not specify the functional form of the random variable X itself; they specify only its distribution $f_X(x)$. Consequently we do not know the momentum for each individual but must be content with talking about the proportion of individuals having a certain momentum. The emphasis on distributions instead of functional forms and individual values marks random variables from the functions characterizing deterministic properties. With the shift in emphasis, *we pass from a fine-grained description of individual elements to a coarse-grained description of groups of elements.* The information lost in the transition to the coarse-grained description suggests the attribute "random."

Constituents, Systems, and Ensembles of Systems

Among the major results of the probability calculus are various limit theorems. Let us examine the structure of the laws of large numbers. Consider the paradigm case of coin tossing. The laws of large numbers involve three kinds of entity: coin toss, sequence of independent tosses, and ensemble of sequences, which I generally call *individual, composite system,* and *ensemble of systems.* They correspond to the micro-, macro-, and supermacrolevels discussed in § 22.

Consider a toss of a coin. The relative weight of heads is p, which would be different from $\frac{1}{2}$ if the coin is bent. This relative weight is a quantity given to the calculus. Next consider a sequence made up of n independent tosses of the coin; if it contains k heads, the proportion of heads is k/n. Finally consider the ensemble consisting of all possible distinct sequences of n independent tosses. The *strong law of large numbers* asserts that if the limit of n equals infinity, the proportion of sequences with $k/n = p$ in the ensemble is 1. In other words, except a few atypical sequences that collectively have zero relative magnitude, in all typical infinite sequences the proportion of heads k/n is equal to p, the original relative weight of heads for a single toss. "All typical sequences" is expressed as "The proportion of such sequences in the ensemble is 1."

As the strong law of large numbers involves three kinds of entity, it invokes the mathematical definition of probability in three places, one for each kind of entity: (1) the relative weight of heads for a single toss of a coin, with value p; (2) the proportion of heads in a sequence of n tosses, with value k/n; (3) the proportion of sequences in the ensemble with $k/n = p$, with value 1 when n goes to infinity. The three occurrences of relative magnitude are common in many results in the probability calculus.

Among the three relative magnitudes, the law of large numbers takes the relative weight of the individual toss for granted and takes as its locus the characteristics of the sequences of tosses. It is similar to theories of composite systems, which assume that the characters of the constituents are known and focus on the characters of the systems. The law of large numbers treats composite systems in a peculiar way. Among all the variations in the sequences of tosses, it proves via the statistics of the ensemble the existence of one feature, a specific value p of the proportion of heads, which is shared by all typical infinite sequences. Admittedly the proportion of heads is a crude description of a sequence, but its importance should not be undervalued, for it supports a *statistically deterministic law* for the sequences. "All typical systems of the kind have a certain character" is the logical form of a statistical law, as "All systems of the kind have a certain character" is the logical form of a universal law. The statistical law about sequences is established from the synthetic view on the ensemble level.

The law of large numbers is not limited to coin tosses. As a mathematical theorem, it is applicable to anything satisfying a few general conditions. The most important condition is that the random variables constituting the composite random variable are *independent.* Independence captures the notion of randomness and is a central concept in the probability calculus. The randomness represented by independence signifies a lack of relation among the constituent random variables; hence it is related to randomness as complexity (§ 32).

Stochastic Processes[5]

A major part of the probability calculus studies various combinations and transformations of random variables. A *stochastic process* or *random process* $X(t)$

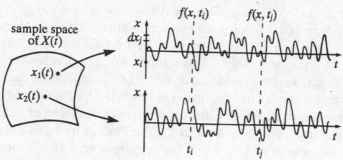

FIGURE 9.2 The sample space of a stochastic process $X(t)$ contains all its possible realizations $x(t)$. A specific realization $x_1(t)$ is a sample sequence that assumes a specific value at each instance of time t. With time fixed at t_i, the stochastic process becomes a random variable with distribution $f(x, t_i)$, which gives the proportion of sample sequences that have values in the range dx around each x at t_i. The two-time joint distribution $f(x_i, t_i; x_j, t_j)$ gives the proportion of sequences with values around x_i at t_i and around x_j at t_2.

is an infinite sequence of random variables X indexed by a parameter t called time, which can be discrete or continuous. In the discrete case, $X(t) \equiv \{X(t_1), X(t_2), \ldots \}$, so that each constituent random variable $X(t_i)$ represents a stage of the process. A value of $X(t)$ is a *sample sequence*, which is a sequence of values of the constituent Xs, $x(t) \equiv \{x(t_1), x(t_2), \ldots \}$. A specific sample sequence $x_1(t)$ is a particular realization of the process. The collection of all possible sample sequences is the *sample space* of the stochastic process $X(t)$ (Fig. 9.2).

For example, let X be the random variable associated with a toss of a coin. X has two possible values: $x = 1$ for heads and 0 for tails. Infinite sequences of coin tosses constitute a random process $X(t)$ with discrete time. A sample sequence $x_1(t)$ is the result of a particular sequence of coin tosses, say $10011101\ldots$. The sample space of the stochastic process, consisting of all such possible sequences, is enormous, simply because of the infinite ways of combining the values of the constituent random variables. It is reminiscent of the Γ-space in statistical mechanics.

Suppose that each of the random variables X that make up the stochastic process has a distribution $f(x)$. Looking at the stochastic process at the instant t_i, we find a random variable $X(t_i)$ with *distribution* $f(x, t_i)$, which gives the proportion of sample sequences that pass through the infinitesimal window of values dx around x at time t_i. The distribution $f(x, t_i)$ is insufficient to characterize the whole stochastic process. We also want to know, for example, the proportion of sample sequences that pass near x_i at t_i and x_j at t_j, which is represented by the two-time *joint distribution* $f(x_i, t_i; x_j, t_j)$. There are higher-order joint distributions involving more time instants. If we consider a set of times $t_1 \ldots t_n$ in the interval in which the process is defined, then an n-time joint distribution is required to characterize the process completely. There are many well-studied processes with familiar joint distributions, for example, the Gaussian process or the Poisson process.

A stochastic process is *stationary* if its joint distribution is not changed by a uniform shift in the time index, so that it is the same if the sequence of measurements is performed today instead of yesterday. Stationarity implies that the dynamic mechanism that generates the stochastic process does not change, not that the mechanism is deterministic. Many processes that find application in the sciences are either stationary or approximately treated as so.

Multitime joint distributions are difficult to calculate. Most applications are satisfied with partial characterizations. The quantities most commonly sought are the averages, especially the *time-dependent mean value* $\langle X(t) \rangle$ and the *auto-correlation* $\langle X(t_i) X(t_j) \rangle$. The autocorrelation, which gives the relation between the process's statistical behaviors at different times t_i and t_j, becomes the *time-dependent variance* if $t_i = t_j$. For stationary processes the autocorrelation depends only on the time lapse $t_j - t_i$. In many applications, the autocorrelation vanishes if the time lapse is larger than a certain value, which is called the *correlation time*.

A specific sample sequence, say $x_1(t)$, looks like the path of a deterministic dynamic system. It may well be, but we do not care. We do *not* consider a single sequence in isolation. Even if a single sequence is produced in an experiment, it is regarded as a *sample* that must be analyzed in the context of the sample space. We do not ask why or how the sample sequence attains a specific value at a specific time; instead, we try to extract from it statistical information about the mean value, correlation, and distribution. These quantities are categorically different from those we seek in the theory of deterministic dynamics.

Generally, the n-time joint distribution of a stochastic process cannot be factorized into a product of single-time distribution functions. It can be so factorized only if the random variables at various times are statistically independent. *Statistical independence* captures the intuitive idea that the values of the random variables have no influence on each other; for instance, the outcome of a toss of a coin is independent of the outcome of its predecessors or successors. Factorizable processes are the simplest sort, for their behaviors can be specified by single-time distributions alone. If the random variables constituting a stochastic process are all statistically independent and have identical distributions, then the process is a *white noise process*.

If the random variables of a stochastic process are statistically independent, then the autocorrelations of the process vanish except at equal times. However, the reverse is not always true; zero correlation is a weaker condition than independence. Physicists call a process with zero correlation time *white noise* even if it is not statistically independent. For instance, the Brownian motion to be discussed shortly is driven by a white-noise process with a normal (Gaussian) joint distribution. White-noise processes in both senses find wide application in science and engineering, where the name originated.

The next complicated class of stochastic process is the *Markov process*, which can be completely specified by the two-time joint distribution. The two-time joint distribution can be expressed as the product of the single-time distribution $f(x_i, t_i)$ and a transition function giving the proportion of sample sequences with value x_i at t_i that take on the value x_j at a later time t_j. The statistical features of the process at n times can be expressed as the sum of the products of the transition functions between two successive times.

A white-noise process lives only in the present, so to speak; its future depends on neither the present nor the past. A Markov process forgets its history; its future depends only on its present state and not on how it is attained. A process is Markovian only if its correlation time is short compared to the interval of characterization; otherwise it would not be able to shake off the influence of history. Many physical processes, including Brownian motion, can be treated as Markov processes only if time is sufficiently coarse-grained.

A well-known example of Markov processes is the random walk, in which the direction and span of each step are uncorrelated to those of previous steps. Another example is the branching process widely used in biological and sociological population models. The process was first developed in 1874 to address the question whether the extinction of upper-class names in England was due to infertility or stochastic factors. It features several initial individuals, all having the same statistical reproductive characteristics and each starting a family of descendants. The transition function represents the proportion of individuals that have a certain number of offspring. The process tracks the population over time.

White-noise and Markov processes are usually deemed random. At the other extreme, there are *statistically deterministic processes*, the behaviors of which are predictable with probability 1 if their infinitely long records are known. *Deterministic* here qualifies a special kind of stochastic process and should not be confused with the deterministic dynamics discussed in the preceding chapter. Statistically deterministic processes generally do not involve dynamic rules.

Systems Driven by Stochastic Processes: Brownian Motion[6]

In many applications, a system is under some fluctuating external influence, which can be represented as a stochastic process. Systems driven by stochastic processes are themselves stochastic processes. We try to find the behavior of a stochastic process dependent on another stochastic process the statistical characteristic of which is assumed to be given. Examples abound in the sciences. Evolution by natural selection can be regarded as a stochastic process driven by the fluctuating environment (§ 37). Cyclical economic booms and recessions are often treated as a stochastic process driven by exogenous factors such as the unpredictable changes in taste and technology (§ 36). Here we will briefly consider Brownian motion, which is the prototype of physical systems driven by random forces. It is simple and easily visualized, yet its characterization embodies many concepts that are central to the theories of more complex systems.

Brownian motion, first observed by the botanist Robert Brown in 1828, is the irregular movement of a heavy microscopic particle in a fluid of light molecules. Brown argued that the movement cannot be attributed to innate vitality because it is executed by organic and inorganic particles alike. The first theory of Brownian motion was developed in 1905 by Einstein, who used the erratic movement of tiny particles in fluids to argue for the atomistic approach to thermodynamics. Soon it was firmly established that the Brownian particle zigs and zags randomly because it is being incessantly bombarded by the molecules that constitute the fluid. Like a person making his way through a crowd, it cannot travel a straight path.

Brownian motion can be characterized in several levels of refinement. On the macroscopic scale, it is represented as a diffusion process. A group of Brownian particles initially concentrated in a tiny region spread out and eventually distribute themselves uniformly throughout the fluid. The locations of the Brownian particles are statistically summarized in a density distribution of the group, the evolution of which is governed by a differential equation called the diffusion equation. Given an initial density distribution, the subsequent distributions are uniquely determined. On a finer level, Brownian motion is described as a Markov process, in which a Brownian particle takes a random walk and its position and velocity jump haphazardly.

Microscopically, the Brownian particle obeys the mechanical equation of motion. Its correlation with its initial position and velocity is lost only after suffering many collisions with the molecules. Thus both the diffusive and Markovian descriptions are valid only in a sufficiently *coarse-grained time scale* in which the characteristic interval between observations is large compared to the typical interval between collisions. This is an example of the importance of time scales discussed in § 38.

The Brownian motion can also be characterized by a *stochastic equation of motion*, for instance, Langevin's equation. Langevin's equation starts with Newton's equation but deviates from it by including a random force, which represents the impacts of the molecules. Formally, Langevin's equation for Brownian motion still says force equals mass times acceleration. Its formal solution for the Brownian particle's velocity and position can be written down readily. The formal solution is empty, however, because the form of the force is not given. Unlike ordinary forces, the random force is a stochastic process specified only by its statistical properties. Because of the random force, Langevin's equation becomes stochastic and no longer yields a unique path for a given initial condition. Its solutions describe the responses of the Brownian particle to the random force in statistical terms. The random force is usually taken to be a Gaussian white-noise process with a certain variance. From this information we can take the proper average to find the means and autocorrelations of the particle velocity and position. Theoretically, the average is taken over all possible sample paths of the Brownian process. Physically, the sample paths can be realized by an ensemble of Brownian particles, which is what is usually measured in actual experiments.

36.   Deterministic and Stochastic Models of Business Cycles

We now turn to how the calculus of probability and stochastic processes are applied to the empirical sciences, and how they compare to other methods. Often a phenomenon can be modeled as either a stochastic or a deterministic process. Different models answer different kinds of question about the phenomenon. For complicated processes, the more practical questions are addressed by stochastic models.

This section examines business-cycle models intended to explain the fluctuations in economic activities and ideally to forecast them or to recommend ways to smooth out the bumpiness. Growth and fluctuation are the two major dynamic phenomena studied in theoretical economics. Economic growth is the most striking fact in the history of industrial capitalist economies.[7] Growth, however, is far from smooth; fluctuation in output and employment is a chronic ailment of the capitalist economy. A *recession* is a period when most major sectors of the economy are in decline; the onset of the period is signaled by two consecutive quarters of shrinking gross domestic product (GDP). Recessions began in 1816, 1826, 1837, 1848, 1857, 1866, 1873, 1890, 1907, 1914, 1921, 1929, and 1937. For reasons still debated, the matured economies in the post–World War II era are less volatile.[8] There was a rare prolonged period of prosperity in the 1960s, and subsequent recessions are shorter and less severe. The 1974–5 and 1981–2 slumps introduced the novel phenomenon of stagflation, in which high inflation coexisted with high unemployment.

Business cycles or successive booms and slumps are macroeconomic phenomena. A cycle is manifested in the nearly simultaneous expansion of several macroeconomic variables, followed by their concerted contraction and subsequent recovery, which initiate a new cycle. The chief economic variable for the definition of business cycles is the *output* in terms of GDP, which can be decomposed into several variables: consumption, investment, and government spending. Other variables characterizing the economy are employment, price level, and interest rate. The variables fluctuate with their own amplitudes, some larger than others, as investment is more volatile than consumption. However, their movements are statistically correlated to each other throughout a business cycle. A variable need not move exactly in phase with the output; it can be leading or lagging. The *lead* or *lag* is manifested by the correlation between the variable at one quarter and the output at later or earlier quarters. The government regularly publishes data on about ten leading economic variables: manufacturers' new orders, building permits for new houses, vendor inventory, initial claims for unemployment insurance, change in prices of basic materials, change in money supply, and the index for stock-market prices. In contrast, the total unemployment rate is a lagging variable whose declines and recoveries occur later than those of the output.

Unlike the motions of physical oscillators, business cycles are recurrent but not periodic, occurring in irregular intervals with various amplitudes and durations that cannot be analyzed as a superposition of periodic oscillations. Although individually irregular, the cycles exhibit certain statistical regularities. The duration of business cycle varies from a year to 10 or even 20 years, but both the prewar and postwar cycles have an average duration of 3.5 years. The economy does not jump from booms to slumps randomly; its performance in one quarter is statistically related to its performance in the previous quarters, so that the quarterly figures show strong serial correlation.

Business-cycle theories try to explain the typical duration and amplitude of cycles, and the covariation, relative phases, and relative amplitudes of various economic variables through the cycles. There are three kinds of business-cycle models. Models of the first kind try to forecast the behaviors of the economy solely on the basis of statistical data, without making any assumption about how the economy works. Models of the second kind regard the economy as a nonlinear deterministic system with chaotic dynamics. Models of the third kind treat business cycles as stochastic processes driven by random external forces. Although stochastic models are by far the most popular among economists, the existence of competitors warns against unwarranted philosophical generalization.

Nonexplanatory Models[9]

The first approach to the study of business cycles can be called *noneconomic*, because it can proceed almost without any knowledge of the economy and without any use of economic concepts. All it needs from economics is a list of relevant variables. Given that, it uses the purely statistical techniques of time-series analysis. A *time series* is a sample sequence of a stochastic process with data points taken at regular intervals, for example, the values of the Dow Jones Industrial Average at market close on successive days. As a sophisticated method to extrapolate data, time-series analysis tries to extract from sample sequences regularities describable in general terms such as moving averages and strength ratios, without appealing to the causal mechanisms of the underlying stochastic processes.

The noneconomic approach looks only at the time series and ignores the underlying economic processes altogether. It employs the notion of Granger causation, which is a case of probabilistic causation discussed in § 39. A variable X *Granger causes* a variable Y if the time-series data of X before time t contain information that helps predict Y at t, and the information is not contained in the time series of other variables. The information is purely statistical; it has nothing to do with the working of the economy.

Proponents of the noneconomic approach argue that their model of business cycles has the epistemological advantage of eschewing a priori theories and hypotheses. Critics argue that the approach explains nothing because it is not at all concerned about the structure and operation of the economy. What it offers is not theoretical prediction but data extrapolation. Theoretical prediction tries to gain some knowledge and understanding of the underlying mechanism; data extrapolation does not.

Deterministic Models[10]

A class of models regard fluctuations as essentially the results of the internal dynamics of the economy. An important subdivision of the class is based on deterministic dynamics. Deterministic models represent the economy by a set of coupled difference equations relating various macroeconomic variables with suitable time lags. For instance, the consumption in one period is determined by the output of the previous period, and the investment by the difference in the outputs of the two previous periods. Some early models are linear; oscillations arise from the inclusion of second-order differences. Other early models include nonlinearities such as floors and ceilings that limit the values of the variables. These models yield only limit cycles that are periodic. They cannot reproduce the irregularity that is characteristic of business cycles. Effort along this line dwindled soon after 1950.

The rise of chaotic dynamics in the past two decades reinvigorated deterministic models. Deterministic systems represented by nonlinear dynamic equations can generate bounded and highly irregular oscillations that appear to be stochastic in a coarse-grained view. Some economists argue that the general behaviors of economic time series can be mimicked by deterministic models with slight chaotic instabilities. In this view the economy is not unlike the weather, which vacillates irregularly between rain and shine but can be represented as a hydrodynamic system governed by a set of coupled deterministic nonlinear equations.

Recent years have seen a plethora of nonlinear deterministic models. In a simple model for economic growth, capital accumulation is modeled by a logistic map representing capital accumulation, with parameters relating to population growth, interest rate, and level of pollution. The capital stock grows initially but is eventually checked by pollution or other resource constraints. The model shows that small perturbations of the parameters can steer the economy along the path of chaos. Proponents of deterministic chaos argue that if a model as simple as the logistic map is capable of generating irregularity, there is no reason to believe that nonlinearity is unimportant in economic modeling. Careful studies of economic time-series data have shown that they do contain nonlinearity. However, so far the deterministic models are very abstract, with only a dubious connection with the real working of the economy. It is not difficult to produce irregular paths that resemble some empirical time series qualitatively. The difficulty is to explain the comovements of several variables, and the chaotic models still have a long way to go.

Stochastic Linear Response

The rest of this section is devoted to stochastic models, which originated in the 1930s and constitute the majority of effort in the study of business cycles. In these models, fluctuations arise mainly from external disturbances, without which the economy settles in a steady state. The models hinge on two basic ideas, *shock* and *propagation*. Like the Brownian particle suffering molecular bombardments, the economy is subjected to shocks originating from the noneconomic sector of the society. The exogenous shock can be a single event, or it can be a stochastic process. The effects of the shock propagate deterministically throughout the economy according to the economy's structure and dynamics, generating the correlated movements of economic variables.

All stochastic models of business cycles share a broad conceptual framework, regardless of the specifics of shocks and propagation mechanisms they posit. Abstracting from the economic content, similar frameworks are found in physics and engineering and are usually called *stochastic linear response theories*. I will first present the general framework, then fill in the economic content.

The general problem is to study how a system responds to an external disturbance that varies with time, perhaps randomly. As a simple example, imagine that we give a damped pendulum a kick. The pendulum initially swings and eventually returns to its equilibrium resting position. The effect of the kick is transient, but it lingers for a finite time, which can be quite long if the damping is weak. Now suppose that an oscillator is modulated continuously by a field whose frequency changes slowly with time. The oscillator almost ignores the field when the frequency of the field is way off its natural frequency. It becomes more responsive when the field frequency approaches its natural frequency, vibrating with increasing amplitudes as it absorbs energy from the nearly resonant field.

The example is trivial; we know the dynamic equation for the oscillator and the form of the external force, so we can directly solve for the oscillator's motion. Most cases are not so simple. The dynamic equations are difficult to solve. More important, often the disturbance is a stochastic process, and we know only its statistical characteristics but not its exact form. The behavior of a system driven by a stochastic process is itself a stochastic process. However, the randomness arises solely from the disturbance; the system structure remains deterministic; a pendulum kicked randomly still obeys its equation of motion. In such cases it is better to break down the problem so that the deterministic and stochastic aspects are treated separately.

A stochastic linear response theory consists of two steps. The first works out the deterministic response of the system to a short and constant impulse; the second takes account of the random variation of prolonged disturbances. However, before the problem can be analyzed into two parts, some basic assumptions must be stated.

The major assumption is that the system responds *linearly* to disturbances, a characteristic rejected by chaotic dynamics. For a linear system, the magnitude of the response is directly proportional to the magnitude of the disturbance, so that doubling a disturbance produces twice as much effect. The secret to the tractability of linear systems is that their responses are *resultant*; the effects produced by several disturbances can be added to yield the total system response.

Consider a time-varying disturbance impinging on a linear system. We can regard the disturbance as a sequence of short impulses, each of which is approximately constant in its duration. The effect of an impulse may linger in the system long after the impulse is over and another impulse is in action. Linearity assures us that the system's response at any time can be obtained by adding the effects of impulses past and present. Thus the problem of finding the system's response to the time-varying disturbance is analyzed into two steps.

We first find out the system *response function* or "propagator," which represents the temporal evolution of the system in response to a single short impulse, or the propagation of the impulse in the system. If the system is characterized by many variables, the response function tells us how each variable evolves. This step addresses the structure and dynamics of the responding system, which behaves deterministically. It contains most of the economic content of the model, describing not only the propagating mechanism but also the impact point of the shock.

The economy is subjected to all kinds of shocks, big and small, singular and indistinct. Its response function, if known, is in principle sufficient to characterize the way it reacts to sharp and singular shocks. The ideas of shock, response, and gradual return to equilibrium are implicit in the description of many historical periods, for instance, the adjustment to the OPEC oil-price hike or to the Federal Reserve's tight-money policy of the late 1970s.

In most situations, the economy is continuously subjected to shocks, and economic fluctuations are the cumulative consequences of the shocks. In such cases, we need to take the second step of the linear response theory, which can take the deterministic or the stochastic route, depending on the nature of the shock. A deterministic linear response theory posits a function specifying the magnitude of the disturbance at each time. The system's response is similarly deterministic. Its state at each time is obtained by integrating over its responses to present and past disturbances.

In stochastic linear response theories, the disturbance is random and represented statistically, such as a random walk or a white-noise process. The behavior of the system is similarly stochastic. We are unable to obtain a path that determines the state of the system at each time, because we do not know the state of the disturbance at each time. However, we can average over the disturbance and response to obtain the statistical characteristic of the system's behavior that conforms to the statistical characteristic of the disturbance. The results of the theory will be given in terms of statistical quantities such as averages, variances, or, as physicists say, line widths.

The Economy to Be Shocked[11]

Economists were so complacent in the prosperous 1960s that they held a conference in 1969 entitled Is the Business Cycle Obsolete? The recessions of 1974–5 and 1981–2 answered with a resounding no. When interest in business cycles revived in the 1970s, the debate revolved around what the proper shocks and propagation mechanisms are. Two broad types of shock are distinguished, monetary and real. A monetary shock is a fluctuation in the supply of money with rippling effects in commodity prices. *Real* in economics means not related to money but concerning such basic economic factors as preferences, technology, resources, government policy, and investment decisions, which influence the demand and supply of goods. Various *real business-cycle models* invoke various kinds of real shock such as government actions and changes in technology and taste.

The propagation mechanisms are more contentious, for they reflect the working of the economy in response to the shocks. New classical economics favors *equilibrium business-cycle models*, in which the market clears in a time short compared to the period of observation, so that it is in equilibrium in each period. New Keynesian economics repudiates the assumption of instantaneous market clearing and argues for propagation mechanisms with various forms of *wage and price stickiness*.

Let us consider the propagation mechanism and response function first. As is common in macroeconomics, the economy is characterized in terms of macrovariables: output, capital stock, investment, consumption, quantity of money, labor supply. The variables are related by the structural equations. In dynamic models, all variables, related as they are, change with time. Under the assumption of homogeneous individuals (§ 25), the economy is conceived as the arena of two big players, Household and Firm, both of which live indefinitely. In some models, Household is decomposed into Cohorts to account for overlapping generations, but we will not consider that complication here.

Household owns labor and capital and derives income by letting them out to the Firm. Now that Household has a temporal horizon, it must take into account both present and future utilities to maximize its expected lifelong utility. Consumption in a period increases the utility of that period but decreases the utility of future periods, for saving and investment increase capital and hence future income. The expected lifelong utility is obtained by summing the utilities of all periods, with future utilities weighted by proper discount factors. Household is Rational; it has a correct model of the economy and accurately foresees the wages, prices, and demand of labor and capital in all future periods. With its perfect foresight, it maximizes its expected lifelong utility under the constraint that its budget is balanced in each period. Thus it decides for each period what amount of labor to supply for the going wages, what portion of its income to consume at the going prices, and what portion to invest. Firm produces output by hiring labor and renting capital at the going wages and prices. It is characterized by a technology factor and a production function, which determine the output for a certain input of capital and labor. Its decision problem is to maximize profit.

When the economy suffers a shock, Household and Firm contribute to the propagation of the shock by their intertemporal decisions. Anticipating fluctuations, Household temporarily changes its behavior to distribute consumption and leisure more smoothly over time. Firm smooths its production by controlling its inventory; when business is brisk, inventory is depleted even with production running at full capacity, and restocking the inventory means high output in the succeeding period also. Thus they carry the effect of a shock into subsequent periods. Their behaviors contribute to various structural equations of the economy.

The dynamic equation of the economy is essentially one of capital accumulation: The capital stock at period $t + 1$ is equal to the properly depreciated capital stock plus the investment at period t. Given the capital stock, the structural equations yield the output and other variables at the same period. Thus the temporal variation of the capital stock is passed on to all variables.

Much depends on the details of the structural equations, which are subject to constant study and debate. Here we consider only a temporal factor. A dynamic equation implicitly assumes a time scale pertinent to the system it represents. Thus the infinitesimal time lapse dt in a differential equation can represent absolute durations as disparate as picoseconds and millions of years, depending on whether the subject matter is molecular motion or organic evolution. Similar scales apply to the duration of periods in discrete time equations. To study business cycles, the time scale must be short compared to the average duration of cycles, which is several years. Usually economic data are reported every quarter of a year, which is also the typical period in business cycle models.

Scales and orders of magnitude cut directly into the objective content of scientific theories. A major point of contention among economists is whether the time scale of the dynamic equations is long compared to the characteristic time for prices and wages to adjust to their market-clearing values. The new classicists answer yes; prices and wages are flexible enough to harmonize the desires of Household and Firm within each quarter. The new Keynesians answer no; the economy contains many institutional structures such as staggered multiyear contracts, which impede the rapid adjustment of prices and wages.

If the characteristic time of price and wage adjustment is shorter than a quarter, the market clears and the economy is at equilibrium in each quarter; consequently its dynamics is a string of successive equilibria. In physicists' parlance, the economy evolves quasi-statically. Whether a process is quasi-static is an objective and not an ideological question. The "price-wage relaxation time to market equilibrium," as physicists would call it, is an objective quantity describing the structure of the economy and the propagation mechanism of perturbations. Its magnitude should eventually be decided empirically and not by ideological principles such as the perfection of the free market.

Technology Takes a Random Walk

The structural and dynamic equations characterizing the economy are deterministic. However, they depend on the technological factor, which has yet to be specified. Besides labor and capital, the output at each period depends on the state of technology in that period, which determines the productivity of the input factors. *Technology* is an umbrella term that includes not only the ability to manipulate things but also the skills to organize humans and channel capital; not only the knowledge of science, engineering, management, and finance, but also the application of the knowledge in concrete situations. Knowledge accumulates, but the application of more advanced knowledge may not always increase productivity. The real world is much more complicated than the idealization of the sciences, so that applications are always made with incomplete knowledge. We may have misjudged the situation in deciding what to apply or neglected some possible deleterious side effects. Pollution resulting from applying some knowledge is an outstanding example, and the effort to clean it up diverts resource from other ends. Certain shifts in the pattern of supply are also collected under the umbrella of technology; supply shifts affect productivity because they may favor certain modes of production that do not match the existing facilities.

On the basis of these considerations, economists argue that technology or productivity is best represented as a stochastic process. However, productivity does not fluctuate like a white noise; time-series analysis shows a strong statistical correlation between the productivity of one period and that of previous periods. Productivity is more appropriately a biased random walk; its value in the period t is equal to its value in $t-1$ plus a constant drift factor and a random shock. In other words, productivity *growth* is a drift plus a white-noise process composed of uncorrelated shocks. The drift accounts for the growth trend in the output, and the shocks account for fluctuations. A positive shock increases productivity growth in one period and productivity permanently.

Economists numerically simulate the random shocks and the consequent economic variables, extract their statistical behaviors, and compare them to the cyclical characteristics of empirical time series. To do so they must first ascertain the variance of the technological shocks. It is found that a random walk with a substantial variance is required to produce the amplitudes of fluctuation characteristic of the U.S. postwar economy. Is it reasonable to assume that technology fluctuates so much?

In a review article, Bennett McCallum said that a "fundamental issue concerns the *nature* of the unobserved random effects that the RBC [real business cycle] literature refers to as 'technology shocks.' If the interpretation is that these are truly technological disturbances, then it is difficult to understand how there could be so much variability at the aggregate level. Different goods are produced by means of different technology processes. . . . Shocks in different sectors should presumably be nearly independent, so economy-wide level variability should be small in comparison to that of any single industry." The standard answer is to invoke events with economywide consequences such as the oil price hikes of 1973 and 1979. However, McCallum argued; "Observations are available on oil prices, for example, so there is no need to treat their changes as unobservable."[12]

Stochastic models depict idealized economies and describe their general responses to random shocks. They answer questions such as whether certain variance of shocks can produce the observed standard deviation of output from trend, the variability of investment and consumption, or the correlation between production and employment. These questions do not apply to a single cycle or a single period, for which stochastic models and statistical explanations are inappropriate. Consequently, the stochastic models do not provide recipes that enable the government to fine tune the economy and smooth out fluctuations.

37. The Survival of the Fittest or the Luckiest?

Wright distinguished three types of factors for evolution: deterministic, stochastic, and unique events.[13] Deterministic processes include recurrent mutation, recurrent migration, and, most important, natural selection. Stochastic processes include fluctuations in the rates of mutation, migration, and selection and genetic drift caused by the random sampling of genetic material in reproduction. Unique events such as the conjectured extinction of dinosaurs partly caused by a meteor are not susceptible to theoretical generalization; they are better addressed in narratives (§ 41).

This section examines various deterministic and stochastic models proposed to explain the observed genetic diversity in natural populations. Since the 1960s, experimental techniques have been available to analyze the structures of proteins and DNA. The data reveal a surprisingly large degree of variation in the molecular level. Genetic variation can be measured by the proportion of polymorphic loci in a population. (A genetic locus is *polymorphic* if it has two or more alternative alleles. An example of polymorphism is the variation in human blood type.) Polymorphism ranges from 10 to 50 percent in natural populations; it is 28 percent for humans. The question then arises; How is this large genetic variation maintained? Why are there so many human blood types?

Genetic variation is generated by mutation, but mutation is not sufficient for variation. To contribute to variation, the mutant must spread in the population but does not completely take over. Mutation occurs infrequently. If the mutant is deleterious, it may be promptly eliminated by natural selection. Even if it is beneficial, it must run the gauntlet of random factors to survive and spread. And if it is successful, it may drive the old allele with lower fitness to extinction and fix itself as the new sole occupier of a genetic locus. All these factors tend to reduce variation. What then causes the observed genetic variation? Does the variation merely represent the overlapping paths of various alleles toward extinction or fixation? Does this explanation agree with the characteristic rates of mutation and molecular evolution? *Molecular evolution* is usually defined as the replacement of a sole allele in a genetic locus by another sole allele. Its rate can be estimated by several methods, such as comparing the genetic makeup of various species. If mutations are like random impulses on populations, what are the propagation mechanisms of the mutants that account for, among other data, both the observed molecular evolutionary rate and genetic variation?

Explanation by Deterministic Natural Selection

Most models of natural selection are deterministic, representing a population by a phenotype, genotype, or allele distribution (§ 17). Given an initial distribution for a population and certain rate constants such as the mutation rate and the fitness coefficients, a dynamic equation determines the evolution of the distribution uniquely. The phenotype, genotype, and allele distributions are definite predicates of the population. They have no probabilistic connotation because they only count heads but do not care whose heads are being counted. The deterministic evolution of the distributions reminds us that the theory is concerned only with the evolution of the population or the species. Individual organisms have accidents; individual mutations occur randomly. However, the theory is not concerned with individuals and takes account of their behaviors summarily in terms of a few rate constants. This is why Wright counted recurrent mutation as a deterministic process. A similar case occurs in the rate equation for the temporal variation of the composition of a large chunk of radioactive material with a constant decay rate or half-life. Both describe the deterministic evolution of statistically characterized composite systems. Neither says anything about the behaviors of the individual constituents of the system, which may be totally random.

There are two schools on the nature of selection. The *classical school* argues that natural selection is mainly purifying. If so, then there would be little genetic variation, for only the fittest allele for each locus eventually survives. The observed degree of polymorphism is far too large to be explained by the passage of the fittest toward fixation. The *balance school* argues there are various mechanisms by which natural selection maintains genetic variation. A once-popular mechanism is heterozygote superiority; if heterozygotes whose two alleles are different have higher fitness than homozygotes whose two alleles are the same, then selection would preserve both alleles. Biologists now agree that heterozygote superiority is rare and contributes little to genetic variation. Other selective mechanisms favoring variety are linkage between genes and diversity of geographic conditions. Variety can also be maintained if the fitness of an allele depends on its frequency in the population, as the fitness of protective mimicry drops when the number of mimics increases. However, it is doubtful whether these mechanisms are strong enough to account for the degree and prevalence of the observed polymorphism.

Stochastic Effects in Natural Selection[14]

In deterministic models, the fitness coefficients are constants or definite functions of environmental parameters. In principle, the environmental parameters change over time, but in practice, the fitness coefficients in most models are independent of time. Thus the deterministic models are valid only for time scales long compared to the rapid fluctuation of weather so that the vicissitude can be averaged out, but short compared to the slow drift of climates so that the environment can be considered fixed.

Drought and flood alternate in a matter of years or decades. In such time scales, natural selection must be treated as a stochastic process. Environmental fluctuations are important for evolution because they help to maintain the variability of the population and safeguard it from being too specialized. If the environment remains the same for a long long time, all except those characters most adapted to it will be eliminated by natural selection. A population so narrowly adapted to a specific environment becomes highly vulnerable to eventual changes. A fluctuating environment favoring some characters in one year and others in the next preserves variation because natural selection acts slowly and seldom exterminates a character in such a short time. The variation will become the base of evolution if the environment shifts permanently.

In stochastic models of selection, the fitness coefficients are treated as stochastic processes to reflect random environmental variations. Consequently the changes in the frequencies of various characters also become stochastic processes. They are driven by the stochastic fitness like Brownian particles being pushed about by molecules. The stochastic models show that environmental fluctuation alone is not sufficient to maintain variation; sexual reproduction is also required. For populations of asexually reproductive organisms, the allele frequencies behave as random walks if the fitness is a white-noise process, and eventually the allele with the maximum geometric mean value of fitness dominates. However, for sexually reproducing organisms, a stationary distribution with finite variance can be maintained. Although sexual reproduction adds complications, stochastic selection shares the basic idea of a stochastic process driven by a random force as discussed in the previous two sections.

For some reason, stochastic models for natural selection driven by a fluctuating environment are not popular among biologists. John Gillespie points out: "This is an area of population genetics that has received relatively little attention over the years."[15]

Genetic Drift and the Neutral Theory of Evolution[16]

In the late 1960s, Motoo Kimura and others argued that the mechanisms considered in mainstream population genetics, notably balancing selection, cannot account for the observed prevalence of polymorphism. Balancing selection also fails to explain the rapid rate of molecular evolution, the constancy of evolutionary rates across species, and the difference in the evolutionary rates for functionally important and unimportant genes. Kimura agreed that natural selection eliminates deleterious mutants and determines the evolutionary course of adaptive mutants; however, he contended that only a minute fraction of molecular changes are adaptive. The vast majority of nondeleterious changes are selectively neutral or nearly neutral, making little if any difference to the performance of the organism. Evolution at the molecular level results mainly from the random fixation of one or another of the selectively equivalent alleles through genetic drift. The bulk of the observed protein and DNA polymorphism is maintained not by balancing of selection but by the balance between random mutation and random extinction. This is the *neutral theory of evolution*, which has stimulated much controversy and much research.

The neutral theory makes *genetic drift* the major mechanism of molecular evolution. Genetic drift occurs because the genetic makeup of the offspring generation is a *random sample* of the possible genetic combinations of the parents. An organism produces many sperm or eggs, each of which contains half of its genes. Only a tiny fraction of the sperm or eggs will fertilize or be fertilized to produce offspring. Suppose that a genetic locus has two alternative alleles A and a, and that the proportion of A is p and that of a is $1 - p$ in the parental generation. Whether an offspring belongs to the genotype AA or Aa or aa is determined in a way similar to drawing two balls from a well-shaken urn containing a fraction of p black balls and the balance white balls. Furthermore, successive drawings are independent; which alleles are inherited by an offspring does not influence which alleles are inherited by its siblings.

Random sampling is well studied in the probability calculus. For infinitely large populations, the law of large numbers implies that, with probability 1, the proportion of each allele in the offspring generation is equal to the proportion of the allele in the parental generation. This is the *Hardy–Weinberg law* asserting a steady-state genetic composition in the absence of selection. It is statistically deterministic if the size of the population is infinite and holds with probability close to 1 if the population is large enough.

The matters are different when the population is not very large, or when a large population is segregated into small reproductive pockets. For a sequence of two billion tosses of a fair coin, the probability is almost 1 that the proportion of heads is 0.5, but in a sequence of twenty tosses, the probability is noticeably less than 1 that the proportion of heads is 0.5. That difference in probability is the source of genetic drift. In small populations, even without selective pressure, the proportions of alleles in succeeding generations are not stable but drift in the manner of a random walk. Eventually one of the alleles meets with the gambler's ruin and becomes extinct, and the other is fixed in the population. The characteristic time of ruin increases with the size of the population; the gambler's pocket becomes deeper, so to speak.

The neutral theory asserts that genetic drift, not selection, is mainly responsible for molecular evolution and the maintenance of genetic variation. It poses a strong challenge to selectionism, one of whose central tenets is that evolution proceeds by the accumulation of minute improvements, because mutations that introduce significant improvements are rare. Neutralists argue that if the improvements are really minute, they will be overwhelmed by random drift. A sculptor who chisels away slowly at a huge rock is unlikely to achieve anything significant when the rock is constantly exposed to sandstorms that polish it randomly. Thus selection's role is either eliminating deleterious alleles or preserving alleles with significant advantages. The image of a petty tinkerer is untenable.

The Limit of Empirical Resolution

A gene is a segment of a DNA strand, which is a sequence of nucleotides. With a few exceptions, each gene encodes the information for a particular protein. A protein is a string of amino acids, the sequence of which is determined by the sequence of nucleotides in its encoding gene. A triplet of nucleotides is a *codon*, or a word in the genetic code. There are four types of nucleotides, making $4^3 = 64$ possible codons. There are 20 types of amino acids, which are coded by 61 of the 64 codons. Since the same amino acid is coded for by several codons, the substitution of one nucleotide in a DNA strand does not necessarily change the amino acid in the protein. Thus the sites in the DNA strand can be divided into *silent sites* that do not change proteins and *replacement sites* that do.

The neutral theory predicts that the silent sites in the DNA evolve several times more rapidly than the replacement sites. The prediction is confirmed by experiments in DNA sequencing. Experiments also confirm that stabilizing selection is operative in the replacement sites, for identical amino acid sequences are produced despite the variation in the silent sites.

The more interesting question concerns the mechanisms responsible for the directional evolution of proteins and replacement sites in DNA. Proteins are essential to body building and perform all sorts of functions in organisms. For instance, the protein hemoglobin carries oxygen in the red blood cell, and a battery of proteins called enzymes serve as catalysts for metabolism. Proteins directly influence the performance of organisms. They belong to the phenotypes of organisms, in terms of which adaptation models are usually framed. They are better for research than other phenotypes because of their susceptibility to more precise definition and experimental study.

A protein is a huge molecule, some parts of which are more important to its function than others. Experimental data show that the functionally less important parts evolve much faster. However, unlike in the case for DNA, here selectionists have a ready explanation. They argue that the rapid evolution in the functionally less important parts is evidence for natural selection's fine-tuning of the adaptedness of proteins. Neutralists countered that the force of fine-tuning is no match for that of random drift.

To gauge the relative importance of selection and genetic drift, contenders try to calculate the rates of evolution according to various processes. Results of calculations are only as good as the parameters that go into the calculations. Stochastic models of neutral evolution depend on parameters such as the average rate of mutation and the effective size of the reproductive population. Deterministic models of selection depend on the average rate of selectively advantageous mutation and the magnitudes of the relative fitness of various amino acids. Unfortunately, the processes at issue – mutation, drift, selection – are all extremely weak. They produce significant results because their effects are accumulated over millions of years. To resolve the conflict between neutralism and selectionism, the fitness coefficients must be determined to an accuracy of one part in a thousand or better. Such accuracies are difficult to obtain in the laboratory and almost impossible in natural environments. Without adequate data, the conflicting hypotheses cannot be tested.

Some doctrines are not testable because they are too vague; for instance, the doctrine that God exists is not testable because anything can be interpreted as supporting evidence. Such doctrines are rightly excluded from the sciences. Some scientific theories are not testable because they run up against the limit of empirical resolution that we are capable of mustering. Such cases occur in all sciences, including physics.[17] In evolutionary biology, the effect of the empirical limitation is magnified by theoretical weakness. Recent studies show that many models with natural selection yield the same general patterns as does the simple model of genetic drift.[18] Thus even a theoretical resolution of the controversy between selectionism and neutralism will need more sophisticated models than are presently available.

According to Lewontin, "Population genetics is not an *empirically sufficient* theory.... Built into both deterministic and stochastic theories are parameters and combinations of parameters that are not measurable to the degree of accuracy required." Twenty-one years after Lewontin's remark, Gillespie said in a review article: "It is stunning how little we have learned from statistical studies of electrophoretically detectable protein variation after 25 years of work. Innumerable attempts to find patterns that give insights into the mechanisms responsible for variation have been almost entirely uninformative. Why is this so?" His answer is similar to Lewontin's: The resolution of the mechanisms requires an accuracy unattainable by the statistical data.

Lewontin also reaffirmed his assessment recently, noting that "the evidence marshalled by these schools [selectionist and neutralist] is virtually the same, for its ambiguity is sufficient to allow both parties to claim it."[19]

38. Causality and Randomness in Statistical Mechanics

We have examined how random elements function in the explanations of economic and biological phenomena. In physics, we will take a broader perspective that encompasses not merely specific kinds of phenomenon but laws covering many kinds of phenomenon. We investigate the role of statistical and stochastic concepts in the derivation and justification of dynamic equations, which are often called laws of physics.

Nonequilibrium statistical mechanics connects the evolution of large composite systems with the motion of their microscopic constituents. The temporal dimension makes its task more difficult than that of equilibrium statistical mechanics, for it must relate not only momentary microstates and macrostates but also the microscopic and macroscopic equations of motion. To achieve the task, physicists have developed an array of dynamic equations connecting microdynamics and macrodynamics.

The vision of nonequilibrium statistical mechanics includes dynamic equations at the micro-, meso-, and macrolevels: *mechanic, kinetic,* and *hydrodynamic.* The hydrodynamic equations are derived from the kinetic equations, which are derived from the equations of mechanics. The derivations are not exact; they contain many conceptual gaps and independent assumptions, including the postulate of stochasticity. The postulates are often justified by heuristic physical arguments such as order-of-magnitude estimations of the speeds of various physical processes. We will examine the role of randomness, the importance of various spatial and temporal scales, the meaning of equilibrium and the relaxation to equilibrium, and the contraction of information as we proceed from mechanics to hydrodynamics.

The dynamic equations in the mechanic, kinetic, and hydrodynamic levels exhibit two salient differences. The two general characteristics of mechanics, *time reversibility* and *determinacy,* are not both satisfied by the dynamics in the other levels. The kinetic and hydrodynamic equations are generally not time reversible. Where and how does temporal irreversibility enter? This question will be considered in § 40. Here we concentrate on the second difference: The hydrodynamic equations are as deterministic as the equations of mechanics, but some equations on the kinetic level are stochastic. Nonequilibrium statistical mechanics presents an intermediate layer of description that requires the notion of randomness, sandwiched between the deterministic descriptions on the micro- and macrolevels. Thus some people say that the motions of microscopic classical particles are governed by deterministic laws but somehow become random on a larger scale. Other people say that macroscopic systems obey deterministic rules but somehow exhibit random phenomena on the smaller scale. How does randomness appear as we go from mechanics to kinetics, and how does it disappear as we proceed to hydrodynamics?

Temporal and Spatial Scales[20]

Although mechanic, kinetic, and hydrodynamic equations apply to the same physical system, they describe different types of behavior. The distinction among them depends on the existence of physical phenomena of disparate scales. A physical interaction has its peculiar spatial scale or characteristic range, a physical process, its peculiar temporal scale or characteristic time. The spatial and temporal scales are not apparent in mathematics. Mathematically, all differential equations describe infinitesimal changes. Physically, an infinitesimal duration in the hydrodynamic regime is an age in the microscopic regime. The difference in magnitudes is as crucial in physical theories as it is in economics and biology.

The microscopic scale is that of particle mechanics. Its characteristic length is the typical range of interparticle interaction, which is on the order of 10^{-8} centimeter. For a gas under standard conditions, the mean particle speed is approximately a few hundred meters per second, so that the characteristic time during which two particles are within interaction range is of the order of 10^{-12} second. In these scales, all information about the particles is necessary to specify their motion, and only mechanics is up to the job. Thus the equations of mechanics are needed to study fast and microscopic phenomena such as the scattering of neutrons and laser light from fluids.

A single collision changes an N-particle system but only slightly. Significant changes of the system occur in a characteristic time that is long compared to the time for particles to be within interaction range. The characteristic length of the mesoscopic scale is the *mean free path* of the particles, which is the average distance a particle can travel without entering the interaction range of another. For a dilute gas, the mean free path is of the order of 10^{-5} centimeter. The time for a particle with average speed to traverse the distance, usually called the *collision time*, is about 10^{-9} second. The kinetic equations, valid only in the scale of collision times, are satisfactory for phenomena with relatively low frequencies and long wavelengths, for instance, the propagation of sound waves.

For the macroscopic scale, the characteristic length exceeds 1 millimeter and the characteristic time for a gas under standard conditions exceeds 10^{-3} second. Descriptions in this gross scale are so crude the behaviors of the system can be characterized in terms of only a handful of macrovariables such as the density, temperature, and velocity of macroscopic fluid motion. This is the scale of hydrodynamics, which is usually sufficient for slow and extended phenomena such as the flow patterns of fluids.

Dynamic Equations in the Mechanic, Kinetic, and Hydrodynamic Levels[21]

Let me describe briefly the general natures of the dynamic equations on the micro-, meso-, and macrolevels before turning to the relations among them. The microscopic equations of motion are well known from mechanics, classical or quantum. We will consider only the classical case; quantum characteristics do not make much difference to the statistical structure, which is distinct from the peculiar quantum indeterminacy. An *equation in mechanics* describes the evolution of the microstate of a composite system by specifying the motion of each constituent particle.

There are several classes of equation in the mesolevel. I often collectively call them *kinetic equations. Kinetic equation* also has a narrower sense that refers to a special class of equation in the mesolevel, the oldest and most commonly used of which is Boltzmann's equation, which describes the evolution of an *N*-particle system in terms of a single-particle distribution and considers only the effect of binary collisions between the particles. The information on the microscopic collision mechanism is encapsulated in a collision cross section, which is calculated in mechanics and handed over to the kinetic equation. The derivation of Boltzmann's equation will be discussed in more detail shortly.

Another class of kinetic equation, called the *stochastic equation*, decomposes the force in mechanics into a slowly varying term and a rapidly fluctuating term. The slow term usually represents some damping mechanism. The fluctuating term represents a random force, which can arise from either the external environment or the internal dynamics of the system. For example, Langevin's equation describes, among other phenomena, Brownian motion and the noise in electrical circuits (§ 35).

A third class of kinetic equation, called the *master equation*, treats the evolution of composite systems as Markov processes. A master equation describes a system in terms of a set of mesostates and the transition rates between two mesostates. The mesostates are obtained from the microstate space of the system by a proper coarse graining, and the transition rates from the equation of mechanics with suitable approximations. The master equation accounts for the evolution of the system in terms of all possible transitions into and out of each mesostate at each period.

More than one kind of kinetic equation is applicable to a certain physical system, but often one kind is more useful than the others. Frequently, certain physical parameters assume small numerical values, which facilitate certain approximations. Two common parameters that can have small values are the density of the system and the relative strength of coupling. Boltzmann's equation is most appropriate for systems with low density, and its generalizations often treat the density as an expansion parameter. Master equations are more suitable for systems with weakly coupled constituents, so that the coupling, which causes the transition from one mesostate to another, can be treated as a small perturbation.

There are several *hydrodynamic equations* and conservation laws suitable for various physical conditions, for example, the Navier–Stokes equations for compressible viscous fluids. Generally, the hydrodynamic equations are partial differential equations with respect to the spatial and temporal variables. They describe the evolution of a composite system in terms of macrovariables such as the average density, average momentum, and average energy, leaving the effect of fluctuations to be calculated on a more refined level. In the hydrodynamic equations, the relevant microscopic information for the specific systems under study is sealed into transport coefficients such as the viscosity, thermal conductivity, and diffusion coefficient. The transport coefficients can be experimentally measured, theoretically calculated in kinetic theories, or for many practical purposes, found in standard tables. In any case they are considered to be parameters given to the hydrodynamic equations.

The hydrodynamic equations contain transport coefficients, which can be calculated from the kinetic equations. The kinetic equations contain scattering cross sections or transition rates, which can be calculated in mechanics. The connection between the mechanic, kinetic, and hydrodynamic equations maintains the conceptual unity among various levels of description and affirms that they are about the same physical system. On the other hand, the fact that an enormous amount of information on one level becomes on the coarser level a few parameters directly measurable in appropriate experiments confers a significant degree of autonomy on the levels.

Randomization in the Derivation of Boltzmann's Equation[22]

To see the meanings of randomness more clearly, let us examine the derivation and solution of Boltzmann's equation; the derivation links it to mechanics, the solution to hydrodynamics. The topic of Boltzmann's equation is the evolution of an N-particle system, where N is something like 10^{23} particles per cubic centimeter. The evolution is in principle governed by the laws of mechanics. Boltzmann's equation distills the information relevant to a wide range of phenomena and represents it in a concise and manageable form. In the process it disregards much of the microstructure of the system, and the approximation is known as randomization.

The difference between mechanics and kinetics is best illustrated in their respective state spaces. The state space of mechanics is the Γ-space, of kinetics the μ-space (see Fig. 3.1). As discussed in § 11, the Γ-space is the microstate space of an N-particle system; each point in the Γ-space is a possible microstate specified by $6N$ dynamic variables for each particle's momentary position and velocity. Given an initial microstate, the $6N$ coupled differential equations of mechanics in principle determine a unique path in the Γ-space representing the evolution of the system. The path is microscopically jagged because of particle collisions. The solution of the $6N$ coupled equations is practically impossible.

The μ-space is the state space of a single particle; each point in it is specified by the 6 dynamic variables for a particle's instantaneous position and velocity. Counting the number of particles with position and velocity within infinitesimal ranges of the position $\mathbf{x}$ and velocity $\mathbf{v}$, we obtain a *single-particle distribution* $f(\mathbf{x}, \mathbf{v}, t)$ for the N-particle system at the instant t. The single-particle distribution gives a statistical description of the N-particle system. Boltzmann's equation governs its evolution, given an initial distribution.

Obviously the single-particle distribution that depends on only six dynamic variables contains much less information than the microstate that depends on $6N$ variables; it neglects much of the structural correlation among the particles. Mathematically, we can concentrate on the behaviors of a few particles or even a single particle in a system by integrating over the variables of all other particles. The question is whether the mathematical procedure is justified by the physical phenomenon under study. They often are not.

Is the representation of an N-particle system by a single-particle distribution justified? Both internal and external factors contribute to the evolution of N-particle systems. Optional external forces such as electric and magnetic fields generate no difficulty, because they act on the particles individually. Internal mechanisms, notably collisions and particle streaming between collisions, are more difficult, for a collision involves at least two particles. Collisions cannot be neglected even if they are infrequent in dilute systems, for they are essential to the transfer of energy and momentum among particles and to the establishment of equilibrium. Only binary collisions are important in dilute systems, for occasions in which three or more particles simultaneously come within the interaction range of each other are much rarer. Myriad binary collisions make the evolution of N-particle systems complicated, just as myriad binary relations make equilibrium composite systems difficult to study.

Boltzmann analyzed dilute systems with binary collisions into two aspects, *mechanical* and *statistical*. He treated the encounter of two particles according to the laws of mechanics, which produce a scattering cross section as a function of the velocities of the incoming particles and the angle of collision. As far as the kinetic theory is concerned, the functional form of the cross section is ready made. It is more concerned with the statistical task, which is to count the number of collisions with each value of the cross section, and to integrate over all collisions to find their effects on the evolution of the system. Since a binary collision involves the relative positions and velocities of two particles, physical considerations require the inclusion of the two-particle joint distribution giving the numbers of pairs of particles, each with various positions and velocities.

The major approximation Boltzmann made is to factorize the two-particle joint distribution into the product of single-particle distributions. The factorization assumes that the two single-particle distributions are *statistically independent*. Boltzmann's *Stosszahlansatz* (assumption on numbers of collision)[23] essentially neglects the statistical correlation between the candidates of collision for all time. Statistical independence is the chief meaning of *randomness* in the probability calculus, so that Boltzmann's ansatz is often called a randomization approximation. Here randomization is a simplification; the system is randomized in the sense that some information on its microstructure and some relations among its constituents are ignored in an approximate theoretical treatment. Randomness appears in coarse-grained descriptions, a result we have encountered in the more mathematical context of deterministic systems (§ 32).

Boltzmann's approximations are partly justified by the distinction between the microscopic and mesoscopic scales discussed earlier. In the scales of kinetic theories, collisions can be treated as occurring instantaneously and collision cross sections as given. Furthermore, the particles are within interaction range for only a minute fraction of the time; mostly they are far apart and exert almost no influence on the behavior of each other.

The Stosszahlansatz has been much criticized because it has no basis in mechanics; it is a statistical assumption posited independently. One criticism is that even if the particles are statistically independent before a collision, they cannot be so after it, for their characters are correlated by various conservation laws in mechanics that govern collisions. The answer to the criticism is that the two particles move apart after a collision and the next collision a particle suffers is with a third particle and not its former partner. Successive collisions with different partners build up multiparticle correlations, which is small in low-density systems. Thus for binary scattering, the statistical distributions of the participating pairs can be assumed to be uncorrelated, at least in a first-order approximation. The approximation is most successful for low-density systems. There are ways to generalize Boltzmann's equation to high-density systems. They all depend on some randomization assumptions that are the generalizations of the Stosszahlansatz.

There are other ways to derive Boltzmann's equation, some of which avoid the Stosszahlansatz but build the randomness into the initial condition. It has been shown rigorously that under certain limiting conditions, if initially the particles are uncorrelated so that the system can be rigorously represented by a single-particle distribution, then the system evolves as described by Boltzmann's equation, at least for a short time. The assumption of statistical independence is crucial for the derivation, although it enters in a different place.[24]

Similar randomizing and coarse-graining approximations are involved in the derivation of other classes of kinetic equations. For instance, the master equations implicitly depend on the factorization of the n-time joint distribution into the product of the one-time distribution and the transition between two times. As Joel Lebowitz said in an article on the foundations of kinetic theories; "To make rigorous the arguments leading from the microscopic evolution to the macroscopic one requires, at present, some amount of stochasticity in the microscopic evolution."[25]

Aggregation in the Derivation of Hydrodynamic Equations[26]

Boltzmann's equation is a nonlinear integrodifferential equation for the single-particle distribution. The nonlinearity and integration come from the collision term, which contains the product of two single-particle distributions and the integration over the velocity and angle of collision. The collision term makes the Boltzmann equation difficult to solve. Exact solutions are almost impossible, but many approximate methods have been developed. The solution describes the evolution of an N-particle system in terms of its single-particle distribution.

Often our main concern is not the single-particle distribution itself but the macroscopic equations and quantities – the hydrodynamic equations and transport coefficients – that follow from it. Thus we try to derive the hydrodynamic equations by directly aggregating over Boltzmann's equation without first solving it. The aggregation transforms the kinetic equation for a distribution to the hydrodynamic equations for several macroquantities: mass density, average momentum, average energy, temperature, pressure, heat flux. Information is again discarded in the process. Unlike in the derivation of the kinetic equations from mechanics, here the loss of information is more appropriately interpreted not as randomization but as averaging. The averaging wipes out whatever randomness is present in the kinetic equations. Thus viewed from mechanics, the randomness of the kinetic description is a simplification and a coarse-graining arising from overlooking certain information on microstructures. From hydrodynamic viewpoints, the randomness of kinetics appears as a complication arising from taking account of more microstructures in a finer-grained description. In either case, the source of randomness is our way of reckoning, not a mysterious force called chance.

The general forms of the hydrodynamic equations turn out to be insensitive to the microscopic details of the system, which can be swept into a few transport coefficients. Two general features of physical systems underwrite the successful derivation and the universal forms of the hydrodynamic equations: the conservation and additivity of certain key dynamic quantities, especially the additivity of the dynamic energy that is crucial in the equations of mechanics.

Consider the quantities characteristic of individual particles. A quantity is *conserved* in a collision if its sum over the incoming particles is equal to its sum over the outgoing particles. According to mechanics, the three prominent conserved quantities are particle number, energy, and momentum. Furthermore, these quantities are additive or *resultant*. Thus the aggregate momentum of the system can be obtained by integrating over individual particle momentum weighted by the distribution giving the number of particles with each value of momentum. The aggregate energy and particle number are obtained similarly. Properly normalized, the aggregate quantities yield the average quantities featured in the hydrodynamic equations. Here we see the importance of resultant properties in linking descriptions on different organization levels.

The totally aggregated conserved quantities for an isolated system in its entirety are not interesting, for they are constants. What are interesting are the *local* aggregate quantities, which are the quantities aggregated over a region in which local equilibrium obtains. The hydrodynamic equations governing the change of a local aggregate quantity can be derived by integrating the kinetic equation weighted by the corresponding individual quantity. The collision term of the kinetic equation, which makes Boltzmann's equation so difficult to solve, either drops out in the integration because the quantity is conserved in each collision or becomes definable as a closed-form transport coefficient.

Besides simplifying the problem, the disposal of the collision term has another important consequence. The microinformation on particle interaction that it carries is now encapsulated in the transport coefficients. Consequently the general form of the hydrodynamic equations depends only on some general features of the underlying equations of mechanics, for example, their additivity. It does not depend on the details of the microscopic interactive structures of the particles. The same hydrodynamic equations govern systems with widely disparate microscopic structures, the difference among which is mainly represented by their different transport coefficients. We have noted in § 23 that many microdetails are irrelevant to the self-organization of large systems. The same holds for their dynamics.

From Local Equilibria to Global Equilibrium

As statistical independence is the important concept in connecting the mechanic and kinetic equations, *local conservation laws* and *local equilibria* are the important concepts linking the kinetic and hydrodynamic equations. A major topic of nonequilibrium statistical mechanics is how a nonequilibrium system relaxes to equilibrium. We have seen in § 9 that equilibrium is not absolute but depends on the scale of the specific description. An isolated system is in equilibrium if its appropriately characterized state is unchanging in the chosen temporal scale. Thus equilibrium fluctuation or quasi-static process is not a contradiction. In equilibrium fluctuation, the microstate of a system fluctuates rapidly in the time scale in which the macrostate is stationary. In quasi-static processes, a system is in equilibrium in the temporal scale of measurement but changes in a longer temporal scale.

Consider a bar of steel heated up at one end and then thermally isolated. The bar is not in equilibrium because it exhibits a temperature gradient. However, temperature is an equilibrium concept; the temperature of a system of particles is defined only in terms of an equilibrium distribution. The apparent contradiction in the temperature gradient is resolved by various spatial scales.

The basic mechanism responsible for the system's relaxation to equilibrium is collision, by which energy and momentum are transferred from one particle to another. In a tiny region with a radius not much larger than the mean free path, a few collisions for each particle can usually bring the system into a *local equilibrium*. The system within the local region is in equilibrium, because it can be characterized by an equilibrium velocity distribution and other equilibrium parameters such as the temperature and pressure, which are approximately constant within the region in times longer than the collision time but short in the hydrodynamic time scale. The equilibrium is local because its parameters differ from those of its neighbors. The approach to local equilibrium falls within the domain of kinetic theories.

Many more collisions and particle streaming are required to transport energy across macroscopic distances to achieve *global equilibrium*. Since mean free paths are on the order of 10^{-5} centimeter, a system a few centimeters across can be divided into numerous local regions. Local equilibrium parameters such as local temperature and pressure and local average quantities such as density, energy, and momentum become the chief variables for the hydrodynamic equations. Hydrodynamics accounts for the way the spatial variations of local parameters are smoothed out in a time scale much longer than the collision time.

Proceeding from mechanics to kinetics to hydrodynamics, the description becomes more and more coarse-grained. The rapidly varying features at the microscopic level of mechanics are averaged out or approximately treated in one way or another, so that only the slower changes apparent on a coarse scale remain in the kinetic equations. The approximations invariably involve some kind of randomization procedure, and there stochastic notions play their vital role. Despite the coarse graining, the kinetic equations still contain a large amount of detail in terms of single-particle distributions. The hydrodynamic equations, valid at even longer time scales, are derived by averaging various dynamic variables over the kinetic distributions. There is a successive contraction of information, as details irrelevant to macrophenomena are filtered out and only gross features are retained. The dynamics of complex systems shares with equilibrium models the basic logic of screening for important macroinformation, as repeatedly emphasized in Part I.

39.   Probability but Not Tychism

The flood of scientific and philosophical books on the applications of the probability calculus that include the word *chance* in their titles indicates how widely the applications are interpreted as the "laws of chance." A scholarly book by six philosophers and historians on "how probability changed science and everyday life" opens with the declaration "Fortuna, the fickle, wheel-toting goddess of chance . . . already ruled a large and growing empire in the sciences."[27]

The probability calculus did originate from the study of the games of chance, and it was once called the doctrine of chance – the *mathematical* doctrine of chance. However, the spread of its application is not the proliferation of casinos. Mathematics typically detaches itself from the substantive content of its origin as it develops; the probability calculus is no exception. The probability calculus is abstract, and the meaning of its applications must be assigned anew depending on the topic to which it is applied. Thus the general interpretation of the sciences that employ the probability calculus as the empire of chance governed by the laws of chance suggests a *metaphysical* doctrine of chance, which Charles Peirce called *tychism*. This section examines two central elements of tychism: probabilistic causation and the interpretation of probability as propensity or objective chance. I argue that neither element has any justifiable claim on the scientific applications of the probability calculus.

I interpret the spread in the application of the probability calculus not as the erosion of determinism by tychism but as the necessity of more than one way of thinking when human knowledge extends to cover increasingly complex phenomena. The general concept of possibility, inherent in the idea of changing things, is shared by the probability calculus and deterministic dynamics in the form of the state space. The probability calculus introduces additional means to partition possibilities into groups and to weigh them, hence to grasp the gross patterns of complex processes. However, it does not apply to situations of genuine uncertainty, the possibilities of which we cannot fathom. It does not shed light on chance, which belongs to the detailed information it overlooks. It does not clarify the ordinary meaning of probability, the vagueness of which tychism exploits. For complex and uncertain phenomena for which details are demanded, we have other ways of thinking such as historical narratives. The limitation of the probability calculus and the ambiguity in the meaning of probability urge the rejection of the metaphysical doctrine of chance.

Probabilistic and Statistical Concepts

I begin the section by briefly reflecting on the meaning of probability and the peculiar logical form of probabilistic assertions, which I have discussed more fully in an earlier work.[28] As discussed in § 35, the only meaning of probability in the probability calculus is a relative magnitude ascribed to a group of individuals sharing a common character, for instance, the relative magnitude of the group of poker hands that are flushes. When the character at issue varies systematically over the value of a character type, the corresponding relative magnitude generates a distribution. The technical meaning is used in all of the statistical theories we have examined. I have avoided calling relative magnitudes "probability" because that usage differs from the ordinary sense of the word.

"Of all possible poker hands, 0.198 percent are flushes" is an objective statement about the relative magnitude of the class of poker hand that is a flush; it conveys no sense of risk or chance. In real life, we are not so much concerned with classes of poker hand as with *individual* hands. We are interested in the probability that *a* poker hand or *this* poker hand is a flush. *Probability* in ordinary usage is about particular cases, not classes of cases; it is on particular cases that we risk our bets and take our chances. It is well to think of all possibilities, but each of us has only one life, and its particularity is what finally counts. We can and do use the values of the relative magnitudes provided by the probability calculus to assess risk, saying that the probability of getting a flush for this hand is 0.00198. However, the utilization of the values in the probabilistic assertion is an additional step of reasoning that is not generally transparent or correct, as when the cards are stacked.

"The proportion of possible poker hands that are flushes is 0.198 percent" is a statistical proposition. "The probability that this poker hand is a flush is 0.198 percent" is a probabilistic proposition. Both propositions invoke the classification of individuals into groups according to various universals such as "being a flush" or "being a straight." However, statistical and probabilistic propositions differ in subject matter, logical form, and empirical verifiability. In the statistical statement, *that* introduces an adjective phrase "being flush" qualifying the subject "poker hands." In the probabilistic statement, *that* introduces a complete proposition, "This poker hand is a flush."

A *statistical statement* is a subject–predicate proposition about a special kind of composite individual. Its subject is the group made up of individuals picked out by a universal and treated as a unit, for example, the group comprising all roses in the garden that are red. Since the universal functions as the independent variable, it is sometimes mistaken for an individual, for instance, as the immortal gene in the vulgarization of population genetics (§ 12). No, the proper subject of a statistical proposition is not a universal but a particular, the particular group of individuals singled out by the universal and treated as an individual. The predicate ascribed to the subject of a statistical statement is not an ordinary universal but a relative magnitude such as 0.198 percent. Since the magnitudes are relative, their only meaningful ascription is to parts of wholes. Statistical propositions are empirically verifiable, for instance, by actual counting. The basic form of statistical statements is preserved in distributions, which are systematic summaries of many statistical statements.

Probabilistic propositions are not as clear-cut, for the meaning of probability is obscure. Since my purpose here is to distinguish them from statistical propositions, I will concentrate on the form that is closest to statistics. Unlike a statistical proposition, a *probabilistic proposition* is no ordinary subject–predicate statement, because it has *two* independent variables: a particular such as "this poker hand" and a universal such as "being a flush." A probabilistic proposition appraises how likely the particular is to be part of the group picked out by the universal from a whole, or, more briefly, how likely the particular is to be an instance of the universal. Probabilistic propositions are not empirically testable, even in principle. When the cards are dealt, the poker hand is either a flush or not, without any sign of probability. When we examine the result of many hands, we have quietly switched from probability to statistics and changed our subject matter. The untestability greatly limits the utility of probabilistic propositions in scientific theories.

The ordinary meaning of probability in relation to a particular individual, on which the idea of chance rests, is used in statistical inference, which tries to determine the characters of a large system by sampling small parts of it. A sample is a particular. Statistical inference belongs to the experimental side of science. On the theoretical side, probability is sometimes used in the sciences we examine, but mostly in interim steps or heuristic reasoning. For probability concerns individuals, and explanations about individual behaviors are precisely what the theories employing the probability calculus eschew. The theories do not attempt to answer why a particular Brownian particle is in a certain position, why a particular allele is extinct in ten million years, or why a particular economic downturn has a certain severity. Instead, they raise a different sort of question and opt to explain patterns of behaviors of many individuals: the spatial distribution of many Brownian particles, the relative abundance of many alleles, or the general characteristics of business cycles. These explanations are statistical and not probabilistic, and they do not involve the idea of chance.

In the sciences we investigate, the probability calculus is often used to partition and quantify the state space of a system. The idea of possibility provides some link to the ordinary meaning of probability. However, the link does not imply the idea of chance, For possibility is a general concept in our categorical framework and functions equally centrally in deterministic dynamics. Furthermore, the probability calculus applied to state spaces yields statistical assertions; we are still talking about the relative magnitude of classes of possible states, not individual possible states.

Statistical theories employing the probability calculus give coarse-grained descriptions of composite systems that overlook much information about individual behaviors. Probabilistic propositions express our desire to know some of the missing information. Since the information is unavailable from statistical theories, probabilistic statements appear as an *appraisal* of propositions such as "This hand is a flush." Unlike statistical propositions, probabilistic propositions do not belong to the object language, in which we talk about objects and the world. They belong to the metalanguage, in which we talk about propositions and theories.

Tychism

The word *Tychism* was coined by Charles Peirce in 1892 for the doctrine of absolute chance, a foundational element of his evolutionary cosmogony. Two of Peirce's central ideas are relevant to the interpretation of the applications of the probability calculus and are developed in various ways by later philosophers, although they do not always refer to Peirce or tychism. The first is Peirce's rejection of causal determinacy in the fundamental sense. Empirical results never match exactly the corresponding results derived from causal laws. Trace the causes of the discrepancies as far back as you please, Peirce contended, "and you will be forced to admit they are always due to arbitrary determination, or chance." Similar ideas underlie the later theories of *probabilistic causation* and *probabilistic explanation*, in which causes are represented solely in terms of probability. Another aspect of tychism is the interpretation of probability as an innate property of things. Peirce defined the meaning of the statement that a die thrown from a dice box has a certain probability of showing a specific number: "The statement means that the die has a certain 'would be'; and to say that a die has a 'would be' is to say that it has a property, quite analogous to any *habit* that a man might have."[29] The interpretation of probability as the *propensity* or the dispositional property of "chance-setups" is winning an increasing number of adherents.

Allow me to frame tychism in terms of the concepts discussed in § 32. Deterministic dynamic systems can be described in two levels of detail: a fine-grained account of state succession according to dynamic rules and a coarse-grained statistical account of types of states. Not all dynamic systems have dynamic rules. Many, notably those undergoing stochastic processes, have only statistical descriptions. Most scientists acknowledge that some detailed information is missing from the statistical descriptions, although the information need not pertain to deterministic dynamics; we simply do not know what we have missed. In contrast, tychism asserts that the statistical descriptions are complete and fundamental, for they account for the working of objective chance as the propensity, tendency, or dispositional property of the systems. There can be no deterministic dynamics underlying the statistical distributions. The dynamic rules we happen to have are superficial and crude descriptions or theoretical apparatus that help us to organize the distributions mentally. Generalizing these ideas to the metaphysical level on the same footing as determinism, tychism asserts that causation is not definite but probabilistic.

Probabilistic Causes and Indications of Causes[30]

Probabilistic causation was proposed by several authors. One version, the Granger cause, has been discussed in the context of noneconomic models of business cycles (§ 36). Suppes developed a probabilistic metaphysics, which contains many interesting and I think correct propositions. However, it also contains two propositions I find objectionable: "Causality is probabilistic, not deterministic, in character." "The fundamental laws of natural phenomena are essentially probabilistic rather than deterministic in character."[31] As metaphysical propositions, they come close to tychism.

Suppes defined *probabilistic causation* as follows: Event B is a prima facie cause of event A if B precedes A and the conditional probability of A given B is greater than the unconditioned probability of A. Prima facie causes can be spurious; it is more likely to rain if the barometer falls, but falling barometers do not cause rain. To test for spuriousness, the causally relevant situation is maximally and exhaustively partitioned into mutually independent factors $\{C_i\}$. A prima facie cause B is genuine if and only if for each i, the conditional probability of A given B and C_i is greater than the conditional probability of A given C_i alone. For example, there is statistical evidence that the probability of an inoculated person's escaping an attack of cholera exceeds the probability of a not-inoculated person's escaping it. Therefore inoculation is the prima facie cause of immunity from cholera. Spuriousness can be tested by exhaustively listing all factors relevant to the sickness: health, age, diet, hygiene, and so on. If the statistical analysis of these data does not refute the statistical correlation between inoculation and immunity, then inoculation is accepted as a genuine cause of immunity.[32]

Let us put aside for a while the meaning of probability in the definition. The important feature of the definition is that causes are defined solely in terms of probabilities or statistics without invoking underlying mechanisms or dynamics. This feature is shared by all versions of probabilistic causation and distinguishes them from the ordinary notion of causation. An ordinary causal account of immunity tries to identify the processes and the interactions among vaccine, body, antibody, and bacterium. The body contains millions of kinds of antibody, whose job is to destroy alien organisms. It takes a specific kind of antibody to fight a specific kind of invader. Immunity works because the vaccine stimulates the profuse replication of the right kind of antibody for defense, and these antibodies remain in the body for years, ready to respond immediately to the next invasion. These mechanisms do not figure in probabilistic causation.

Why are inoculated persons immune to cholera? I think most people would find unsatisfactory the answer "because inoculated persons have a higher probability of escaping the disease than not-inoculated persons." They would insist, That we know, but *why*? The "because" in the explanation in terms of probabilistic causation sounds like a fraud; what it introduces is not an explanation but an indication of what needs a causal explanation.

Probabilistic causation is sometimes used in econometric analysis. However, economists are careful to call it "Granger cause," for they agree there is a more legitimate notion of causation. Of the three approaches in modeling business cycles presented in § 36 – in terms of Granger cause, stochastic processes, and deterministic dynamics – Granger causation alone yields no information on how the economy works or how to modify economical behaviors. Granger causation is not fundamental.

Medical cases, the most commonly cited examples of probabilistic causation, are complicated situations where details matter and the particular is paramount. Here underlying processes are often poorly understood, and statistical data are often effective in indicating possible causes. Responsible scientists and physicians who make causal claims based solely on statistics are as careful as if they are walking on breaking ice in a spring thaw. When they say something is a probable cause, *probable* refers to their own hesitant attitude, which is different from the probability in probabilistic causation. More often than not, the claims of statistics and probable cause are contested by other experts. Witness the debate on whether smoking causes lung cancer, which is one of the less controversial cases.

Proponents of probabilistic causation argue that deterministic causal accounts are infeasible or impractical for complicated cases. True, but the same complexity also forestalls the feasibility of genuine probabilistic causal accounts. Prima facie probabilistic causes have little weight. To find genuine probabilistic causes, we must determine the exhaustive causal background of the effects and maximally partition it. Even granting that the delineation of the background has not already presupposed some ordinary notion of deterministic causation, the exhaustive delineation and partition of causal backgrounds are still assumptions no less speculative than underlying dynamic rules. For what make the situation complicated are the interaction and interference of the factors. Without proper tests for spuriousness, it is all too easy to mistake spurious probabilistic causes for genuine, for instance, to conclude from the statistics of intelligence tests that the race of a person is the cause of inferior intelligence. If the situation is so complex we can only manage to see indications, it is better to acknowledge them as indications and refrain from making causal claims.

That a causal theory uses the probability calculus does not imply that it supports probabilistic causation. The central question is whether the causal statements made by the theory have the logical form of probabilistic causation. Our investigation of the sciences shows that they mostly do not. The statistical notions in many scientific examples describe ensembles of systems or perhaps classes of possible states of a system, whereas the probability in probabilistic causation pertains to a single system or a single state of the system. Theories of stochastic processes overlook the detailed evolution of individuals and address a set of questions that do not apply to individual paths. How can the conditional probabilities for various states of individuals be derived from answers to questions that have omitted them?

When many causal factors contribute to a complex situation, it is not enough to pick out a factor that probabilistically causes, or that has something to do with, an event. We also want to know how strongly the cause influences the event, as compared to other competing or complementary causes. Probabilistic causation dictates that causal analysis can only be made statistically. Statistical analysis is extensively used in biometric, econometric, and other social sciences. If there is a common conclusion, it is the difficulty of teasing out statistically the contributions of various causal factors. The protracted controversy on the relative importance of natural selection and genetic drift in organic evolution is a testimony to the difficulty (§ 37). In an article on statistical methods in biological research, Lewontin explained the ineptitude of the methods in separating causes. He gave detailed examples to illustrate how the statistical analysis of variance gives completely erroneous pictures of the causal relations among genotype, phenotype, and environment. And he warned about conceptual confusions, "What has happened in attempting to solve the problem of the analysis of causes by using the analysis of variation is that a totally different object has been substituted as the object of investigation, almost without our noticing it."[33]

Chaotic dynamics is often cited to support the need for probabilistic causation. Chaotic dynamics cuts both ways. Bernoulli systems produce results in the coarse-grained description phenomenologically indistinguishable from those generated by roulette wheels (§ 32). This can support the argument that deterministic systems display random behaviors; it can also support the argument that apparently random phenomena obey underlying deterministic rules. The underlying rules are not necessarily empirically inaccessible, for we are not living within a logical black box and receiving numbers through a slot. We live in the world and know that we are experimenting with a billiard ball and not a roulette wheel. The contextual knowledge includes much information not found in the abstract formalism of Bernoulli systems. It enables us to connect the parameters of models, which contain much physical content, to other factors. We can change the parameters of the billiard to produce statistically measurable results. Such experiments exemplify the importance of the dynamic rule.

Probabilistic Explanation[34]

Probabilistic causation is closely associated with probabilistic explanation, which illustrates more clearly the deep conceptual confusion involved in probabilistic statements. The most famous philosophical theory of scientific explanation is the *covering-law model of explanation*, which also plays a significant role in the argument for reductionism and microreductionism. Even in its besieged state today, it still serves as the starting point and target of attack for its many rivals, none of which has enough strength to claim to supersede it. Identifying explanation with prediction and logical argument, the covering-law model asserts that to explain a proposition is to derive it from certain universal laws and statements about certain antecedent conditions. For instance, from Newton's law and an antecedent position of a planet we can explain or derive or predict its subsequent positions.

The ideas of probabilistic explanation and causation arise naturally in the empiricist philosophy of science. Dynamic and statistical descriptions have their own notions of determinacy. For one, deterministic means evolving according to a dynamic rule; for the other, it means statistically deterministic or predictable with probability 1 if given the entire past statistical record (§ 35). Since causal laws in the Humean tradition mean no more than the regularity of succession, both versions of determinacy can be interpreted as laws.[35] Both are good for prediction and hence explanation. Once statistical determinacy stands on its own as causal laws, the way is paved for the transition to probabilistic causation and explanation that relax the requirement that the regularity holds with probability 1, for the conceptual structure of statistics and probability is already in place.

Carl Hempel, who developed the theory of explanation by universal laws, followed it with a theory of probabilistic explanation by statistical laws. A *probabilistic explanation* has two premises: First, a statistical law asserts that under certain conditions, the proportion of events of type A in events of type B is r, a value very close to 1; second, a is a case of B. The two premises explain with a probability r that a is a case of A. The meaning of probability here is conjectural; it represents the credibility of the conclusion that a is A.[36]

I distinguish two kinds of probabilistic explanation and call Hempel's definition the first kind. Because of the ambiguity in the meanings of "statistical law" and "probability," it is often confused with a second kind of probabilistic explanation, which is far more prevalent and closely related to probabilistic causation. As discussed in § 35, many theorems of the probability calculus, for instance, the laws of large numbers, involve three kinds of entity – individual, system of individuals, and ensemble of systems – each with its own technical probability. *Probabilistic explanations of the first kind* invoke the statistical regularity of an ensemble to explain the behaviors of its systems. *Probabilistic explanations of the second kind* invoke the statistical regularity of a system to explain the behaviors of its individuals. The first kind of explanation is supported by the major results of the probability calculus; the second kind is not.

The strong law of large numbers uses the statistical feature of the ensemble to make a statistically deterministic statement about a certain character of infinitely large systems (§ 35). It inspires and supports probabilistic explanations of the first kind, in which the explanatory statement must hold with probability close to 1. This kind of probabilistic explanation does have an important role in scientific reasoning, mainly in the establishment and justification of statistically deterministic laws. For instance, it is used in the justification of the second law of thermodynamics. Here the second law of thermodynamics is explained, and the statistical law invoked in the premise of the explanation is the mathematical law of large numbers applied to an abstract construction of hypothetical ensembles. The law of large numbers ensures that the statement of the second law holds with probability 1 for infinitely large composite systems, and the statistical determinacy justifies the status of the second law as a statistical law (§ 40). Hempel argued that the probability involved in the explanation refers to the credibility of the explained law. Credibility is not part of objective statements; rather it is the evaluation of such statements. Thus probabilistic explanations of the first kind are more appropriately viewed not as accounts of causal factors but as statistical justifications of laws or of approximate conceptualizations of regularities.

There are few examples of probabilistic explanations of the first kind, for they apply only to regularities of large systems that can aspire to the status of laws, and laws are scarce. Most examples of probabilistic explanations offered in the philosophical literature, including Hempel's own writing, are of a different kind. Probabilistic explanations of the second kind explain not laws or lawlike regularities but singular events. Their explaining statistical laws are usually not theoretical but empirically established statistical regularities more appropriately called empirical generalizations than laws. They are not specific about statistical determinacy and place no restriction on the value of probability, which is often significantly lower than 1. In one of Hempel's examples, a "law" asserts that a large percentage of persons exposed to measles contract the disease; Jones is exposed; therefore it is highly probable that Jones will have measles. The probability calculus sheds no light on the probabilities of individuals, which it takes as premises. Similarly, the statistical sciences seldom if ever explain the behaviors of individuals; their concerns are not individuals but systems. Thus most examples of probabilistic explanations, which belong to the second kind, are not explanations given within the sciences. Instead, they borrow scientific results on systems to make educated guesses about individual events, for instance, Jones's measles. These examples greatly confuse the definition and meaning of probabilistic explanation.

Wesley Salmon relaxed the restriction on statistical determinacy and introduced conditional probability, which was used earlier by Hans Reichenbach in the context of causation, into the definition of probabilistic explanations. The event B is an explanation of an event A if the conditional probability of A given B is not equal to the probability of A. The move closes the gap between probabilistic explanation and probabilistic causation. Both are problematic, as the preceding analysis shows.

The Meanings of Probability[37]

The preceding discussion presents the logical structures of probabilistic causation and explanation but leaves the meaning of probability murky. Without a clear explication of the meaning of probability, the logical theories are like castles built on sand.

There are a dozen or so conflicting interpretations of probability. A major school of interpretation, called the frequentist, identifies the concept of probability with relative frequency in various ways. The frequentist interpretations differ crucially from the technical definition of probability as a relative magnitude. Consider sequences of coin tosses. The proportion of heads varies from sequence to sequence and the probability calculus takes account of the statistical variation of all possible sequences. The variation is not admissible in frequentist interpretations, for which the probability of heads is a fixed value. Thus frequentists often identify probability with the limiting frequency for infinite sequences and hence open a whole new can of worms regarding infinity. The interpretation leaves in obscurity the meaning of probability in most practical usages related to finite or singular cases. It renders meaningless the common question we ask the doctor, What is the probability of my catching the flu? I reject the frequentist interpretations because they conflate statistical and probabilistic concepts and violate the ordinary meaning of probability. Proportion or relative magnitude is a clear and objective concept; there is no reason to muddle it with the name of probability.

As is typical with mathematics, the applicability of the probability calculus depends on the subject matter and modeling. Nothing in principle prevents its application to abstract quantities such as the magnitudes of patterns of beliefs or the magnitudes of evidential supports for hypotheses. Indeed it has been so applied, for instance, in the attempt to build an inductive logic. In these applications, the meaning of *probability* is strictly speaking statistical, because it is a relative magnitude. But as the magnitude of the strength of belief or evidence, it is naturally interpreted in terms of an ordinary meaning of probability. *Probable* and *probably* are commonly used to express the uncertainty or the tentativeness of judgments, as we say, "It is probably true" or "It is probable that such an event happened." Probability quantifies the tentativeness. In common usages such as "The probability that he will recover from cancer is 70 percent" or "The probability that the second law of thermodynamic is valid is 1," probability is a metatheoretic notion that evaluates a theoretical proposition about the objective world. This is the sense of probability that I have been using, as distinct from relative magnitudes.[38]

As a metatheoretic concept, probability operates on a different logical plane from objective concepts that represent the objective world and does not mingle with them. The interpretation puts probabilistic causation in a curious position, for causation should be objective and ascribed to the world. Suppes said that probability in probabilistic causation can assume one of at least three meanings: as a theoretical notion, as relative frequency, and as degrees of belief.[39] In my presentation and criticism of probabilistic causation, I have kept to the technical definition of probability as relative magnitude, which I think is the most advantageous to probabilistic causation. The objection to defining causation in terms of probability would be stronger if probability is the degree of belief. It smacks of faith healing.

The Propensity Interpretation of Probability[40]

Probabilistic causation seems to demand an interpretation in which probability represents some realistic feature of the world. Such a demand is met by the propensity interpretation, which was adopted by Salmon to increase the explanatory power of probabilistic causation. The idea of factors that incline but do not necessitate was an important part of the medieval debate on divine predetermination and the possibility of evil and free will. The idea was revived by Peirce and vigorously promoted by Popper, who interpreted probability as propensity. A system's *propensity* is its inherent property that generates the characteristic statistical patterns of outcomes when the system is engaged in repetitive experiments. For instance, the propensity of a fair coin is physically responsible for the pattern of roughly 50 percent heads when the coin is repeatedly tossed. Propensitists scorn identifying probability with the results of infinitely repetitive experiments, which is maintained by their archrivals the frequentists. They want to ascribe probability to a single throw of a die. Hence they articulate propensity in counterfactual statements, virtual experimental results, or results in possible worlds. There is no lack of mathematical models and axiomatizations of propensity. What is lacking is the connection between the hypothetical world or worlds in which propensity is manifested and the actual world. The lack is fatal because propensity is not a mere idea but is posited as a physical property of things.

The idea of propensity has an anthropomorphic appeal that is apparent in Peirce's metaphor of men's habits. Beyond that its meaning is obscure. The propensity of a coin is something above and beyond its uniform density, its flat and parallel faces, and the ways it interacts with other physical elements. What is that something? Propensitists cannot even agree to the entity for which propensity is a property, the coin alone, or the coin and whatever tosses it, or the coin and its surroundings. They also disagree on what counts as "chance set-ups" that possess propensities, for instance, whether they include deterministic systems. Many tychists argue that propensity is absent in deterministic systems, although deterministic systems have the strongest "habit" of yielding the characteristic results in repeated experiments.

Propensity is posited as a dispositional property, but it differs essentially from usual dispositional properties. Dispositional qualities such as soluble and flammable stand apart from occurrent qualities such as green or massive. Ordinarily, a dispositional quality is a shorthand expression of the idea that a system has certain possible occurrent states under certain conditions, as solubility refers to the possible state of being dissolved, which can be realized under certain conditions. This interpretation does not apply to propensity, which does not refer to any possible state or collection of states of the system that can be realized. It does not refer to outcomes of either single or repeated experiments, for it is a fixed quality, whereas actual outcomes vary.

Dispositional qualities for physical systems usually have deeper explanations in terms of the structures of the systems, as the solubility of salt is explained in terms of its atomic structure. Propensity has no such explanation. Although propensity is supposed to generate actual statistical patterns, there is no account whatsoever of how it generates them. A tychist may even argue that an account of how propensity works is impossible in principle. For if the account uses probabilistic concepts, then we must further explain how the propensity represented by the concepts works, and we land in an infinite regress. If the account does not use probabilistic concepts, then the mechanism is deterministic and cannot be responsible for the working of propensity. If the hand of chance is susceptible to deterministic explanation, it is not chance anymore. Since propensity is never realized and no account of any mechanism links it to observable effects, it cannot be experimentally tested. Like the existence of God, the existence of propensity is not falsifiable. Like the idea of God, the idea of propensity can be deleted from our scientific theories without making any difference in the explanation of natural phenomena.

Of What One Cannot Speak, One Must Remain Silent

The vague commonsense idea of chance can be incorporated into our conception of the complicated world or even complex deterministic systems without positing mysterious properties or causes. We often call unpredictable events chancy, and unpredictability is compatible with deterministic dynamics. Even the behavior of an ideally isolated deterministic system can be unpredictable if the system is chaotic.

Deterministic dynamics is possible only for ideally isolated or carefully controlled semiclosed systems in which the idea of "total cause" as discussed in § 34 makes sense. However, in the real world, isolation is often invalid or a poor approximation. The impossibility of isolation and the inevitable influence of contextual factors are other sources of randomness and the idea of chance. The Brownian particle moves erratically not because of its innate disposition but because it is pushed by surrounding molecules.

Our theories and models, however comprehensive, represent only small fragments of the world. If a model is not to be worthless, the factors it excludes must be minor and intrude only infrequently. In linear systems where the magnitudes of causes and effects are proportional, the effects of the neglected minor causes are similarly minor and can be overlooked. In nonlinear or unstable systems, minor causes can have large effects, and an occasional intrusion of minor external factors can wreak havoc in a system. We habitually call the intruding factors contingent, regardless of the feasibility of expanding the model to encompass them deterministically. Consider a pendulum with its bob resting at the vertically upward direction. The equilibrium state is unstable, and a slight perturbation will send the bob swinging down. We say it is a matter of chance whether it will swing to the right or the left, because the perturbation is not included in the dynamic theory. In this simple case the theory can be expanded to include the perturbation; there are many theories of instability. Of course the expansion is often infeasible in more complicated cases, but its possibility in the simple case suggests that the contingent is not absolute but is relative to our models.

The idea of chance is also relevant in ideally isolated systems. Consider the mechanic, kinetic, and hydrodynamic descriptions of a many-body system (§ 38). The evolution of the system is the result of a large number of minute causal factors pulling in all directions. The physical systems are homogeneous and stable enough that the probability calculus and its statistical concepts enable us to average over fine details and distill gross patterns of change. Since the statistical descriptions are coarse-grained and pertain to systems as wholes, they leave unanswered many questions about the features and behaviors of individual constituents. We call the individual behaviors that are left out of the description random or chancy.

In sum, what we usually call chance is what is left in the dark by current or anticipated theories. Science cannot be expected to explain everything in all levels of organizational detail, not even everything in physics. The sciences answer only those questions that can be framed within their respective theories, letting many questions fall through the cracks separating the theories. It is well to have the vague idea of chance to remind us that our scientific theories are far from complete and probably will never totally explain the world in all its detail. This is the valuable insight of Suppes's probabilistic metaphysics. Suppes rightly argued that the certainty of knowledge is not achievable, the sciences are pluralistic, and scientific theories do not converge on a fixed result giving us complete knowledge of the universe.[41]

Like test pilots, philosophy pushes the outside of the envelope in its search for fundamental concepts; like them, it crashes if it pushes too far. The idea of chance should be allowed to remain vague because it is where scientific theories fall silent. Tychism with its laws of chance and innate propensities goes too far and lands in metaphysical illusion. Theories employing the probability calculus illuminate a type of phenomenon different from that illuminated by deterministic dynamics. They provide coarse-grained explanations about gross patterns in complex systems and preclude questions about many details. They do not illuminate the questions falling into the cracks vaguely known as chance. For these questions probabilistic causation and probabilistic explanation do not help either.

A great deal of knowledge and many substantive concepts are required for a model to employ the probability calculus successfully. The requirement cannot be met for many complicated and uncertain cases; for that reason many people disagree that the covering-law model of explanation, including probabilistic explanation, is applicable to history (§ 41). Keynes explained that uncertainty is not merely a distinction between what is known for certain and what is only probable. The game of roulette is not uncertain, nor are situations where distinct possibilities can be posited and probabilities assigned to the possibilities. Uncertain knowledge refers to matters such as the prospect of a European war. "About these matters there is no scientific basis on which to form any calculable probability whatsoever. We simply do not know."[42]

Quantum mechanics, the favorite of the metaphysicians of chance, is actually an example of the impotence of probabilistic explanation. Briefly, quantum mechanics contains two kinds of descriptions linked by the idea of probability: a deterministic description of a single quantum system and a statistical description of the classical quantities resulting from the measurement of an ensemble of systems. The classical description is necessary because we human beings can only measure classical quantities and without experimental support quantum mechanics would be empty. The bulk of quantum mechanics concerns the quantum description. Steven Weinberg said: "Quantum mechanics in fact provides a completely deterministic view of the evolution of physical states. In quantum mechanics, if you know the state at one instant, and of course if you know the machinery of the system, you can then calculate with no lack of precision what the state is in any subsequent instant. The indeterminacy of quantum mechanics comes in when human beings try to make measurements."[43]

Quantum quantities such as the wavefunction, propagator, coupling strength, scattering amplitude, and path integral, the calculation of which absorbs most effort of quantum physicists, are deterministic. Quantum physicists would be helpless without some proficiency in differential equations, but they can do research comfortably without any training in the probability calculus, for quantum mechanics invokes the notion of probability in the most peripheral and naive manner. The probability of a single system enters as a simple definition and is immediately converted into the statistics of classical data on an ensemble of systems. The meaning of probability in quantum mechanics is as obscure as the phrase "the collapse of the wavefunction." The cause of the "collapse of the wavefunction" is usually associated with observation, which is not covered by quantum mechanics. What is left in the dark by quantum mechanics is not illuminated by probabilistic causation or explanation; the probability calculus is explicitly shown to break down here. What happens in the measurements that forces us to invoke the idea of probability, we simply do not know.

Quantum mechanics contains a conceptual gap between the quantum and classical descriptions of the world. Perhaps the gap will be filled by some future solution to the quantum measurement problem that explains how the measured classical quantities materialize. Meanwhile, quantum mechanics is like Oriental paintings, where an artist does not try to fill every inch of the canvas but leaves blanks as an integral structure of the art. Similar blanks are found everywhere in our knowledge. Philosophy would do well to refrain from filling in the substantive concepts left out by the sciences with metaphysical speculations. We have to learn to live without the illusion of a complete substantive account of the world, which deterministic dynamics plus the probability calculus cannot supply. Instead of building substantive models for our theories, it is more important that our categorical framework be broad enough to include the general idea of the incompleteness and to permit the intellectual frankness expressed in Feynman's remark "I think I can safely say that nobody understands quantum mechanics."[44]

Directionality, History, Expectation

40. Causality and Temporal Asymmetry

Mechanical processes lack a preferred temporal direction. Directionality appears in thermodynamic processes, which nevertheless have neither past nor future. Evolution is a historical process in which past changes are frozen into the structures of present organisms. Economic agents use the capital accumulated from the past and determine their action according to their expectations of the future. As dynamic processes become more complicated, so do the corresponding concepts of time.

There are hardly any phenomena more obvious than the temporal asymmetry of almost all processes we live through and observe. Moving things slow because of friction but do not spontaneously accelerate; we have memories of the past but only expectations for the future. Does the temporal asymmetry merely reflect our own life process and the processes familiar to us, or does it have a more objective grounding?

This section considers the objectivity of a global temporal direction. A particular kind of irreversible process has a temporal direction, but there are many kinds of irreversible processes. Physicists alone talk about at least four "arrows of time": quantum, thermodynamic, electromagnetic, cosmic. More arrows occur in other sciences. The relations among the various temporal directions are not clear. Can they be harmonized objectively and not merely conventionally into a global temporal direction, which can be briefly called the direction of time? Failing that, can we find some kind of pervasive and irreversible process to serve as a standard direction? The result is not encouraging. Most elementary physical processes are temporally symmetric, and the second law of thermodynamics asserting the monotonic increase of entropy contains too many loopholes to do the job. We may have to admit that the temporal direction with which we comprehend the processes in the world is a convention, albeit a reasoned one.

Causality, Common Cause, and Temporal Asymmetry

Temporal asymmetry and directionality have attracted much philosophical attention.[1] Often a substantive time over and above temporal phenomena is assumed, so that people distinguish between the asymmetry of time and the asymmetry in time. I reject the substantival time existing independently of things, events, and processes (§ 26). Therefore I will not talk about the asymmetry of time itself but will consider only the temporal symmetry of various kinds of processes and more generally of causality.

The best way to establish a temporal direction is to base it on a general concept, so that we need not go into the substantive details of various processes. The most likely candidate for such a general concept is causality. If we can find a notion of causation that is independent of the concept of time, then we can establish a general temporal direction by identifying it with the direction of causation. The task is not easy. Since Hume's critique, causes and effects are usually regarded as instances of certain lawlike regularities, and causes are analyzed in terms of various combinations of necessary and sufficient conditions. Determining conditions are temporally symmetric. The assertion that A is necessary for the occurrence of B places no restriction on whether A occurs earlier or later than B. This point is made explicit in deterministic dynamics. For invertible dynamic systems, a state determines its successor state just as it is determined by the successor; that is why dynamic rules are equally good for prediction and retrodiction. Determining conditions are insufficient to distinguish causes from effects; to do so the temporal precedence of causes must be posited independently; for that reason Hume included in his definition the criterion that a cause precedes its effect.

Many people who want to reduce time to causation tried to find criteria of causes without resorting to their temporal precedence. The most popular criterion is the principle of common cause. Suppose the occurrences of two types of event A and B are statistically correlated. Then there must be a third type of event C such that the statistics of A and B conditioned on C are independent. C is defined solely in terms of its ability to "screen off" the statistical correlation between A and B. The event C is inferred as the *common cause* of the events A and B and asserted to be temporally prior to them. For instance, the occurrences of thunder and lightning are statistically correlated but become independent when conditioned on the occurrence of electric discharges. Thus a discharge is the common cause of lightning and thunder and hence precedes them.[2]

The principle of common cause aims to bypass our ordinary notion of causation that presupposes the temporal priority of causes. It tries to discern causes solely by looking at the statistics of events and by crunching numbers. It is a version of probabilistic causation and shares its weakness. Critics argue that many empirical statistical correlations are accidental; physically related events need not have a common cause according to the statistical definition; even if several effects have a common cause, their statistical correlation is not necessarily screened off by the common cause.[3] Here I discuss only the objections for basing temporal precedence on the idea of screening off.

The statistical relation among the types of events A, B, and C does not rule out the inference that C is the effect of the joint causes A and B. Proponents of screening off argue that an event of type C is interpreted as the cause rather than the effect because genuine coincidences are rare. It is much more likely that a coherent cause has diverse effects than that a diversity of causes has a coherent effect. Coincidences are rare, but not so rare to qualify as the metaphysical ground for so general a concept as temporal priority. Consider how many coincidences must have occurred for Lee Harvey Oswald to succeed in assassinating President Kennedy singlehandedly. Must the coincidences be caused by a conspiracy?

The principle of common cause has tacitly assumed that causes are always more compact and neatly defined than effects. Perhaps the assumption is rooted in the coarse-grained view that separates causes from standing conditions, but it is false. Causes are not always compact. Proponents of common cause have neglected self-organization, which is studied in many sciences. When mobs gather or speculators panic, the effects of riots or market crashes are clear, but the causes are diffuse and often elusive to investigators. The idea of uncoordinated actions producing unintended coherent results, captured by the metaphor of the invisible hand, is most important in the social sciences. If we look solely at the statistics of such phenomena, the common causes we extract are more likely to be "teleological causes" that occur later than the statistically correlated events.

Spontaneous organization abounds in the natural sciences. The laser is the coherent effect of the motion of countless atoms. Billions of spins align themselves in magnetization and the phase transition is described statistically in terms of the correlation and fluctuation of the spins (§ 23). The correlation produces the coherent effect of a permanent magnet, but it is difficult if not impossible to capture the conductor directing the concerted motion of the spins; for that reason the magnetization is called spontaneous. Temperature may be cited as a cause in the macrodescription of magnetization, but not in the microdescription of spin correlation. The general spin–spin coupling obtains in all temperatures; hence it cannot be counted as the cause of the long-range spin correlation that occurs only during phase transition. Thus the spin correlation is simply described, with no attempt to seek a further common cause. If the spin correlation could be statistically screened off, it would be screened off by the coherent effect of magnetization.

Philosophers do not succeed in finding a time-independent notion of causation. Causality and temporal directionality are intertwined in our basic conceptual structure and are responsible for making intelligible individual processes and causal chains. Attempts to reduce one to the other prove futile.

In Search of a Global Temporal Direction

Unable to base a global temporal direction on the general concept of causation, we turn to investigating substantive processes. As in the case for the global planes of simultaneity, we again begin with individual processes and things, which are clearly defined (§ 27).

A temporal direction is inherent in a *particular* process or a *particular* enduring thing. The stages of a process or the states of a thing undergoing the process are individuated and identified by the temporal parameter t, which is an inalienable part of the concept of a process or a thing. The parameter is represented by the real number system; hence it carries at least four characteristics of the system: an *order* intrinsic to the ordering of numbers, a *direction* inherent in increasing numbers, an *interval* due to the additivity of numbers, and an *origin* $t = 0$. All four characteristics have objective meaning for the representation of *individual* processes. The ordering and succession of stages are crucial to the definition of a process, the interval represents its extension, and the origin denotes its beginning.

Can the temporal parameter t with all its characteristics be generalized to the global structure of the world? It applies to the cosmological process, but the cosmological process is only a particular process that describes the evolution of the universe on the largest scale and the crudest level. The cosmic time scale is so gross most processes are instantaneous in it. The global temporal structure must include the cosmological process and countless other processes on various scales and organizational levels. It must harmonize the temporal features of all the processes and exclude a feature if no agreement can be reached objectively among all or a great majority of the processes. It is questionable whether agreement on any one of the four temporal characteristics of individual processes can be reached objectively without arbitrary conventions imposed by us. Relativistic physics has shown that a global ordering in the sense of a sequence of global planes of simultaneity is not tenable. Global origin and interval are also spurious because they depend on global simultaneity. Can a global temporal direction be established?

Each individual process proceeds in its own direction, as represented by the "past" and "future" light cones in relativistic physics. Can the temporal directions of various processes be harmonized objectively? The harmonization may be impossible as a result of certain general features of the world's global spatiotemporal structure. For instance, it is impossible to define a direction consistently on a Möbius strip because of the twist in the strip. An arrow pointing upward will become downward pointing if it is transported parallel to itself all the way around the Möbius strip; hence there is no consistent definition for the way the arrow points; the strip is nonorientable. Similarly, the spatiotemporal structure of the universe may turn out to be spatially or temporally nonorientable. If so, then a global temporal direction would be ruled out. However, let us assume that the universe or at least a large region of it is temporally orientable so that the problem of a temporal direction can unfold.

A definite global temporal orientation requires much more than the temporal orientability of the universe. It requires at least a kind of pervasive process representable in a universal law so that we can uphold the direction of the processes as the norm for the world. The hope of fulfilling this requirement is quickly dashed by fundamental physics, in which almost all processes are temporally symmetric.

The Time-reversal Symmetry of Processes[4]

An *individual* process always has its temporal direction, but a *kind* of process does not always have a distinct temporal direction. An individual is more specific than a kind, and its peculiarity may be lost in the generalization to the kind. That is why dynamic theories distinguish between dynamic rules and initial conditions; the dynamic rule represents what is common to all processes in a kind and the initial conditions the peculiarities of individual processes. The dynamic rule is the place to look for the global temporal direction.

On the basis of the symmetry properties of dynamic rules, various kinds of process can be divided into two broad classes: reversible and irreversible. A process is *reversible or temporally symmetric* if its dynamic rule or equation of motion is unchanged under *time reversal*, which means the replacement of the time parameter t by $-t$. A symmetry generally erases a peculiarity in a representation, and time-reversal symmetry erases the direction of increase of the real-number representation. A process is *irreversible or temporally asymmetric* if its dynamic rule changes under time reversal. To define reversibility correctly, care must be taken to reverse all occurrences of the time parameter, even if it is hidden in the definitions of quantities such as the electric current or magnetic field. Velocity, which is the change of position with respect to time, must be reversed under time reversal. Since velocity and hence motion reverse with the time parameter, motion reversal becomes the surrogate of time reversal in experiments. A dynamic rule that is invariant under time reversal does not specify a definite temporal direction.

Time reversal symmetry has a spatial analogue in space inversion symmetry. A circle is symmetric under space inversion; it looks the same in a mirror. In contrast, a hand is asymmetric under space inversion; a right hand in a mirror looks like a left hand. To visualize a time-reversed process, we take a moving picture of the process and run it backward. The process is temporally symmetric if we cannot tell whether we are viewing the movie forward or backward, just as we cannot tell whether we are seeing a circle or its mirror image. Temporally symmetric processes have no temporal direction since they can go both ways. In contrast, temporally asymmetric processes have a temporal direction because they go only one way. For instance, if things in a kind of temporally asymmetric process always grow bigger, than we get a temporal direction by identifying later with bigger.

Imagine seeing a brief segment of a movie of two billiard balls approaching each other, colliding, then separating. The process is temporally symmetric; if the movie is run backward, we see a similar process with the velocities of the balls reversed. The temporally symmetric episode is typical of mechanics. Now suppose we are shown a more extended segment of the movie that includes the balls slowing and finally stopping. Now we can at once tell whether the movie is run forward or backward, for we have learned from experience that balls at rest do not start moving on their own accord. Moving balls slow and stop because their energy is dissipated by friction. The extended process is temporally asymmetric. When our reasoning includes friction and the like, we have passed from the realm of mechanics into the realm of thermodynamics.

Almost all processes we encounter in our daily life are irreversible, but that is because we are complex beings operating on a level of complex organizations. If we consider the composition of systems from elementary particles upward, we find that temporal asymmetry emerges in thermodynamics and becomes almost the rule for processes of larger and more complex systems. Below thermodynamics, all processes covered by classical and quantum mechanics and elementary particle, atomic, and molecular physics are temporally symmetric, with the possible exception of a few very feeble processes in the weak nuclear interaction.

The reversibility of elementary processes has a great impact on the idea of a global temporal direction. Since motion and reversed motion are equally allowed by physical laws, neither can be the basis of the global temporal direction. Furthermore, the temporally symmetric processes are the constituents of all macroscopic processes. Thus any asymmetric macroscopic processes must reconcile themselves with the temporal symmetry of the motions of their constituents. The last point becomes a major issue in the development of statistical mechanics.

Entropy Increase and the Emergence of Irreversibility[5]

Irreversibility is an emergent character of macroscopic processes. The breaking of the time-reversal symmetry of the laws of mechanics by thermodynamic processes illustrates the relation and tension between the micro- and macrodescriptions of large composite systems.

Consider a large system composed of N structureless classical particles, where N is a huge number. The system can be described both microscopically and macroscopically. In the microdescription, we write down the dynamic rule of mechanics that determines the trajectory of the *microstate* of the system in its $6N$-dimensional microstate space. The trajectory specifies at each time the position and velocity of each and every particle. The dynamic rule is symmetric under time reversal as mechanics demands. If the velocity of every particle is *exactly* reversed, the entire system retraces its path in the microstate space. The time-reversal symmetry of the evolution of the microstates of large composite systems is demonstrated in computer simulations and in spin echo experiments.[6]

The macrodescription of the same composite system gives a different picture. Here the *macrostate* of the system is characterized in terms of a density function giving the statistical distribution of the particles or a few macrovariables such as the temperature or pressure. The macroscopic dynamic rules governing the evolution of the macrostates are usually not symmetric under time reversal. The particles in the system disperse but never congregate. Heat flows from hot spots to cold spots of an isolated system but never the other way around. All these phenomena are summed up in the *second law of thermodynamics*, which asserts that the entropy of an *isolated* system always increases. Isolation is crucial to the second law. Vacuum cleaners gather dust particles in a bag and refrigerators pump heat from cold to hot places, but utility bills remind us that the appliances require external energy to perform their tasks.

The entropy of a composite system is a macrovariable that measures the system's disorder manifested in its macrostate. Strictly speaking, the notions of order and disorder do not apply to microstates per se, for each microstate precisely determines the positions and velocities of all particles. Order and disorder become significant when the microstates are classified according to macrostates. Each macrostate is compatible with a host of microstates. The *relative disorder* of a macrostate is measured by the relative number of microstates compatible with it. An ordered macrostate is satisfied by a smaller number of microstates than a disordered macrostate, because it has more macroscopic structures and imposes more restrictive conditions. Technically, the disorder of a macrostate is represented by the volume W of the portion of the $6N$-dimensional microstate space that includes all microstates satisfying the conditions of the macrostate. Boltzmann defined the *entropy* of a macrostate as proportional to the logarithm of the volume W of the microstate space compatible with it. The definition of entropy links macro- and microphysics.

How does the temporally asymmetric increase of entropy reconcile with the temporally symmetric evolution of the system's microstate? Consider a gas of N particles initially confined to the right-side half of a box. When the confining partition is removed, the gas is in a nonequilibrium state. As the gas evolves toward equilibrium, the particles disperse until they fill the box uniformly. In the microdescription, the process is represented by a jiggling trajectory of the microstate in the $6N$-dimensional microstate space of the gas, as illustrated in Fig. 10.1. The trajectory is reversible.

To represent the increase in entropy, we partition the microstate space into cells according to the macrostates of the gas, in this case according to the fraction of the box that is successively occupied by the spreading particles. The volumes of the cells corresponding to more diffused macrostates are larger, because each particle has a larger possible range of positions, leading to larger numbers of possible microstates. Let W_i denote the volume of the cell in the microstate space corresponding to the initial nonequilibrium macrostate in which only half the box is filled, and W_f the volume of the cell corresponding to the final equilibrium macrostate in which the box is totally filled. Since the possible position of each particle is doubled in the final equilibrium state, the number of possible microstates increases 2^N-fold. Thus $W_f/W_i = 2^N$, where

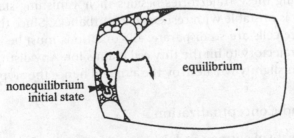

equilibrium

nonequilibrium
initial state

FIGURE 10.1 The evolution of the microstate of a large composite system is represented by a reversible trajectory in the microstate space. In the macrodescription, the microstate space is partitioned into cells distinguished by different macrostates. The volume of the cell corresponding to the equilibrium macrostate is unimaginably larger than the volumes of the cells corresponding to nonequilibrium macrostates (the cells are not drawn to proportion). The typical trajectory starting in a tiny cell proceeds to cells with increasing volume and hence macrostates with increasing entropy.

$N \approx 10^{24}$ for a small box of ordinary air. The cell of the equilibrium macrostate takes up almost the entire microstate space. The initial cell is so unimaginably tiny it cannot be found by the most powerful microscope if the microstate space is drawn to scale.

Because the number of particles N is large and the volume W of a cell in the microstate space increases exponentially with N, a minute increase in the possible position of each particle leads to a huge increase in W. Therefore the cell corresponding to a slightly more diffused macrostate is overwhelmingly larger than the cells corresponding to less diffused macrostates. Once the trajectory of the gas wanders into a larger cell, it has vanishing probability of finding its way into a smaller cell again, and it has probability close to 1 of going into an even larger cell. Thus the gas almost always spreads. Except a few atypical trajectories with vanishing statistical weight, any *typical* trajectory will proceed with probability very close to 1 to macrostates with increasing microstate-space volume and hence increasing entropy, until it enters and settles down in the equilibrium macrostate that practically occupies the entire microstate space.

This is the *probabilistic justification* of the second law of thermodynamics based on the statistics of the number of possible microstates conforming to various macrostates. It invokes the notion of probability because it addresses the evolution of *single* composite systems instead of ensembles of systems. The law of large numbers ensures that the second law holds with probability very very close to 1 for systems of macroscopic size. The large size of the system, to which "large numbers" refers, is crucial. If N is a mere handful, then the particles are quite likely to be found congregated in a corner of the box, for the volumes of the state-space cells would be comparable, and the trajectory could wander into any one of them.

Entropy increase is a statistical law that applies to macrodescriptions. It ignores the atypical cases that have zero statistical weight. Examples of atypical trajectories are those with the velocities of the gas particles exactly reversed, so that the gas particles collect themselves in one side of the box. There is another justification for ignoring these trajectories besides their vanishing statistical weight: the atypicality is unstable with respect to disturbances. Since the sizes of the microstate-space cells are so disparate, all conditions must be exactly right for the atypical trajectory to hit the tiny cell it aims for. Any disturbance that foils the conditions slightly will destroy the aim and hence the atypicality.

Self-organization and Conceptualization

Can the incessant increase of entropy and the second law of thermodynamics be made into the definition of a global direction of the world's temporal structure? Can they be the ground for our intuitive idea of the past and future? Can they be a substantive replacement for the general concept of temporal priority presupposed in the notion of causality? I am afraid not. The trouble is not that the second law of thermodynamics is only a statistical law; the drawback is not serious since we have to wait for times longer than the age of the universe to see exceptions. There are more serious objections.

As the critics of Boltzmann pointed out and as Boltzmann and later physicists conceded, the reasoning that the microstate trajectory goes into larger and larger cells is equally valid for retrodicting instead of predicting the behavior of the system. Suppose we are shown a box of gas with all molecules congregating to one side, told that the gas has been isolated for a long time, and asked to explain how it got itself into such a state. Our first reaction would be incredulity; we would protest that it is impossible and the gas must have been just tampered with. However, if we are persuaded that the gas indeed was not pumped into the corner by external means and we are indeed lucky enough to witness one of those extremely rare fluctuations, then the same statistical accounting of the number of possible microstates would lead us to the conclusion that the gas was previously in a more diffused and hence more probable state. It is like finding something in an incredibly tight spot and judging that it got there from a roomier place instead of from an even tighter place.

The increase of entropy is peculiar; it holds only for isolated systems, but not for systems isolated for too long. In typical illustrations of the law, the process exhibiting entropy increase is defined to start right after the system is put into a nonequilibrium low-entropy state by some external interference, for instance, the removal of a constraining wall in a box of gas. The question then becomes why nature has so many initially ordered and nonequilibrium systems for the second law to be so commonly manifested. The answer to the question illuminates at once the general nature of the world and the general nature of our conceptualization of the world.

Our minds are far too small to study the world in all its richness. We cut out bits and pieces for investigation. Isolated systems are our idealizations. Permanently isolated systems are beyond our empirical knowledge because to know them we must directly or indirectly interact with them. We approximately ignore weak relations when we conceptually carve nature at its joints and acknowledge individuals for their internal coherence. Low-entropy systems, being more ordered, have more prominent macroscopic structures. Hence they are more likely to be singled out and individuated as things in our conceptualization of the world.

In nature, order and structure emerge spontaneously in processes of self-organization, of which phase transitions and the growth of organisms are examples. Self-organization does not violate the second law of thermodynamics, for the thing that spontaneously emerges from the organization is only a *partial* product of a larger system, part of which becomes even more disordered to compensate for the order in the self-organized part. In our rather crude partition of the world into individuals, we conceptually isolate the part with prominent structures as our system and sweep the disordered part into the shadow of randomness. Thus nature's self-organization and our conceptualization jointly produce the nonequilibrium low-entropy initial states for the second law of thermodynamics to be ubiquitously manifested.

The preceding explanation of the prevalence of nonequilibrium systems has at least two problems that further undermine the argument to identify the global temporal direction with entropy increase. The first problem concerns the role of conceptualization. If conceptualization is allowed to play an important role in isolating ordered systems whose entropy increases subsequently, there is no justification for ruling out other conceptualizations that yield processes whose temporal directions have nothing to do with entropy increases.

Entropy is a subtle quantity and its rigorous definition requires the conjunction of micro- and macrodescriptions. Many kinds of processes studied in the sciences, and certainly many encountered in daily life, are characterized in only one level. Their entropy is difficult to define, but they are nonetheless temporally oriented in their own way. We have theories for open and dissipative processes that exhibit increasing order. The lives of organisms are far from equilibrium processes that do not approach equilibrium. These processes, although conforming to the second law of thermodynamics if reformulated as isolated systems that include their surroundings, are clearly defined by themselves and do not manifest the increase of entropy or disorder. Consequently, we cannot establish a common temporal direction by comparing the orderliness of the stages of different kinds of processes. The successive stages of some processes become increasingly structured while those of others become increasingly disordered. We do not have an objective notion of global temporal direction just as we do not have objective global simultaneity. We have to be satisfied with conventional ones, basing the temporal direction on the direction of the processes that we ourselves are undergoing, on our memory and expectation, our past and future.

The Ultimate Source of Order: The Initial Condition of the Universe

Self-organization and conceptualization only succeed in pushing back one step the question why there is so much order for the second law to operate. The formation of self-organizing structures is possible only by exporting disorder into the environment. To absorb the disorder, the environment must be relatively ordered. Why is the system plus the environment in a relatively ordered state in the first place? If the second law of thermodynamics holds not only for local isolated systems but also for the universe as a whole on the cosmic scale, then we must give some account of the ultimate source of order. When the question is pressed, we are invariably led to the initial condition of the universe.

Starting from home, the major source of order is the temperature disparity between the burning sun and the surrounding space at the freezing temperature of $-270°C$. The energy the earth receives from the sun, in the form of high-energy photons, is rather coherent and ordered. All the energy the earth receives is eventually returned to space as heat, but the heat is disordered, for it is composed of a much larger number of low-energy photons. Thus entropy is increased and structure destroyed in the complete process. This large process drives numerous smaller processes, including many self-organizations and structure formations in the earth's meteorological, oceanological, and biological systems. About a tenth of a percent of the sun's energy falling on earth passes into the biosphere and supports dissipative systems known as life.

Why are there suns and stars? How does the disequilibrium between them and the surrounding space come about? We have theories for the formation of galaxies from rotating dust by the force of gravity and the ignition of stars by nuclear fusion. How is the formation of these structures financed by the export of entropy? The answer can only be found in the initial conditions of the universe. It is conjectured that the universe started in an ordered state with low entropy. According to modern cosmology, the universe was born about ten billion years ago in the Big Bang and has been expanding ever since. Energy and matter were distributed uniformly in the primeval fireball. Entropy increases because the universe expands.

Will entropy decrease if the universe eventually stops expanding and re-contracts? The answer is speculative. Roger Penrose hypothesized that the uniform distribution of the early universe was not an equilibrium state, for it did not include the effect of gravity, which pulls matter into clusters, galaxies, stars, and planets. When gravity is taken into account, black holes are closer to equilibrium than the uniform distribution of energy. The entropy of black holes has been estimated, and it is enormous. If the universe ends in a Big Crunch, which is like a gigantic black hole, its final entropy will be much much bigger than the entropy of the Big Bang because of the effect of gravity.[7]

The probability for the realization of a macrostate is proportional to its entropy, because the entropy is equal to the relative number of possible microstates compatible with it. It is estimated that the probability of the low-entropy macrostate in which the universe began is about 1 part in 10^n where $n = 10^{123}$. Why was this state of fantastically small probability realized? Why did the Big Bang create such an ordered system? Finally physics runs out of even speculative answers.

As the question about the source of order and nonequilibrium is pressed, the answer recedes from universal laws to statistical laws to narratives about the evolution of the universe to conjectures about its initial conditions. The question "Why does entropy increase?" ends with the question "Why did the universe begin the way it did?" The answers along the way invoke many physical laws formulated and confirmed in general relativity, elementary particle physics, nuclear physics, and other branches of physics. The narrative in which various laws are judiciously called upon to explain the cosmic evolution includes contingent factors resulting from cosmic phase transitions. Some of the contingencies are expressed in a significant number of constants that, try as physicists do, cannot be determined from existing theories. Above all, the narrative depends critically upon the initial condition of the universe, the unique case for which no law exists. In cosmology, physics meets history.

41. Narratives and Theories in Natural History

Dobzhansky said that evolutionary biology addresses two types of question: The first concerns the actual course of organic evolution leading to the diverse and adept life-forms on earth; the second attempts to find out the causes and mechanisms underlying the historical course of life on earth.[8] By themselves, the two types of question are common to all sciences; physics studies both the principles of planetary motion and the actual orbits of particular planets. In physics, the two types of question are addressed in a unified framework; actual cases are usually regarded as the instances of general laws or models for the causal mechanisms, and the validity of the laws or models is established only by testing against actual cases. The unity is absent in evolutionary biology, which splits into a theoretical and a historical branch for addressing the two types of question.

So far we have considered only the theoretical branch, which has almost monopolized the philosophy of biology. Thoughtful biologists, however, are increasingly emphasizing the historicity of evolution and the role of narrative in explaining evolutionary events. They realize that historical narratives, which differ in aim, method, and conceptual emphasis from theoretical models, are not only more appropriate but almost mandatory for explaining specific events in the evolution of life on earth. Theoretical generalization is powerful, but generalization exacts a heavy price, which may destroy the concreteness of history. Moreover, the power and generality of theoretical models differ greatly. The theoretical apparatus of evolutionary biology is not robust. Most basic concepts, even the units of evolution and selection, are confused and controversial (§ 12). Population-genetic models in terms of fitness overlook so many causal factors they are criticized as trivially circular. Optimization models for adaptation are static, lack proper genetic explanation, and do not explain how the adaptations evolved (§ 17). Potentially important evolutionary mechanisms are brushed aside (§ 24). Theories and experiments are unable to differentiate the relative efficiency of various mechanisms (§ 37). Theoretical models are too skeletal to handle the complexity of concrete evolutionary events, for the understanding of which biologists resort to historical narrative.

Of the three types of evolutionary factors cited by Wright – deterministic, stochastic, and unique events – the first two can be addressed theoretically; the third demands historical treatment. Not only are catastrophes and appearances of conspicuous novelties unique, most if not all concrete historical events are unique, if only because possibilities outnumber actualities by such an unimaginably large factor. The evolutionary course of a species depends on many contending internal and external forces, which are often called accidental because an adequate account is beyond our theoretical capability. Being contingent, historical events cannot be represented deterministically; being rare, they cannot be represented statistically. Why did the dinosaurs become extinct? How did human beings evolve? These and many questions we commonly ask about evolution can only be answered by reconstructing the past itself. They belong to history, and their explanations take the form of narrative rather than theory.

This section explores the historical nature of evolutionary biology. We examine narrative explanations, their efficacy in addressing complex processes, and their relation to the covering-law model of explanation. Finally, we contrast historical narratives with the just-so stories that are fictitious explanations for evolutionary events generated by the abusive use of theoretical models.

Systematics and Phylogenetic Inference

"History" can mean either the course of events or an account of the course. It usually means accounts in methodological discussions such as this. Historians distinguish among chronology, chronicle, and history, none of which is trivial. Just to establish a chronology that properly synchronizes and dates historical events is a difficult task that engaged the minds of people like Newton. Chronicles record events chronologically, suggesting some underlying causes but making no attempt to bring them into relief. Histories, mostly in the form of narrative, try to discern some meaning and give some explanation of the facts recorded in chronicles.

The construction of a *phylogeny*, or a family tree of organic species, is like the construction of a chronology of ancient events for isolated civilizations. Descent with modification is a central Darwinian postulate with the corollary that all species on earth are related. Phylogenetic inference answers questions such as whether humans are more closely related to chimpanzees than to orangutans. As chronologists establish dates by comparing historical records, phylogeneticists establish lineages by comparing existent and fossilized organisms, arguing that evolutionary changes are frozen into the structures of organisms, which can be viewed as some kind of text in which their ancestry is written. Like a chronology, a phylogeny arranges items in an objective order but does not explain the causal relations among the items, although causality is presupposed in its construction. As chronologists appeal to independent records of eclipses and other astronomical events, phylogeneticists make causal assumptions about evolutionary processes such as the rarity of novelties and the preservation of some novelties in descendant species. The construction of both chronology and phylogeny is laborious, with results that are fraught with gaps and ambiguities.

Systematics, which is responsible for describing living and fossilized organisms, classifying them, and constructing their phylogeny, is analogous to a multicivilization chronicle. As chronicles draw on the results of various branches of research, including archaeology and geography, systematics draws on the results of other branches of biology, paleontology, biogeography, and various comparative disciplines. Systematics is fragmented into several schools, which differ in their basic assumptions and methods of comparison.

Historical Narratives in Evolutionary Biology

A historical narrative is more than a chronicle. It does not merely record what happened when but also gives some explanation of why an event happened by relating it to earlier events and highlights its significance by putting it in perspective. Historical narratives are important in addressing macroevolutionary phenomena.

Population biology, which mainly contains theoretical models, concentrates on *microevolutionary* processes, or processes underlying changes within individual species. It pushes to the periphery two of Darwin's central theses, the conversion of intraspecific variations into the diversity of species and the common descent of all life-forms on earth. These postulates concern *macroevolutionary* trends and the patterns observed in the fossil record and described in systematics. The historical record is full of puzzles that befuddle population biology: the emergence of novel characters, the long-term stasis of characters in changing environments, the varying evolutionary rates (§ 24).

In macroevolution, biology faces the problem of explaining conspicuous macroscopic developments as the combined results of numerous microscopic causes. We have seen how the solution of similar problems in physics requires a new whole class of theory accompanied by its own postulates (§ 38). The biological problem is more difficult because biological processes are more heterogeneous and complex. Consequently, biology has neither a macrotheory comparable with the hydrodynamic equations nor a theoretical connection between the macro- and microlevels of description. Macroevolutionary concepts are sparse and macroevolutionary theories speculative. Most population biological models are for equilibrium situations, unsuited to the dynamics of evolution. Under such circumstances, how do evolutionary biologists explain specific historical events? Like historians, they use *narratives* to give coherent accounts of the effects of competing microcauses.

Mayr said: "It might have been evident soon after 1859 that the concept of law is far less helpful in evolutionary biology ... than the concept of historical narratives." "The explanation of the history of a particular evolutionary phenomenon can be given only as a 'historical narrative.'" Mayr's assessment was echoed by Gould: "Many large domains of nature – cosmology, geology, and evolution among them – must be studied with the tools of history. The appropriate methods focus on narrative." Lewontin, one of the most theoretical of biologists, alluded to the historian Leopold von Ranke's motto that history wants to tell what really happened and said that historical evolutionary biology "attempts to tell the story of the common ancestry of organisms *wie es eigentlich gewesen*." He explained that evolutionary biology assumes the simultaneous operation of several forces that include chance perturbations and regards the actual event as the nexus of these forces. Under this assumption it tries to provide a correct narrative of the past events and an account of the causal forces and antecedent conditions leading to that sequence. "The historian may have some difficulty in distinguishing this description of evolutionary theory from, say, the structure of Khaldûn's *Universal History*."[9]

Narrative accounts of descent by modification abound in the evolutionary literature. There is the narrative of how Devonian bony fishes divided into two classes, lobe-finned and ray-finned; how the lobe fins gradually evolved into tetrapod limbs and lobe-finned fishes into amphibians and reptiles that invaded the land; how ray-fins remained the way they were and ray-finned fishes populate the water today. Examples of narratives such as the evolution of the horse or the invertebrate eye can be indefinitely multiplied. Anyone familiar with the literature would agree with Lewontin that "a great deal of the body of biological research and knowledge consists of narrative statements."[10]

Narrative

Narrative is the chief form of presentation in history. For more than fifty years historians have argued that narratives constitute an alternative to the covering-law explanation. Narrative explanations are more suited to account for the concrete details of complex and dynamic historical processes, which tend to get lost in theoretical generalization and covering laws. Since history without qualification usually means human history, much of the argument revolves around the human factor.[11] The debate on narrative explanation is simpler in the context of evolutionary biology because we need not worry about the human factor.

Traditionally, a narrative organizes its material into a coherent account with a central theme told in a chronological sequence, as in Edward Gibbon's *The Decline and Fall of the Roman Empire*. This is probably the form of history familiar to most readers. Since the 1930s, many academic historians have imported ideas from the social sciences, which helped them to make explicit their presuppositions, sharpen their concepts, broaden their scopes, and improve their research tools. They are called "new historians" and their research "scientific history." New history broadens the scope of history from the traditional locus of political processes to cover demographic, social, economic, cultural, and other processes. Its style is more analytic than narrative, more cross-sectional than chronological, making frequent use of statistics. For instance, Peter Laslett's *The World We Have Lost* gives a cross-sectional analysis of preindustrial English society, probing into questions such as whether peasants really starved or whether girls married in their early teens as Shakespeare had us believe. Scientific historians often use sociological and psychological theories. However, their enthusiasm for the theories waned as they found the theories too idealized and remote from reality to illuminate concrete historical events. Lawrence Stone, a pioneer of scientific history, recently noted a return to the more traditional narrative: "The movement to narrative by the 'new historians' marks the end of an era: the end of the attempt to produce a coherent and scientific explanation of change in the past."[12]

The structures of many works that Stone cited as the "revival of narrative" are analytic according to traditional historians, for they separately focus on different aspects of big movements. Since my purpose is to compare and contrast explanations by narrative and by covering law, I will use the term *narrative* even more loosely to include analytic monographs in scientific history before the revival, for example, Edward Thompson's *The Making of the English Working Class*. By *narrative* I refer not to a literary style of presentation but to the peculiar mode of understanding appropriate to the sort of complex subject matter investigated by the historical sciences.

The cross-sectional and thematic account of many monographs often lacks the dynamic flavor of traditional history. I count them as narratives – Joycean narratives, if you please – not only because they are more akin to the description of organismic behaviors in evolutionary biology; in the broad distinction between theory and narrative, they are much closer to traditional narratives than to theoretical models covered by general laws. The structure of a complex process at any specific stage is complicated. A clear view of the configurations of the stages enhances our understanding of how one stage leads to another. The view is provided by Joycean narratives, which cut across the grain of time and examine the structure of the slices in detail. Furthermore, just as many processes go on simultaneously in a system of particles, many processes intermingle in the current of history. Various processes have their own characteristic times, as the political wind blows faster than the glacial transformation of social institutions. Hence some processes may appear to be stationary in the time scale characteristic of the account of other processes. This is the reason for the triple-decker structure of many works in the French *Annales* school of history.

The Inadequacy of Covering-law Explanations

Some philosophers, notably Hempel and Popper, argued that the covering-law model of explanation applies equally to history, if allowance is made for statistical laws and the partiality, sketchiness, and incompleteness of explanations. Hempel gave the recipe of a historical explanation: "if fully and explicitly formulated, it would state certain initial conditions, and certain probability hypotheses, such that the occurrence of the event to be explained is made highly probable by the initial conditions in view of the probability hypotheses." For example, he explained the French Revolution in terms of certain prevailing conditions and the lawlike hypothesis that a revolution is highly likely when there is growing discontent in a large part of the population.[13]

Covering-law explanations are strenuously resisted by historians. Recently, some biologists joined the debate. Mayr complained about those who try "to deny the importance of historical narratives or to axiomatize them in terms of covering laws."[14] Indeed, it is difficult to imagine how the evolution of birds from reptiles can be explained in terms of general hypotheses.

A persistent criticism of the covering-law model is that it does not fit the explanatory practice in existing histories, even leaving aside the human factor. Examples such as that people died in a famine because there was not enough food are trivial and do not count as historical explanations. Historical explanations indicate answers to questions such as what caused the French Revolution, how the United States got into the Vietnam War, or why socialism was weaker in America than in Europe. There are no satisfactory generalizations for revolution, conflict, or mass alignment. The sort of simplistic hypothesis conjured up for covering-law explanations would fail a grade-school history course. Stalinist historians did try to frame historical laws to explain economic development, but by the 1960s, the idea of historical law was abandoned by most Marxist historians.[15]

The more crudely we look at a system, the more likely we are able to discern patterns. If enough distinctions are jettisoned, we can always find some commonality and frame some lawlike statements, even for very complex systems. "What goes up must come down" is a law for the stock market. It is not trivial; what is in the pits does not necessarily rebound. Combined with proper initial conditions, the law can explain every fluctuation, correction, and bear market, but few people would be satisfied, for the explanation fails to answer what we want to know about market movements. For similar reasons, the new historians retreated from sociological theories to revive narratives; the theories are so crude and abstract they destroy the substance of history. In the level of detail we aim for, historical processes are too complex to be susceptible to comprehensive theoretical representation and covering law.

History and Theory

Almost all processes of any complexity are unique. History and theory raise questions on different levels of generality and demand different degrees of detail in description. In theoretical studies, we try our best to abstract, generalize, and formulate idealistic models of complex systems, even if the models are woefully unrealistic. If the theoretical models are applied to particular cases, we allow specific features to yield to the distortion of the general model. In historical studies, we try our best to retain concrete details at the expense of generalization; the details forge the particularity of the system. Both goals are ideal and often unattained; more important, generality and concreteness are not absolute and mutually exclusive. Theoreticians keep an eye on real particulars, which they try to understand and which would confirm or refute their conjectures. Historians must adopt a particular viewpoint in selecting material, and the criteria of relevancy always involve some general ideas. General ideas such as revolution, industrialization, proletarianization, and state formation serve as themes that illuminate the significance of particular situations and guide the historian in his interrogation of the sources. Similarly, Darwin's ideas of variation, proliferation, heredity, and natural selection constitute a general framework that makes sense of a large amount of natural historical data. These general ideas, too vague to function as laws, serve as themes, guiding principles, and centers of organization for historical evidence.

To distinguish further historical processes from nomological processes, let us consider the simple example of a Brownian particle, which is a heavy particle that jiggles in water because it is being randomly bombarded by water molecules (§ 35). In principle, the movements of the particle and all the molecules are summarily covered by the law of mechanics, so that given the initial conditions, we can explain according to the covering-law model why the particle is at a particular position at a particular time. Such explanations are not attempted even in physics.

Physicists treat the movement of the Brownian particle as a diffusion or stochastic process. In doing so they look only at the statistical regularity of a long path of the particle or the regularity exhibited by the movements of a large ensemble of particles. They give up knowledge about the positions of the particle at certain times and how it gets there – they relinquish the history of the particle. The generalization of the statistical law is purchased at the price of knowledge about individual stages of Brownian motion. Physics can pay the price because it is mainly interested in large ensembles of similar Brownian particles.

History refuses to pay the price of generalization, not least because there are no ensembles of historical processes, not in the degree of detail history demands. There is nothing comparable to the French Revolution or the evolution of the bird's wing. And even if we can collect a few similar cases, the number is too small for a theoretical statistical treatment.

Suppose history wants to explain the movement of a particular Brownian particle. How does it do it? The problem can be roughly divided into two parts: the particular encounter between the particle and a given water molecule, and the account of the surrounding molecules.

The free motion of the Brownian particle and its collision with a water molecule are governed by the laws of mechanics. They behave causally or according to some lawlike regularity. The assumption of causality pervades our everyday discourse and historical narrative; just think of the explanations you give in your daily activities. Even for human behaviors, we often assume some regularity for what we call normal and in what we offer as motivating reasons. The regularity promotes mutual understanding and can be loosely interpreted in causal terms. The commonsense notion of causality is reflected in words such as *because, therefore, naturally,* which occur as frequently in historical narratives as in ordinary conversations. It is the valid point of the covering-law model of explanation, except that the causal regularity implicitly assumed is only a minuscule part of historical explanations. Historical narratives can make use of causal laws and theories, perhaps weighty theories. Yet a law or a theoretical model accounts only for a small aspect of the process narrated. It spins a causal thread that has to be woven into the large tapestry of the narrative. The covering-law model of explanation suffers the same defect as its cousin reductionism. It assumes that knowledge of causal threads automatically implies knowledge of the tapestry, just as reductionism assumes that knowledge of individual constituents implies knowledge of the composite system. Both assumptions are naive and false.

Unlike the closed systems studied in theoretical models, the Brownian particle sits in an open context. We know its behavior when it crosses paths with a molecule, but we do not know the paths of the myriad molecules, which move and jostle among themselves. This is a common predicament in history, which is driven by the crossing of many causal paths. Historical events are the nexus of many contending forces. The historian John Bury, who was a strong advocate of history as a science, realized that the confluence of paths is not covered by laws and introduced the idea of contingency: "It was the conflux of coincidence which proved decisive."[16] The contingent conflux of coincidence is similarly decisive in biological evolution.

The paths of the water molecules that have the potential for striking the Brownian particle fall beyond the pale of our causal theory; hence the intersection of paths is called random by physicists and contingent by historians. Because of the randomness, there is no way to tell the trajectory of the Brownian particle except to follow its movement step by step, surveying the surrounding molecules at each step and accounting for each collision that deflects it. We find ourselves giving a *narrative* in which causal laws play but a minor role; the major task is to ascertain the configuration of the surrounding molecules at each step, for which we must rely on empirical evidence.

The movement of the Brownian particle can be compared to the evolution of a population. Population biology features rate equations for the statistical distribution of various characters in a population. Given the relative fitness of the characters and an initial distribution, the rate equations govern the evolution of the population just as the laws of mechanics govern the free motion of the Brownian particle. As we have seen in § 17, the rate equations are the results of numerous approximations and simplifications. There is no theoretical model for the developmental pathway linking genotypes and phenotypes, and the interaction between phenotypes and the environment is obscure. Consequently, the relative fitness is usually determined empirically for each fixed environment. Applied to the evolution of a population in a naturally changing environment, the fitness must be empirically determined for each period, just as the presence of water molecules on collision courses with the Brownian particle are empirically determined. Thus although the rate equations of population genetics have the form of dynamic rules, in practice they are parts of a narrative in which the character distribution of the population is recounted step by step, each requiring new input information on the fitness. This differs dramatically from the kinetic equation of gases, which determines the entire future behavior of a gas given only one initial distribution.

Narrative Explanation

The example of the Brownian particle illustrates the inevitable contingent factors of history. It is nevertheless misleading because it is too simple. Its causal factors are clearly defined; given the configuration of the surrounding molecules at an instant, we have rigorous dynamic rules from which we can derive the motion of the particle for the next step. These conditions are seldom satisfied in historical processes. A historical system not only sits in more varied contexts, it has internal structures whose dynamics drive its development. Its causal background is amorphous and the causal factors must be first delineated, described, and assessed. Historians laboriously sift through records and archaeological finds to discern causal threads. Similarly, evolutionary biologists pour tremendous effort into systematics, which appraises and classifies fossil remains. In the process they use their empathic understanding, commonsense judgment, causal reasoning, and causal laws if available.

When the contingent causal factors at one stage are sketched, the historian cannot simply call on a dynamic rule to deduce what happens next; no such rule exists. The primary sources of history, documents, records, letters, or fossil remains, are small in scope and indicative of tiny causal factors. Many causal factors pull and tug in all directions in a historical process, and the historian must compound them to find the movement of history. I hope that after hundreds of pages on how theoreticians wrestle with compounding causes for relatively simple and homogeneous many-body systems, I need not explain why it is hopeless to find general hypotheses for aggregating the microcauses in historical systems, which are far more complex and heterogeneous. That is why historians cannot predict. They can assess the relative importance of various factors and decide what to include as the major causes in their accounts only because they are helped by hindsight. Hence narratives often contain anticipatory predicates, for instance, describing the shot fired at Sarajevo on June 28, 1914 as the "first shot of World War I."[17] The predicate locates the event in the historical sequence and highlights its significance by alluding to its effects. In evolutionary biology, anticipatory predicates are found in the idea of preadaptation; for instance, lobe-finned fishes used their fins to walk on the bottom of the lakes, and hence their fins were preadapted for subsequent movement on land. To a lesser extent, the historian's predicament is shared by theoreticians, who cannot predict many emergent phenomena but can try to explain them after they are observed.

By exhibiting the contending causal forces in one stage and explaining why certain factors are dominant, the narrative guides us to the succeeding stage of the historical process. However, it sets no determinate direction for the process, for at the next stage new contingencies and new causal factors appear, closing certain possibilities and opening others, channeling the future development of the process. The historian must return to delineating the factors, weighing them, then proceeding to the next stage. The step-by-step reckoning leads to the chronological organization of traditional narratives.

An explanation responds to a question and addresses a block in understanding. Often the obstacle is cleared if we can fit the problematic item into a scheme and see its multiple connections with other things. A theory presents a scheme that is abstract and general; an event is seen as an instance of some generality. A narrative presents a scheme that is concrete and encompassing; an event is seen as a part of the current of history. Both enhance our understanding; both explain.

Historical Narratives Versus Just-so Stories

Gould and Lewontin, who emphasized the importance of historical narratives in evolutionary biology, are strongly critical of "just-so stories" that are symtomic of panadaptationism. Gould said: "Rudyard Kipling asked how the leopard got its spots and rhino its wrinkled skin. He called his answers 'just-so stories.' When evolutionists try to explain form and behavior, they also tell just-so stories." Take the diminutive fore legs of *Tyrannosaurs*, for example; some stories say they are adapted to titillate sexual partners, others to help the dinosaur to get up from a prone position.[18] Some commentators distort Gould and Lewontin's point completely by bundling historical narratives and just-so stories into "storytelling." The natures of narratives and just-so stories are diametrically opposite. In historical narratives, theory is always subordinate to evidence. In just-so stories, it is the other way around.

Historical narratives can invoke theoretical models, but the models are usually regional and their applicability unclear. The burden of the narrative is to weigh competing models and show that one is most appropriate in view of the evidence. It also has to determine the parameters of the chosen model.

Since many parameters change randomly, it has to fall back constantly on evidence. Evidence is not merely raw data; the historian must authenticate the sources and ascertain their relevance to the situation. The scholarly treatment of source and evidence in historical narratives demands respect.

Historians often experience the painful predicament that the evidence for the causes of certain events is destroyed in the course of time. A similar predicament occurs in evolutionary biology; soft tissues are seldom fossilized, and many behaviors leave no physical trace. Thus the evolution of human intelligence and social behaviors, which fascinates many people, remains in obscurity because of the lack of evidence. In such cases conscientious scientists acknowledge their ignorance and refrain from wild speculation. Some people, however, ignore the dearth of evidence, make "bold hypotheses" with adaptation models, and spin just-so stories about the evolution of intelligence and social behaviors.

Just-so stories lack the support of evidence and the careful weighing of evidence that are crucial to historical narratives. They result from the wanton extrapolation of crude adaptation models and perhaps the fabrication of "facts" to substantiate their claims. Just-so stories abound in human sociobiology and new Social Darwinism, which have provoked much criticism. As the philosopher Philip Kitcher said in his patient in-depth survey: "The dispute about human sociobiology is a dispute about evidence."[19]

Take, for example, Wilson's account of indoctrinability, which he claimed to be one of the biologically determined universal human characteristics that can only be changed by incurring unknown costs. Wilson said: "This leads us to the essentially biological question on the evolution of indoctrinability. Human beings are absurdly easy to indoctrinate – they *seek* it." Assuming that indoctrinability results from natural selection, he offered two alternative adaptation models to explain how, both based on the posit of a "conformer gene." In the individual-selection model, the nonconformer gene is eliminated by selection because individuals bearing it are ostracized and repressed by the community. In the group-selection model, the conformer gene is selected because the groups of individuals bearing it fare better than groups of individuals bearing the nonconformer gene.[20]

What is the evidence of the alleged indoctrinability? What is the evidence that it is biologically instead of socially based? Wilson collected a miscellany of examples of conformity ranging from ancient religions to modern politics but ignored the many counterexamples offered by the rich cultures of human history. He did not analyze whether these superficially similar behaviors are homologous or analogous,[21] or whether they result from cultural or biological conditioning, although such analyses are mandatory in historical biology.

Like other just-so stories, the account of indoctrinability belongs not to narrative explanation but to covering-law explanation. Under the general doctrines of panadaptationism and genic selectionism, Wilson offered as hypotheses two adaptation models in which conformity is the optimal behavior and "a gene for" is added to make its evolution biological. The adaptation models are crude, idealistic, appropriate only for equilibrium conditions but not their evolution, empirically tested for only a small number of special cases, and unsupported by an understanding of the underlying genetic and developmental mechanisms. The important questions are not which adaptation model is better but whether the adaptation models are applicable at all and whether the effects they describe are prominent. These questions take much research to decide. Wilson did none and cited only one highly conjectural paper. There is no historical evidence for higher frequency of ancestral nonconformers that disappeared later, no empirical evidence for the relative reproductive success of conformers, and no evidence for the genetic heritability of conformism. Without evidential support, optimization models become tall tales.

History resorts to narratives because historians are keenly aware that grand generalization is infeasible. Science generalizes, but unwarranted generalization is bad science. Just-so stories appear when optimization models for adaptation are generalized to areas beyond biology for easy and sensational results. The models are not strong enough for such generalization; their underlying genetic, developmental, physiological, environmental, and other causal mechanisms are poorly known. Since the simplifications made in the models are so drastic, qualifications for their applicability are numerous. Whether the qualifications are met must be determined by the particular circumstance of each historical event, and the determinants are more important than the model in the explanation of the event. The adaptation models can be used to support piecemeal arguments in a narrative explanation of the historical evolution of specific behaviors, in which case their constraints and parameters must be meticulously based on historical evidence. They do not have the strength of covering laws or general hypotheses. When they are carelessly deployed as hypotheses in covering-law explanations, they lead to just-so stories. Their deployment in human sociobiology and new Social Darwinism is not unlike the Stalinist histories that try to explain historical events by a simplistic grand theory.

42. Dynamic Optimization, Expectation, Uncertainty

Marshall said in 1890, "The element of time is the center of the chief difficulty of almost every economic problem." Similar judgments are voiced by many economists wrestling with time. Recently Peter Diamond quoted Marshall and added: "I share Marshall's view of time as a source of difficulty. The picture of Harold Lloyd conveys my image of a theorist grappling with the modeling of time." Lloyd hanging precariously onto the hands of a clock high above the bustling streets in the film *Safety Last* is featured on the cover of his book.[22]

Time is difficult to model in economic theories for several reasons. The first is common to the study of complex systems. Dynamic models of complex systems are more difficult than static models, and that is why they are also less developed in physics and biology. There are typically several processes with varying characteristic speeds going on simultaneously, making the distinction among the short, intermediate, and long run mandatory. Most important, economics depends on human decisions and actions. Decisions, made in historical contexts, are inevitably influenced by the legacy of the past and the uncertainty of the future. Thus their description requires the concept of tense as *memory and expectation*, which is not necessary in physical or biological theories. However, memory and expectation are intrinsically perspectival and partial, thus not well suited for theoretical representation, which inclines toward neutral and comprehensive views. The proper representation of expectation in decision making is one of the chief obstacles in theoretical economics.

Static and Dynamic

The confusion surrounding statics and dynamics is a manifestation of the difficulty of time in economics. Samuelson said, "Often in the writings of economists the words 'dynamic' and 'static' are used as nothing more than synonyms for good and bad, realistic and unrealistic, simple and complex." Fritz Machlup said: "Typically, 'Statics' was what those benighted opponents have been writing; 'Dynamics' was one's own, vastly superior theory." He also found in a broad survey that *statics* and *dynamics* have a "truly kaleidoscopic variety of meanings" in economics.[23] Why do *static* and *dynamic*, which have intuitively distinctive meanings, become so muddled?

Suppose we adhere to Samuelson's definition and the common usage in other sciences and call *dynamic* those models that represent system behaviors by dynamic equations in which time is the independent variable.[24] Three factors contribute to its blurring with statics. Dynamic systems can settle in stationary states. Several processes with different characteristic speeds may proceed simultaneously in the same system. A dynamic process regarded as a whole is static. All three factors occur in other sciences and have been discussed previously in various contexts.

Most dynamic models are found in macroeconomics. Most economic equations describe what economists call *equilibrium dynamics*, in which demand and supply equilibrate at each time, although the demanded and supplied quantities and other aggregate variables vary over time. Equilibrium dynamics differs from comparative statics in that the varying parameters are not arbitrarily chosen but are governed by temporal rules with causal bases. Nevertheless, the emphasis on equilibrium conveys a sense of statics.

Equilibrium dynamics has at least two presuppositions: First, the economy is essentially stable and returns to demand–supply equilibrium after small perturbations. Second, the return to equilibrium occurs in a time short compared to the characteristic time scale of the dynamic equations in which the aggregate variables change perceptibly. Abstractly, the presuppositions of equilibrium dynamics are similar to the conditions of hydrodynamic equations (§ 38). Each depends on at least two kinds of process. The fast process establishing the temporary or local equilibria – the price mechanism or molecular scattering – is implicitly assumed but not explicitly addressed in the dynamic equations. The evolution of the slowly changing macrovariables characterizing the temporary or local equilibria – for example, the aggregate consumption or the average density – is governed by the macroeconomic or hydrodynamic equations. Furthermore, there may be superslow processes represented

by constant exogenous parameters. Those concerned with the fast or the superslow process would find the dynamic models "static" because their time
scales are either too coarse to resolve the rapid changes or too fine to detect
the glacial movement.

Optimal Control: From the Integral View to the Differential View[25]

Dynamics is not necessarily more comprehensive than statics. From the bottom up, we obtain a dynamic view by stringing together static stages; from the
top down, we obtain the dynamic view by analyzing a broader static view. If
we intellectually grasp all successive states of a system or all stages of a process,
we get a synthetic picture that is static because it has encompassed all changes
within itself. We then analyze the synthetic picture to examine its dynamic
parts. Thus in relativistic physics, the static synthetic framework of the four-
dimensional world underlies the introduction of proper time and dynamic
processes (§ 27). Similarly we can derive dynamic equations by analyzing a
static synthetic framework in economics.

Consider a household deciding how much to save and how much to consume at each time. It has income from wages and interest from its assets.
If it consumes less at a time, the saving adds to its assets, which yield more
income to pay for more consumption at later times. The household takes a
comprehensive view of its lifespan and, with its perfect foresight of future
wages and interest rates, maximizes its lifelong utility, which depends on its
consumption at each time within its lifespan. Thus formulated, the problem is
static; it is conceptually not different from the household surveying a stretch
of land and deciding where to put soil to form a most satisfying landscape.

The first step to derive a dynamic model from the comprehensive static
view is to factorize the lifelong utility function somehow. The most common
way is to assume that it is the sum of instantaneous utilities at various times,
which depend on instantaneous consumption at those times. At each time,
the household computes its current total utility by integrating present and
future instantaneous utilities, discounting future utilities according to its preference of consuming now over consuming later. It then finds the stream of
instantaneous utilities that maximizes the current total utility subjected to
the equilibrium budget constraint, so that at each time the sum of wage and
dividend equals the sum of consumption and saving. The constrained maximization is a standard procedure in optimal control theory, which yields a
dynamic equation for the household's consumption and asset accumulation.

There are several theories that derive a dynamic equation from a static
optimization problem, and they are used in many sciences. The first to appear
is the calculus of variation, which originated in the late seventeenth century
and was developed by Leonhard Euler and Joseph-Louis Lagrange. Optimal
control theory, which generalizes the variational calculus, was developed in
the late 1950s and quickly found applications in economics and many areas
of engineering. It contributed greatly to the *Apollo* mission to the moon, for

instance, in finding the rocket trajectory that maximizes terminal payload by controlling the timing, direction, and magnitude of various thrusts.

Control and variational theories differ from the atemporal optimization theories discussed in § 9 in that they optimize entire temporal paths of dynamic variables instead of instantaneous values. An *optimal control theory* contains the usual elements of optimization – possible states, constraints, and the objective – and introduces several complications. It divides the possible states into control and state variables. More important, it adds the temporal dimension, so that all variables are functions of time. One of the constraints takes the form of a transition equation relating a state variable at two successive times. The objective is a temporal integral function of the control and state variables. The control problem is to choose the time paths of the control variables that maximize or minimize the objective, subjected to the constraints and appropriate initial and terminal conditions. The choice leads to differential equations for the state variables.

The case of household saving has only one state variable (asset) and one control variable (instantaneous consumption). Its transition equation is the equilibrium budget constraint; the saving is the change in assets. Its objective is the current total utility. Optimal control theory yields a differential equation for the rate change of consumption and consequently the rate of asset accumulation. More exactly, the result is an equation form, for it is expressed in terms of the instantaneous utility function and the time preference function discounting future utilities, the forms of which must be supplied for a definite dynamic equation. When the forms are specified, the differential equation gives the rate change of consumption at time t as a function of the variables at t. From it and a given initial condition at t_0, subsequent states can be determined, just as in other dynamic equations. Like that of other dynamic equations, its solution is often difficult if not impossible to obtain.

Control and variational theories are examples of synthetic microanalysis. Starting with an *integral* view that encompasses all states of a system or stages of a process during a finite temporal interval, they derive a *differential* view that follows the development of certain dynamic variables in infinitesimal time increments. The integral view taking account of entire processes provides a unifying synthetic framework; it is static not because it excludes temporal variations but because it includes them all. The differential view with its dynamic equations is more analytic; it enables scientists to delve into the temporal stages of unfolding processes.

The derivation of differential equations for certain dynamic variables from the maximization or minimization of an integral function is common in physics. Newton's second law can be derived in this way, as can other basic equations of motion in mechanics and field theories.[26] Newton's law follows from the principle of least action, which asserts that a particle going from one place to another always takes the path that minimizes the action, which is the temporal integral of the difference between the particle's kinetic and potential energies along the path. The principle of least action is analogous to the principle of maximization of the temporal integral of discounted instantaneous utility. Depending on the nature of the terminal condition, the results of the principles are the Euler–Lagrange equations or their variations, which are the generic names of equations that find application as physical equations of motion or economic dynamic equations.

Although the principle of least action and the principle of maximum current total utility are similar in mathematical structure, their interpretations are different. In physics, the integral view in which least action is minimized is the theoretical view of physicists, who stand outside the theory and do not explain why the action is minimized. In economics, the integral view in which current total utility is maximized is attributed to the household in each moment of its existence and interpreted as its expectation and plan for the future. Expectation is further interpreted as the humanity of economic units, for through it the state of the economy at one time is affected by the states at later times, and that is deemed impossible except through human anticipation and agency. If the integral view of the future adequately represented human expectation in decision making, time would not be such a difficulty in economics. It is not adequate, as many economists realize.

Stochastic Factors and Risk Reckoning[27]

So far we have considered only behaviors under conditions of certainty and perfect information, where each decision yields a unique result. In real life, risk and uncertainty are prevalent. Frank Knight has distinguished between risk and uncertainty: In both cases a decision has many possible outcomes, depending on circumstances beyond the individual's control or foresight. A situation is *risky* if the individual can list the possible outcomes and assign a probability value to each. It is *uncertain* if the individual is unable to enumerate the possible outcomes or to assign probabilities.[28] Knight's distinction is not commonly observed; probability distributions are customarily called uncertain.

The dynamic models of optimal control theory can be extended to include risk reckoning. Most economic models treat some states of the world as stochastic. Economic units do not know beforehand which one of a range of the world's possible states will be actualized at a specific date. However, they know exhaustively the range of future possibilities, and in most cases they are able to assign numerical probabilities to the possibilities. The probabilities are often represented as distributions over the possible states. The interpretation of probability varies. For some it is "objective" like the posted odds of lotteries; for others it is "subjective" and descriptive of the degrees of belief of an economic unit. Objective probabilities are the same for all participating economic units. Subjective probabilities differ from unit to unit; it is implausible that all units have the same degree of belief, even if they all carry the "conformer gene." On the basis of the numerical probabilities, economic units compute the *mathematic expected utility*, which is the average of the utilities in various possible states of the world weighted by the probabilities of the states. They then maximize the mathematic expected utility.

In stochastic control theories, the current mathematic expected utility is often conditioned on the information available at specific times. The optimization problem is analyzed into a sequence of problems, each of which involves two successive times. For instance, in the case of household saving, future interest rates can be a random walk with a stochastic element represented by a certain distribution. Each actualized step of the random walk yields information. At each time an economic unit chooses a consumption plan but is not committed to the chosen future consumption; the plan is revised at later times as more information about the random walk becomes available. The future gains a more definite meaning in these models, although its full meaning is still elusive.

Uncertainty and the Future

Knowledge and action are among the most important elements in the asymmetry of the time in our everyday life. Actions are forward looking, and deliberate actions imply some kind of knowledge and steady expectation about the future. We are not totally ignorant about the future; we can predict the occurrence of the next eclipse more accurately than we can tell what happened in a murder scene last night. The things we use everyday are dependable although not absolutely so; the people we deal with behave more or less consistently, and we can trust their reputation if the stake is not too high. The reliability makes possible intelligible experiences with their notion of temporality. Because of it we can form stable and persistent expectations about the future. We do not know the details of what will happen, but for routine matters in the near future it is not unreasonable to assume that we can roughly enumerate the possibilities or even assign numerical probabilities to them. The routine decision making is quite satisfactorily represented in economic models with risk.

What the economic models miss is that life is more than routines; it is also full of surprises. Surprises increase exponentially with the expanding temporal horizon and the increasing complexity of situations, which are the typical conditions of long-term investment decision making. The importance of uncertainty in Knight's sense was hammered home by Keynes, who argued that we cannot assign probability to future uncertainties, about which we simply do not know. Yet life never stops to wait; business decisions must be made whether or not the desired knowledge is available. Holding off investment in the fear of ignorance is itself a decision, one that may have the dire consequence of generating deficient demand and depression. Consequently many decisions are made with a certain amount of "animal spirits," utter doubt, precariousness, hope, and fear. Plans and expectations may be unstable and subject to sudden and drastic reversals. The volatility of decisions is enhanced by the interaction of opinions; we have to speculate about not only the states of the world but also the expectations of other people. Keynes emphasized that the instability of beliefs due to the "dark forces of time and ignorance which envelop our future" has great economic ramifications.[29]

The uncertainty of the future still eludes theoretical economics. Hahn and Solow said recently:

The longer the effects of a current decision will last, the more that decision is affected by uncertainty, and the larger the number of inferences that the decision maker has to risk until the game is over. Probably this is why durable capital is so hard to handle outside of steady state or other transparent situations, and why either perfect foresight or arbitrary convention is the usual way out for macroeconomics. The problem with both of those artifices is that neither of them really captures the buildup of "Knightian uncertainty" during the lifetime of the fixed capital. It hardly matters whether this sort of uncertainty is essentially nonprobabilistic (as Knight or Keynes thought) or just much too complicated to probabilize. In either case we still lack a clearly satisfactory way to go about modeling decisions and their consequences.[30]

In real life, often we not only are incapable of appraising probabilities, we cannot even enumerate all the future possibilities. The ability to list possibilities decreases with the time horizon and the degree of detail required. Classical contracts that specify the obligations of the parties under all conceivable contingencies have been mostly abandoned in modern times, for issues become so complicated the anticipation of contingencies becomes practically impossible. Managers cannot list the possible merchandise available ten years hence to conclude comprehensive contingent trade contracts, for some future products are yet unheard of.

All the scientific theories we have examined postulate state spaces that encompass all possible states of the systems under investigation. The idea of a state space appears to be inappropriate when the systems become humans with their memory of the past and openness to the future. When state spaces are posited in the integral and theoretical views, as in economic models, human beings are torn out of their past and future and treated as things. In thinking about humans living in their world, the general concept of possibilities remains as vital as ever, but the substantive circumscription of possibilities in a state space is no longer feasible. The future is possibility, but we are also aware that some future possibilities are beyond our wildest dreams now and that our actions will be partly responsible for bringing them about. Anticipation, action, and unfolding novelties contribute to the dynamics of human temporality. This dynamic is lacking in the models in which all contingent courses are spelled out beforehand. As history defies the constraint of a covering law, human beings with their awareness of the past and future defy the representation by a comprehensive state space. State spaces are for closed systems, which are the major topics of science. The human mind is intrinsically open. To understand humans a whole new class of categories is needed.

Epilogue

43. A Look Backward and Forward

Science reveals complexity unfolding in all dimensions and novel features emerging at all scales and organizational levels of the universe. The more we know the more we become aware of how much we do not know. Gone is the image of a clockwork universe. Equally untenable are the image of a clockwork science that claims to comprehend all the diversity by a single method and a single set of laws and the clockwork scientists who are absorbed in deducing the consequences of the laws by applying given algorithms. Scientific research is a highly creative activity. Scientific creativity, however, is not anything-goes arbitrariness. There are general guiding principles, which are discernable across diverse disciplines.

We have examined the general way in which theoretical reason comes to grip with complexity. This synthetic microanalytic approach is not restricted to science. Readers would hear a familiar ring in the following passage from a textbook in computer engineering: "The techniques we teach and draw upon are common to all of engineering design. We control complexity by building abstractions that hide details when appropriate. We control complexity by establishing conventional interfaces that enable us to construct systems by combining standard, well-understood pieces in a 'mix and match' way. We control complexity by establishing new languages for describing a design, each of which emphasizes particular aspects of the design and deemphasizes others."[1]

Scientists lack the engineer's liberty to design clean interfaces and standard modules, because they are constrained by the natural phenomena they try to understand. Nevertheless, they too control the representation of complexity by high-level abstractions that hide details; by approximate models that emphasize particular aspects of natural phenomena; by their best efforts to discern patterns and regularities, carve nature at its joints, and analyze a whole into typical parts that interact only weakly with each other. The nearly decoupled parts come in all sizes and structures, which I called independent individuals (Ch. 4) and collectives (Ch. 5). Modularization confines much of the difficulty of multilateral relations in complex systems to the internal structures of the parts, which are easier to analyze because they tend to be more coherent and smaller in scope. If the parts are still too complicated to understand, at least the modules can be treated as black boxes whose behaviors can be found empirically. The similarity in scientific and engineering approaches is not surprising; scientists and engineers all use the theoretical reason common to humans.

Nature does not always oblige our penchant for analysis by exhibiting joints. Many natural phenomena are not explainable by modularization. Emergent phenomena such as phase transitions involve structures that span entire systems and cannot be obtained by summing the contributions of parts, however they are defined (§ 24). Emergent properties of systems often bear little resemblance to the features of the constituents, for they belong more to the structural aspect than to the material aspect of the systems. Furthermore, their representation often requires a scope of generalization that differs from the scope of generalization for the constituents, for instance, by the inclusion of contingent or accidental factors. Because of the difference in generalization, the concepts characterizing emergent and constituent properties are often "out of joint." Emergent properties are missed in models that straightforwardly extrapolate or aggregate constituent behaviors. Thus structural unemployment is incomprehensible in models that treat the economy as a scaled-up household (§ 25), and the evolution of novelty is difficult to explain in models that consider only the accumulation of minor changes (§ 26). Emergent properties are volatile and hard to control, because the effects of a few constituents can propagate throughout the system in the thoroughly interconnected structure. They may be unwelcome to engineers, but they make nature much more interesting. Many-body theories show that these exotic phenomena result from the large-scale composition of familiar things and simple relations without any intervention of exotic force or substance.

Many-body systems are complex as a result of the large number of constituents and interconnections among the constituents. They are susceptible to theoretical treatment for several reasons. First, they are rather homogeneous; their constituents and interconnections are typical and not specialized. Typicality is conducive to generalization and statistical treatment. Second, some properties essential in their macrodescriptions are resultant and representable in cumulative variables such as total energy, aggregate income, or number of surviving organisms. The resultant properties are like the pillars connecting two floors. The micro- and macrodescriptions can generalize and classify in their own ways like two disparate floor plans, yet they are united by the resultant properties. Third, the systems are ideally isolated or partially isolated, so that the influence of the rest of the world is packaged in exogenous parameters that are unaffected by the dynamics of the system. The controlled isolation enables us to circumscribe relevant causal factors in theoretical models. Fourth, we accept coarse-grained accounts of system behaviors, for instance, the statistical descriptions of dynamic processes that forsake the successive steps of their temporal evolution. We sacrifice detail for generality. Fifth, we often settle for equilibrium models, even if the systems are intrinsically dynamic.

When these conditions are not satisfied, theorization is more difficult. Some of these conditions are ill justified even for some many-body phenomena. They are tolerated on pragmatic grounds; they are the best we have so far. We have seen how theoretical models yield to historical narratives when natural selection, which is a dynamic process wide open to the environment, is taken seriously in evolutionary biology (§ 41).

Among complex systems, many-body systems are simple, at least as they are idealized in the theories. Nevertheless, the lessons we glean from many-body theories are valuable for the understanding of more complicated phenomena. Science will generalize and theorize whenever it can. We can expect that theoretical models for highly complex phenomena will be more partial and regional than those for many-body phenomena. Hence it is even more important to have a broad general conceptual framework that makes sense of these substantive regional models, highlights the kind of question they answer, delimits their significance for the phenomena under study, and prevents us from falling into excessive claims or dogmatic proscriptions.

Consider, for instance, the mind–body problem, which interests many people. Most who reject soul stuff regard mind as a high-level property of humans who are purely physical beings. How do we theoretically represent mind and its relation to body? No scientific theory yet exists, but there are plenty of philosophical doctrines, the most prominent of which are mind–brain identity and functionalism. Mind–brain identity, a brand of reductionism, defines a type of mental state by identification with a type of predefined brain state, as pain is identified with the firing of a particular kind of neuron called a c-fiber. The identification dispenses with the corresponding mental concepts. There are many objections to it, but the "knockout argument" served up by functionalists is its inability to entertain the possibility that other kinds of system such as computers can also have mental properties. This multiple-instantiation argument converted many to functionalism, which defines mental states holistically and abstractly in terms of their functional roles. Functionalist mental states are not analyzable individually and are meaningful only en masse within a certain "program." Furthermore, they are totally abstract and can be "implemented" in any substance, including electronic hardware or soul stuff.[2]

The continuing debate between mind–brain identity and functionalism shows how effort can be wasted when one speculates about theoretical forms without examining relevant existing cases. Scientific theories include two broad types of explanation for composite systems: Macroexplanations elucidate the characters and dynamics of the systems as wholes without mentioning their composition. Microexplanations connect the macrobehaviors to the behaviors of and interactions among the constituents. Functionalism advocates a holistic macroexplanation for mind, mind–brain identity a simplistic microexplanation. They argue at cross purposes.

On the basis of our knowledge of existing theories, we can see the implausibility of eliminating mental concepts in favor of neural concepts. Except aggregate quantities, we are seldom able to recognize independently types of constituents behaviors for identification with system behaviors. Instead, we delineate microtypes only with the help of concepts characterizing system behaviors, for instance, in the definition of ensembles in statistical mechanics (§ 11). This time-tested approach stands the reductive thesis of mind–brain identity on its head.

The fall of mind–brain identity, however, does not imply the triumph of functionalism, whose "knockout argument" packs little punch, because multiple instantiation of high-level properties is a commonplace. It is found, for example, in the universality of scaling laws in phase transition (§ 23), Feigenbaum's parameters in period-doubling bifurcation (§ 30), Kauffman's ahistorical universals for organic evolution (§ 24), and the hydrodynamic equations (§ 38). These macrocharacters are found in systems with the most diverse constituents. Their macroexplanations are ordinary causal theories, which also include constituent-specific features such as the critical temperatures of phase transitions (§ 23) or the viscosity and other transport coefficients of various hydrodynamic systems (§ 38). Nowhere do they call for the extreme abstraction and unanalyzability of functionalism. Furthermore, their microexplanations (§ 23, 38) are nothing like the functionalist "implementation," which demands a miraculous structural isomorphism. Doctrines too simplistic even for many-body problems can hardly be expected to be adequate for the mind–body problem.

To bring the behaviors and micromechanisms of complex systems under the ambit of theories, idealizations and coarse-graining approximations are mandatory. Theories seek the general and the regular, and they pay a price for them. The more coarsely we look at a system, the more likely it is that we can descry patterns and regularities. In making theoretical generalization, we overlook many peculiarities of the microstructures, average over others, and throw still others into the pile labeled *randomness* or *contingency*. We call the suppressed details irrelevant. Relevancy, however, is relative to specific models aimed at investigating certain phenomena at certain degrees of abstractness. Information irrelevant to the microexplanations of system behaviors may be very relevant to someone interested in the behavior of a particular constituent in the system, which statistical theories overlook. Thus scientific theories do not answer all questions, because the information pertinent to many questions falls through the cracks of approximation and modeling. Sometimes the questions left unanswered are the ones many people deem important; people are often more concerned with particular details than universal generalization, for instance, in the evolutionary course of a particular species. In these cases theoretical science should refrain from spinning just-so stories that rhetorically extrapolate crude models beyond their regions of validity. It should acknowledge its limits and yield to other methods of inquiry such as historical narrative, which, although less adept in predicting and controlling, nevertheless abet understanding.

Scientists have developed many powerful conceptual tools to investigate complex phenomena. Users of powerful tools, however, always run the risk of falling prey to them, as those who have lived under the shadow of The Bomb know. Technicality can captivate the mind and turn scientists into calculators. Sometimes a technique gains a mystical aura because people forget that meaning lies not in the technique but in the realistic situations that summon its use. Thus applications of the optimization calculus generate a narrow-minded notion of rationality (§ 13) and the Panglossian misinterpretation of evolutionary biology (§ 17); applications of the probability calculus generate the illusion of a force called chance (§ 39); more generally, applications of mathematics generate the dream of reductionism (§ 5). Sometimes the dominance of a powerful theoretical technique steers the research direction of an entire field and inhibits attention to factors the method cannot handle. Even worse, factors unmanageable by the technique are sometimes stigmatized as "unscientific" and those who worry about them as "superstitious."

To claim much and deserve as much was the "crown of the virtues" to the ancient Greeks.[3] Science can proudly claim the Olympian gold, on the basis of its achievements and contributions. The crown of virtues, however, can degenerate into hubris, the deadly vice of insolence and wantonness. Such degeneration is visible in scientism that promotes the illusion of omniscience and omnipotence, advertises the final theory of everything, urges the corrosion of all meaning, wields "science" as a bludgeon to beat ideas not to its liking into the line of alchemy and witchcraft. Overselling science while betraying the scientific spirit, scientism provokes the antagonism toward science that surprises many scientists.[4] Criticism starts at home. Instead of merely complaining about the diminishing public willingness to support scientific research, perhaps we should examine ourselves to see whether we have given too much over to scientism.

Notes

1. Introduction

1. The three kinds of particle are the electron and the two light quarks. There are twenty-one other kinds of known elementary particle: the neutrino, two sets of copies of electron, neutrino, and quarks with heavier masses, and the antiparticles. However, these particles do not constitute stable matter and are found only in high energy cosmic rays or in particle accelerators. The four kinds of fundamental interaction are gravity, electromagnetism, and the two nuclear interactions. As explained in § 4, the formation and transformation of all medium-sized things are essentially the work of electromagnetism. These particles constitute the known or luminous part of the universe. Evidence from gravity indicates that most matter in the universe is dark, with unknown constituents.

2. Theories of composition have received little philosophical analysis, although there is no lack of doctrines, mostly reductive, that stipulate the forms they should take. Shimony (1987) is exceptional for its descriptive account of the methodology of synthesis.

   Economists are traditionally reflective on the methodology and implications of their research. See, for example, Machlup (1991), Robbins (1984), Shackle (1972), and the many articles collected in Hausman (1994). Some economists have borrowed ideas from evolutionary biology; see Jacquemin (1987) and Nelson and Winter (1982). Mirowski (1989) relates the exchange of ideas between economics and physics in history. The past two decades have seen a revival of interest in the philosophy of social science, especially of economics. See, for example, Blaug (1992) and Hausman (1992). Rosenberg (1992) compares economics with biology. Generally, the philosophical emphasis is on methodological issues such as falsification and paradigm. The assumption of rationality in economic theories is also much discussed. There is a large literature relating economics to social, political, and ethical problems.

   The philosophy of biology is a boom industry in the past twenty years. Attention is concentrated on the conceptual muddles in evolution, the methodological debates in systematics, the spin-off controversies over sociobiology, the relation of evolution to human nature, and the reduction of social science to biology. The philosophy of biology is deeply involved with the substance of the science; philosophers contributed about a third of the articles in a recent dictionary on the keywords in evolutionary biology (Keller and Lloyd, 1992). Some recent books are Depew and Weber (1995), Dupré (1993), Rosenberg (1994), and Sober (1993). They contain extensive bibliographies to the mountain of earlier works. Many biologists reflected on their own field; two examples are Mayr (1988) and Lewontin (1991). Sober (1984) collects many interesting papers.

The foundation of statistical mechanics has been fiercely debated by physicists. O. Penrose (1979) gives a comprehensive review; Lebowitz (1993) is less extensive but more accessible. It has attracted the attention of philosophers, although to a much lesser extent than relativity, quantum mechanics, or biology. The extensive critical review of Sklar (1993) covers the works of physicists and the philosophical concepts involved in their interpretation. As typical in the philosophy of physics, the emphasis is on metaphysical problems such as the directionality of time and technical problems such as the ergodic hypothesis.

Philosophers have produced many theories for the meaning of probability; see Weatherford (1982) for a review of the major positions. The host of recent books, for example Hacking (1990) and Plato (1994), are oriented toward the technical and historical development of the probability calculus. More references can be found in Auyang (1995, Appendix A), where I analyze the structure of the probability calculus and interpret its elements.

Recently, there is much research effort on chaotic dynamics and its applications to physics, biology, and economics. Chaos and the complexity of dynamics have generated an avalanche of popular books. For a philosophical reflection, see Kellert (1993).

3. The most extensive cross currents occur in the Sante Fe Institute for the Science of Complexity. Since 1984, the institute has regularly organized interdisciplinary workshops, lectures, and seminars, the proceedings of which are published by Addison-Wesley; examples are Anderson et al. (1988), Stein (1989), Perelson and Kauffman (1991), Cowan et al. (1994).

4. The Pauli exclusion principle applies to fermions such as electrons and quarks. It does not apply to bosons such as photons, many of which can share identical properties. Only fermions constitute tangible matter; bosons are responsible for the interaction among fermions.

5. Raup (1966).

6. See, for example, Chatin (1975). For an n-digit sequence, the lower end of complexity is on the order of $\log_2 n$ for large n, and the upper bound of complexity is n.

7. More correctly, the computation time varies with the length of the input sequence specifying the problem to the computer. There are many schemes of coding the problem, but the results of computation time are quite independent of the coding schemes. See, for example, Gary and Johnson (1979).

8. Strawson (1959, p. 9).

9. In our times, analytic philosophers argue that the way to analyze thought is to analyze language, in which thought is conveyed and from which it cannot be separated. Despite the linguistic turn, the philosophical goal persists. W.V.O. Quine states, "The quest of a simplest, clearest overall pattern of canonical notation is not to be distinguished from a quest of ultimate categories, a limning of the most general traits of reality" (1960, p. 161).

Analytic philosophers, too, are descriptive or revisionary. Descriptions of natural languages often lead to the conclusion that they are so rich and complex that their structures cannot be formalized in neat systems. Revisionists scorn the messiness of natural languages and try to devise artificial formal languages more susceptible to logical analysis or to regiment natural languages into canonical forms. Quine explains how the revisionary regimentation is achieved: "We fix on the particular functions of the unclear expression that makes it worth troubling, and then devise a substitute, clear and couched in terms to our liking, that fills those functions" (1960, pp. 258–9).

10. Auyang (1995).

11. Kant (1784).

12. There are many books on the history of economic thought and analysis. The classic is Schumpeter (1954), which is still unsurpassed in its breadth of intellectual vision and depth of detail. Blaug (1985) is more technically oriented. Deane (1978) gives a concise history of the major economic ideas. Polanyi (1944) sets the economic ideas in the context of the politico-economical history of the nineteenth century.

13. Smith (1776, p. 532). Smith cited oppression as one of several possible social arrangements. He generally believed that wealth benefits people in all classes.

14. John Galbraith has said, "In the months and years following the stock market crash, the burden of reputable economic advice was invariably on the side of measures that would make things worse" (1972, pp. 187–91). Such harsh judgment is mitigated in some recent accounts of the Great Depression.

15. Keynes (1937, p. 119, his italics). Although some effort in social accounting was made as early as the seventeenth century, the first credible estimate of the British national income was not made until 1911. The modern system of national income accounting was developed in the late 1930s, when Keynesian economics gained influence.

16. The epithet *neoclassical* was first used by Veblen to designate the economics of Marshall. Later it was extended to cover the marginalist school. Hicks and Stigler identified the core of neoclassical economics as consisting of the subjective theory of value, marginal substitutivity, and the emphasis on the self-centered behavior of individuals. As such it is mainly microeconomic. The term was popularized by Samuelson's classic textbook *Economics*. Starting from the third edition of 1955, it presented a "grand neoclassical synthesis" that covers macroeconomics.

17. Blanchard and Fischer (1989, p. 27).

18. Economists have various ways of classifying them. Phelps (1990) distinguishes seven schools according to two criteria: (1) whether they assume real wage to be flexible, sticky, or rigid; (2) whether the expectation they assume is "rational" or not. Leijonhufvud (1992) offers another scheme in which various schools are classified according to their assumptions on the nature of the shocks that disturb the economy and the propagation mechanisms of the shocks. The shock or the propagation mechanism can be either real or nominal.

19. Almost ten books on the history of probability and statistics have appeared in the last decade or so, for example, Porter (1986), Gigerenzer et al. (1989), Hacking (1990), Plato (1994). There are many interpretations of probability; see Weatherford (1982). I have briefly reviewed them while analyzing the structure of the probability calculus (Auyang 1995, Appendix A).

20. Schumpeter (1954, pp. 12–14).

21. Darwin read Quetelet at about the same time he read Malthus and the works of other political economists. He acknowledged his debt to Malthus only, but some interpreters argue that these works had some influence on him (see Bowler, 1989, pp. 172–4; and references cited therein).

    Maxwell read about Quetelet's work in a review by John Herschel when he was nineteen and wrote to Lewis Campbell saying that statistics was the "true Logic for the world." His derivation of the velocity distribution, published ten years later, bears strong similarity to Herschel's argument for the normal distribution. Later he said in explaining the rationale of the kinetic theory, "The modern atomists have therefore adopted a method which is, I believe, new in the department of mathematical physics, though it has long been in use in the section of Statistics" (Brush, 1976, § 5.1; Porter, 1986, Ch. 5).

22. There are many historical accounts of evolutionary biology. Bowler (1989) narrates the development of evolutionary ideas in a broad cultural context. He has written several other informative books on related topics. Mayr (1982) describes the development of evolution, systematics, and genetics colored with interpretation. Depew and Weber (1995) includes recent advancements in the field.

23. Darwin (1859, p. 6).

24. George Simpson made a triple distinction: microevolution for changes within a species, macroevolution for speciation, and megaevolution for the origin of higher taxa. Some biologists use *macroevolution* for changes in taxa above the species level. I combine the two usages.

25. Gould (1983).

26. Brush (1976) narrates the development of the kinetic theory of gases and statistical mechanics up to 1915. Brush (1983) also covers the early application of statistical mechanics to various areas including the quantum theories of solids and liquids.

27. Hoddeson et al. (1992) is an extensive history of solid-state physics up to 1960. The volume on condensed-matter physics in the series *Physics Through the 1990s*, published by the National Academy Press in 1986, gives a good and readable survey of the field.

28. Quantum mechanics generated two new ways of accounting: the Fermi–Dirac statistics for fermions such as electrons, protons, neutrons, and the Bose–Einstein statistics for bosons such as photons. Both can be handled readily by statistical mechanics.

2. Theories of Composite Systems

1. Feynman et al. (1963, Vol. I, p. 3–10). Feynman quoted an unknown poet: "The whole universe is in a glass of wine." Similar ideas are familiar in Asian philosophy: A Taoist script says, "The heaven and earth are manifested in a finger, ten thousand things in a horse"; a Buddhist script states, "A mustard seed contains the universe, a grain of dust the world." These sayings are much older than William Blake's "To See a World in a Grain of Sand."

2. Coase (1937) and Chandler (1977) discuss the causes and history of the emergence of large corporations and administered coordination.

3. Szathmáry and Maynard Smith (1995) discusses the major evolutionary transitions and the emergence of large compound units. Other examples are transition from unlinked gene to chromosome, ribonucleic acid (RNA) to deoxyribonucleic acid (DNA), prokaryote to eukaryote, asexual clone to sexual reproduction, solitary individual to colony.

4. The dual holistic and atomistic nature of quantum field theory is discussed in Auyang (1995).

5. A more general and serious consequence of bare elements is the collapse of the scheme of predication, because predication depends on the relation of similarity. See Auyang (1995, § 26).

6. Consider the way electrons couple to each other via the electromagnetic interaction. In ordinary physical theories, we write three statements: The first two are subject–predicate statements, each describing the kinetic energy of an electron in terms of the electronic mass m. The electrons are treated as if they are isolated and m is called the free electron mass. The third statement is relational; it spells out the electrostatic coupling, which depends on the electronic charge e and the distance between the electrons. The formulation gives the impression of two bare elements and an imposed relation. The impression is wrong; the electronic mass m and charge e represent not characters of bare electrons but intrinsic relations, as shown in quantum electrodynamics, which investigates the fundamental interaction of electromagnetism.

   Electrons interact via the electromagnetic field. Quantum electrodynamics again starts with two subject–predicate statements and a relational statement. This time the subjects are the electromagnetic field and a bare electron characterized by a mass parameter, and the relation between them is characterized by a charge parameter. When physicists tried to solve for the details of the interaction between the bare electron and the electromagnetic field, hell broke loose. The results were all infinities, which make no physical sense at all. The problem haunted physics for twenty years. Its solution, the "renormalization procedure" developed toward the end of the 1940s, has become the cornerstone of elementary particle theories.

   The basic idea behind the complicated mathematics of renormalization is that the bare electron has no physical meaning. The electromagnetic field, which alone enables the electron to couple with other electrically charged particles, is generated by the electronic charge, which is an intrinsic character of the electron. Being the spontaneous source of its own interacting mechanism, the electron is forced to interact with whatever enters within range of its field. Even if there happens to be no other charged particles within range, it must interact with its own field, and the self-interaction energy gives rise to infinities. An electron is fundamentally a part of the electron–electromagnetic field system. The bare electron mass and charge parameters posited initially are at best hypothetical, because they try to isolate characters that cannot be isolated. At some point in the calculation, they must be replaced by parameters that have taken some account of the interaction. The renormalization procedure folds the self-energy into the mass and charge parameters so that the infinities cancel each other, then replaces the result with the "real" free-electron mass m and charge e measured in experiments. These are the mass and charge not of a bare electron but of an electron fully "dressed" in the electromagnetic field, and they ensure that the electron is relatable. Even when an electron is observed far away from any other particles, it is always accompanied by its own electromagnetic field. If it were otherwise, we would not be able to observe it at all. As for the bare parameters that initially appear in the equation, physicists are not sure whether they are infinity or zero. In any case they are insignificant. For more discussion, see Auyang (1995, § 29 and, the reference cited there).

7. Hobbes (1839, Vol. II, p. 109).

8. The contractarian view in political philosophy was influential from the midseventeenth to the early nineteenth century, developed in various forms by Hobbes, Locke, Rousseau, and Kant, and recently revived by John Rawls. In Rawls's meticulously argued theory, the state of nature is replaced by an original position, where representations of free and equal citizens negotiate behind a veil of ignorance for the fair terms of social cooperation (1971, Ch. 3).

9. Careful theories of reduction were developed by positivists, including Rudolf Carnap, Carl Hempel, Ernst Nagel, Paul Oppenheim, and Hilary Putnam. The first reductionist program, which tried to reduce terms about physical objects to terms about sense impressions, failed despite the effort of able logicians. Philosophers then turned to science, as illustrated by Nagel's statement "The objective of the reduction is to show that the laws or general principles of the secondary science are simply logical consequences of the assumptions of the primary science" (Nagel 1960, p. 301). The program is no more successful. Many classic articles are collected in Boyd et al. (1991). Charles and Lennon (1992) and Cornwell (1995) contain more recent articles on the issue.

10. Oppenheim and Putnam (1958) is the best known paper advocating microreductionism. It argues there are six necessary and sufficient reduction levels: social groups, living things, cells, molecules, atoms, and elementary particles. Theories about cells cannot be reduced precipitously to theories about elementary particles; they can only be reduced to theories in the immediately lower level. However, since the levels form a hierarchy, the paper urges as a working hypothesis "the possibility that all science may one day be reduced to microphysics."

    Microreductionism is different from Weinberg's grand reductionism, which is not a methodological doctrine and which stands in contrast to petty reductionism addressing the relations between systems and constituents (Weinberg, 1995). Microreductionism is petty in this view.

11. Lukes (1973) gives a detailed discussion of methodological individualism. Lycan (1990, Part IV) collects some papers on material eliminatism.

12. The term "methodological individualism" was first used by Joseph Schumpeter in distinction to "political individualism." The debate between methodological individualism and holism has flared up time and again. One such debate was between the neoclassical economists and the more holistic German historical school of economics. Another occurred in the 1950s, when methodological individualism was championed by, among others, Friedrich Hayek, Karl Popper, and J.W.N. Watkins. In this debate, the argument for methodology is often mixed up with arguments for political or economical individualism. Blaug (1992) considers the methodology in the context of economics (pp. 42–7, 209–12).

    Methodological individualism should not be confused with the following individualistic doctrines: Individualism includes the ideas of dignity of the person, autonomy, self-development, and privacy. It is promoted by philosophers as diverse as Kant, Mill, and Marx and has been included in various declarations of human rights. Economic individualism calls for private property, free trade, and competition. It embraces what C.B. Macpherson has called "possessive individualism," which is much narrower than individualism, although it may not be incompatible with it. Political individualism advocates a form of government based on the consent of citizens and aims to preserve personal liberty. Ontological individualism, which asserts that society or social institutions comprise persons, is a case of atomism. None of these doctrines implies methodological individualism, although they are often confusedly used to bolster the latter.

13. Anderson (1972); Purcell (1963, p. 21).

14. See Frey (1977, Chs. 2 and 8) for the differences between human and computer chess. An average master-level chess game has 42 moves (84 half-moves or moves of one side). It is estimated that the average number of *good* moves in a given position is only 1.76. Considering only good moves, there are 4.3×10^{20} possible positions for a game of 84 half-moves. Chess grandmasters on the average consider a total of 35 moves, and for important cases, they can think ahead seven half-moves in depth. The top-notch programs use some heuristic rules to prune the search space but rely mainly on advanced hardware technology for fast evaluations of positions. For instance, IBM's Deep Blue searches up to 200 million positions per second.

15. Holistic ideas were first expounded by Parmenides, who argued that all reality is one, continuous, and unchanging, and the integral whole alone is the suitable subject of thought. He criticized people who "name two forms, of which they need must not name so much as one" (in Kirk et al. 1983, pp. 255f).

16. Greek mathematicians used analysis and synthesis as methods of proof. Aristotle discussed them briefly, but not until the beginning of the third century A.D. did the geometer Pappus of Alexandria formulate them clearly. According to Pappus, in analysis, mathematicians seek the proof of a proposition by deducing consequences from it until they reach a consequence that is independently known to be true. In synthesis, they start with a known true proposition and end in the proposition to be demonstrated.

Plato discussed the methods of collection and division, which underlie our "ability to discuss unity and plurality as they exist in the nature of things." Collection means "to take a synoptic view of many scattered particulars and collect them under a single generic term." The generic term, good or bad, at least enables the argument to proceed with clarity and consistency. Division means to "divide a genus into species again, observing the natural articulation" (*Phaedrus*, 265–6). The Platonic methods come closer to the scientist's way of formulating and analyzing physical problems than to the mathematician's way of demonstrating propositions.

Methods of analysis and synthesis became truly dominant with the rise of modern science. Galileo's method is often described as that of "resolution and composition." Hobbes visited Galileo and adopted his method. Descartes regarded them as two steps of a single method, which he gave a prominent place in the *Rules for the Direction of Thought*: "We shall be following this method exactly if we first reduce complicated and obscure propositions step by step to simpler ones, and then, starting with the intuition of the simplest ones of all, try to ascend through the same steps to a knowledge of all the rest. This one Rule covers the most essential points in the whole of human endeavor" (Rule 5). Gerd Buchdahl noted that "analysis" in Descartes's writing can refer to methods of both discovery and proof (1967, pp. 126–30). Rule 5 is ambiguous about this distinction.

Newton's formulation of analysis and synthesis is unambiguously about empirical inquiry: "By this way of analysis we may proceed from compounds to ingredients, and from motions to the forces producing them; and in general, from effects to their causes. . . . And the synthesis consists in assuming the causes discovered, and established as principles, and by them explaining the phenomena proceeding from them, and proving the explanations" (1704, Query 31).

17. Bechtel and Richardson (1993) discuss many historical examples in physiological and psychological research. They argue that the general research procedure can be analyzed in the steps of decomposition and localization: The former resolves the behavior of the

system into various functions; the latter locates a function in a component.

18. The doctrine was suggested by Pierre Duhem and elaborated by Quine. In Quine's words, "Our statements about the external world face the tribunal of sense experience not individually but only as a corporate body" (1951, p. 41).

19. Author unknown, *APS News*, November 1995, p. 3.

20. A major topic in the philosophy of science is the ideal form of scientific theories. The two major ideal forms are, respectively, patterned after the proof theory and the model theory of mathematical logic. In the *syntactic view*, the ideal theory is a formal system, some of whose terms, called observation terms, are given explicit definitions by some correspondence rules. The privilege of the observation terms and the nature of correspondence rules are much disputed. In the currently popular *semantic view*, the ideal theory is a class of models, which are abstract definitions that make no reference to the world. The models are supposed to be isomorphic to the world, but the meaning of the isomorphism is unclear.

21. Keynes, letter to Roy Harrod, July 4, 1938, in Hausman (1994, pp. 286–7).

22. Hermeneutics originally applied to the interpretation of texts, but in contemporary Continental philosophy it is generalized to cover the general structure of human understanding. Original discussions of the hermeneutic circle are found in the works of Martin Heidegger, Hans-Georg Gadamer, and Paul Ricoeur.

23. Kant (1781, p. 304). Einstein said, "It gets nearer to the grand aim of all science, which is to cover the greatest possible number of empirical facts by logical deduction from the smallest possible number of hypotheses or axioms" (1954, p. 282). Notice the difference between the coverage of "the greatest possible number of empirical facts" and the absolute claim of "the theory of everything."

24. See, for example, Cornwell (1995), Galison and Stump (1996).

25. Auyang (1995, § 15).

26. A familiar definition of convergence in mathematics posits the limit x_0 in advance. Let $\{x_n\}$ be a real sequence. The sequence *converges to* x_0 if and only if for each $\varepsilon > 0$, there corresponds some positive integer N such that $|x_n - x_0| < \varepsilon$ if $n > N$. The definition is borrowed by some philosophers to suggest that the convergence of scientific theories implies a bare reality toward which the theories converge. They have forgotten the Cauchy convergence condition that does not invoke the limit. In Cauchy's version, a sequence $\{x_n\}$ is *convergent* if and only if for every positive ε there is a positive integer N such that $|x_n - x_m| < \varepsilon$ whenever $n \geq N$ and $m \geq N$. It does not posit in advance a x_0 toward which the sequence converges. This more general notion of convergence is adopted in topology. It is also a more appropriate analogy for the convergence of scientific theories.

27. Differential geometry contains two main ideas. The first is the patchiness of the local coordinate systems (idea 1 in the main text). The second is symmetry. Ideas 2, 3, and 4 in the text refer to the three elements in a symmetry: conventional coordinates T_1 and T_2, transformations $T_2 \cdot T_1^{-1}$, and the invariant object x. A transformation establishes the equivalence of representations in two coordinate systems, and the object is defined as that which remains unchanged under all transformations among all possible representations. The invariance signifies the object's independence of the representations. The objects are also the loci that glue the local coordinate patches into an integral manifold. For further explanation, see Auyang (1995, esp. the discussion of Figures 5.1, p. 96, and 6.4, p. 153).

28. Feynman (1965, pp. 124f). Feynman was not the only fundamental physicist with an antireductionist attitude; see, for example, the articles by Freeman Dyson and Roger Penrose in Cornwell (1995). For the rejoinder by a reductionist particle physicist, see Weinberg (1995).

29. Gould and Lewontin (1978) criticize the adaptation program in evolutionary biology as the "Panglossian paradigm" (§ 17). Hahn and Solow (1995) criticize some economics models: "That is the economics of Dr. Pangloss, and it bears little relation to the world" (p. 2).

30. Lewontin (1974, p. 269, his italic), Futuyma (1992), Maynard Smith (1982, p. 8). According to Lewontin, the first problem of population genetics is its empirical inadequacy:

Built into the theory are parameters that are not measurable to the required degree of accuracy (see § 37 for more discussion).

31. Aristotle argued that if heavenly motion is the measure of all motions, if the measure is the minimum, if the minimum motion is the swiftest, and if the swiftest motion is that which follows the shortest paths, which are circular for closed orbits, then planetary orbits must be circular (*De Caelo*, 287a).

Leibniz distinguished four kinds of why questions. Some questions, such as "Why does the earth revolve around the sun?" can be answered by referring to mechanical causes. More general questions, such as "Why do the equation of motion and the law of gravitation hold?" can only be answered by referring to God's having chosen this world above all possible worlds on the basis of what He thinks is best. Leibniz insisted that the standard of the best is objective and interpreted "divine perfection" to be "the simplicity of the means counterbalances the richness of the effects" (*Discourse on Metaphysics*, 5). For more on Leibniz's principle of the best, see Mates (1986, pp. 166–9).

32. For an accessible exposition of the principle of least action, see Feynman et al. (1964, Vol. II, Ch. 19).

3. Individuals: Constituents and Systems

1. Auyang (1995, § 15).
2. Plato, *Phaedrus*, 265–6.
3. Zhuang Zhou (sometimes translated as Chuang Tsu), "The Preservation of Life." This is the third of seven authentic chapters ascribed to Zhuang, who was one of the founders of Chinese Taoism.
4. The criterion of objective individuals is more stringent than the criterion of entities in logic. Individuals in the world are causally connected to each other. For the individuation of things, clearly recognizable boundaries are insufficient. If what is enclosed in a boundary violates a pervasive causal pattern, we would deny it the status of a thing. For example, the spot made on the wall by a spotlight has a clear boundary and moves continuously as the spotlight turns. Logically the moving spot can stand as the subject of a sentence. However, it cannot be counted as an individual thing, for it can violate fundamental physical laws. If the wall is far enough away and the beam can remain collimated all the way, the spot can "move" faster than the speed of light although the light source, which is properly a thing, moves slowly. Thus the moving spot is an illusion; it is not a thing and does not travel. The phenomenon we see is the succession of different light quanta scattering off different areas of the wall. The example is not a fantasy. Pulsars or rotating neutron stars are analogous to rotating spotlights. One of the most swiftly rotating pulsars sits in the Crab Nebula some 6,500 light years away. Besides the usual radio wave, it also emits a beam of visible light that shines on earth with a period of 0.033 second. The "spot" on earth gives the illusion of moving at 1.2×10^{24} cm/s, which is 40 trillion times the speed of light.
5. Auyang (1995, §§ 19 and 20). There I investigate the idea of individuals in field theories. According to modern physics, the fundamental ontology of the world is a set of fields. A field is a continuum, the popular image of which is an amorphous whole. The image is wrong. A field is rigorously a set of discrete field points or Events with definite characteristics. The systematic individuation of the Events is achieved with two ideas: The first is a state space individually assigned to each Event. Since the state spaces of all Events of a field are identical in structure, they embody part of the idea of a kind. The second idea is the spatiotemporal structure of the field, which comprises the numerical identities of all the Events.
6. Aristotle, *Metaphysics*, Books Zeta and Eta.
7. The association of a kind with a state space should not be confused with the definition by extension, in which the kind bird is defined by the collection of all birds. The state space accounts for possible states, many of which may not be realized in actual individuals.

Moreover, the state space is not merely a collection of possible states; it is the result of rules and contains many structures.

8. See, for example, Huang (1987, §§ 3.1 and 4.2).

9. A lucid explanation of ergodic theory and its relation to statistical mechanics is given by Lebowitz and O. Penrose (1973). O. Penrose (1979) is a more technical review of the issues concerning the foundations of statistical mechanics. Sklar (1993) provides detailed accounts of the historical development of ideas and physical and philosophical contentions.

10. Jancel (1963, p. xxx); O. Penrose (1979, p. 1938).

11. The other dispute concerns the general criterion of what makes a group of organisms a species. Like other classificatory terms, "species" has a type usage and a token usage, as the particular species *Homo sapiens* is a token of the type species. Biologists call a type a *category* and a token a *taxon*. The debate over the definition of the species category is mainly a contention among various schools of taxonomy. Many papers on both controversies over species are collected in Ereshefsky (1992).

12. Ghiselin (1974), Hull (1977). The 1989 issues of *Biology and Philosophy* contain more contributions to the debate. Many articles are collected in Ereshefsky (1992).

13. The following are the two major arguments against species as natural kinds: (1) The logical role of a species in the theory of evolution is an entity. As argued in the following main text, the idea of species as an individual does not exclude the idea of species as a kind. (2) The notion of kinds is suitable only for inanimate things and not for life forms because organisms vary widely and species evolve.

Organisms are substantively much more complex than inanimate things, but some biologists overstate the difference when they argue that organisms are peculiar not only substantively but also logically. As discussed in § 5, a type of character has a certain range of values and the type-value duality is general. Not only organisms, but electrons assume various states and are always described by distributions. Inanimate things also change, atoms decay radioactively, and the creation and annihilation of elementary particles are routine. It is tremendously difficult to spell out the criteria whereby an organism belongs to a species, but the difficulties are shared to a lesser degree by inanimate kinds. Physical things too have great variety. Polymorphism is not unique to the life world; diamond and graphite are polymorphs of carbon. Many mineralogy kinds are roughly defined; granite is approximately 75 percent SiO_2, 15 percent Al_2O_3, plus other trace chemicals. The assertion that typology is possible only if there are "unbridgeable gaps" in the features of various kinds is unwarranted. The compositions of alloys vary continuously; for instance, $Hg_{1-x}Cd_xTe$ denotes a kind of alloy with a fraction x of cadmium.

The technical difficulty in defining species is not alleviated by a change in the ontological status. The problem only shifts from delineating kinds to delineating individuals. Mayr's "biological species concept" invokes "potential interbreeding" of organisms. The potential is based on organismic properties. The physiological mechanism that produces scent or other signs in the mating season, the behavior of courtship, the mechanics of copulation, the cell chemical characteristics required for fertilization and zygote viability, the vigor of hybrids, all are properties of organisms. These properties can be used to differentiate species as kinds as well as individuals.

14. Dobzhansky (1970, p. 23).

15. There are two prominent relations in set theory: *membership*, which obtains between a set and its elements, and *part of* or conversely *inclusion*, which obtains between a set and its parts or subsets. Strictly speaking, an element is not a part of a set; what is a part is the singleton subset that consists of only that element. When we talk about the composition of the set, as we do in most of this work, what we mean by "element" is actually the singleton subset. That is why I always refer to them as *constituents* and not *members* of the system; members belong to a kind.

The relations of inclusion and membership lead to two general interpretations of sets. In terms of inclusion, a set can be interpreted as a whole comprising parts or subsets; similarly a subset is a smaller whole comprising smaller parts. The smallest parts are the singleton subsets. The wholes, big or small, can be regarded as particulars and individuals.

In terms of membership, a set is often interpreted as a universal or a common concept by philosophers. This is most apparent in the definition of sets by abstraction; for instance, the set "electron" is the abstraction from all electrons. In this case the set abstractly represents the universal "electron" and a member of the set is an instantiation of the concept. The two notions of sets underlie the notions of species as individuals and as natural kinds.

16. Wright (1931, pp. 97f.), Fisher (1922, pp. 321–2). The statistical nature of evolutionary biology has little to do with Mayr's "population thinking," despite the nuance of the word *population*. The clearest definition of "population thinking" given by Mayr is "Population thinkers stress the uniqueness of everything in the organic world. What is important for them is the individual, not the type" (1982, p. 46). Thus population thinking is the opposite of statistical thinking, which overlooks the individuality of particular items and counts only the number of items with certain typical traits. Perhaps that is why Mayr disputed the importance of the statistical theories of Fisher, Haldane, and Wright in the modern synthesis. I am unable to figure out the role population thinking plays in biological and evolutionary theories.

17. The metaphor that electrons die and are born at collisions is not outrageous. In the quantum mechanical description of collisions, it is generally impossible to tell which outgoing electron corresponds to which incoming one. The incoming electrons lose their identities because of the phase entanglement during impact. In the modern formulation of collisions of classical particles, the point of impact is a singularity, which is removed for analysis. The removal breaks the temporal continuity of the electrons. In both cases, the outgoing electrons are regarded as newly created entities because they are not individually identifiable with the incoming electrons.

18. Lewontin (1974, pp. 12–16).

19. For instance, consider an initial genotype distribution such that half of the population is in the genotype *Aa*, and a quarter each in *AA* and *aa*, so that the allele distribution is half *A* and half a. Suppose both homozygotes are lethal, so that all organisms with *AA* and *aa* die immaturely; selection is strong. However, the homozygotes are replenished in the next generation by the reproduction of the surviving heterozygotes. The genotype distribution remains constant from generation to generation; no evolution occurs in the situation of stable equilibria.

20. Darwin acknowledged that natural selection is not necessary for evolution: "I am convinced that Natural Selection has been the most important, but not the exclusive, means of modification" (1859, p. 7).

21. Many essays on the controversy over the unit of selection are collected in Brandon and Burian (1984) and Sober (1984). Maynard Smith (1987) and Sober (1993, Ch. 4) assess the debate from different standpoints.

22. Lewontin (1970).

23. Dawkins (1989, p. 40). The doctrine of genic selectionism originated in the work of George Williams (1966) and was popularized by Dawkins. Dawkins introduced the notions of replicator and vehicle, Hull the idea of interactor. Some philosophers have said that replicator and interactor are generalizations of genotype and phenotype; the assertion exhibits a basic logical confusion between universal and particular, type and token. Genotype and phenotype are universals; replicator and interactor are particulars.

24. Maynard Smith (1987). Ordinarily, during the reproductive segregation of chromosomes, each allele at a locus has half the chance of being passed on to the offspring. Meiotic drive occurs when an allele is consistently passed on with greater than 50 percent probability. The aggressive allele spreads if there is no countereffect on higher levels of complexity. Meiotic drive is rare; moreover, its effect can be counteracted by effects on the organismic level.

25. Varien (1992, Chs. 2 and 7).

26. Mill (1836, p. 52). A currently popular answer has been given by Lionel Robbins: "Economics is the science which studies human behaviors as a relationship between ends and scarce means which have alternative uses" (1984, p. 16). The definition is both too narrow and too broad. It is too narrow because it excludes most of macroeconomics,

which, operating on the aggregate level, does not invoke individual behaviors. It is too broad because finitude is a universal human condition and all our actions involve choice. The study of choice under scarcity in general is more descriptive of existential philosophy than of economics.

27. See, for example, Heap (1989, and references cited therein).

28. The Greek word that is translated as "reason," *logos*, also means "word," "speech," and "account." Plato explained in the *Republic* that to have knowledge of something is to be able to give an account of it, to say clearly what the matter is about. Generalizing the idea, an action is deemed rational if some acceptable reason or account can be given for it. One of Leibniz's major principles is that God does nothing without sufficient reason. The reasons we mortals give may not be watertight, but at least a rational person should be largely consistent in judgment, open to questions, responsive to evidence, critical of himself, and capable of mustering some defense for his opinions and actions.

29. For discussion of the plurality of utility, see Sen (1980).

30. Arrow (1987), Nelson and Winter (1982). Many bounded rationality models are found in the two volumes of Simon's collected works; see also Simon (1983).

31. In one type of experiment, subjects are asked to choose between two lotteries with the same expected return, one with a high probability of winning a small prize, the other with a lower probability of winning a bigger prize. Later, the same subjects are asked to name the amount they are willing to pay for the lottery tickets. The majority of subjects consistently prefer the lottery with the higher the probability of winning but pay more for the lottery with the higher prize. Since one prefers more money to less, the willingness to pay more for a less favored lottery is a preference reversal that contradicts the preference theory basic to economics. Most empirical studies are carried out by psychologists. The few experiments by economists yield the same result. Hausman (1992, Ch. 13) gives an interesting account of the results of the experiments and the reactions of economists, together with ample references. Blaug (1992, Chs. 6–8) gives a methodological appraisal of the theories of firms and of consumers and of the general equilibrium theory, with emphasis on empirical evidence.

32. Cyert and March (1963).

33. Friedman (1953). I wonder how well economics, which is not famous for its predictive power, fares under the positivist doctrine. Hutchison (1992) reviews the current trend in economics, showing the gap between theory and forecast. In 1987, there was an estimated total of 130,000 economists in the United States, of whom 52 percent worked in industry, 23 percent in government, and 20 percent in academia. Most of those working in industry and government were responsible for forecasting. The value of such forecasts is being increasingly challenged or even denounced by economics professors (pp. 67–9).

34. Kuznets (1963, pp. 44, 47f.).

4. Situated Individuals and the Situation

1. More discussion on methodological individualism is found in note 12 in Ch. 2. Steven Lukes (1973, pp. 118–22) roughly classified individual predicates according to the degree of social reference. On the one end are pure physiological and nonsocial notions such as brain states, on the other fully social notions such as voting and writing checks. He persuasively argued that explanations of social phenomena in terms of nonsocial or weakly social predicates are either incredible or partial. Almost all explanations in terms of concrete individuals involve social predicates. Thus social concepts have not been eliminated as demanded by methodological individualism; they have only been swept under the rug.

2. The material covered in this section is standard in many solid-state textbooks. For the free-electron model, see, for example, Ashcroft and Mermin (1976, Ch. 1).

3. Anderson (1963, § 2B), Ashcroft and Mermin (1976, Ch. 8).

4. The valence bands are usually quite full. When they are not completely filled, we meet a totally new kind of entity, a hole. A *hole* is a vacancy in the valence band. It is treated

exactly as a positively charged particle with its character described by the valence band structure and its behavior measured in experiments. Holes are manipulated as particles and play crucial roles in semiconductor devices such as transistors.

Dynamic processes such as electric conduction require the transfer of electrons from one state to another. Electrons in an energy band that is completely filled usually do not contribute to dynamic processes, because there is no readily accessible empty state for them to enter. These electrons are like people blocked from career advancement because all desirable positions in the company are filled. Their only opportunity is to be promoted to a higher band with vacancies, but that process often requires energies not readily available.

A material is an *insulator* if its valence bands are completely filled, its conduction bands completely empty, and its energy band gap large compared to the thermal energy. It is a *metal* if there are many electrons in the conduction band; these electrons are mobile and contribute to dynamic processes, with the result that metals are good conductors. If a material has only a few electrons in the conduction band, it is an *n-type semiconductor*. If it has only a few holes in the valence band, it is a *p-type semiconductor*. A transistor is made up of a piece of *n*-type semiconductor sandwiched between two pieces of *p*-type semiconductor. *Many* means more than 10^{21} particles per cubic centimeter; *a few* means 10^{12}–10^{16} particles per cubic centimeter. The particle type and concentration can be controlled by introducing impurities that accept or donate electrons.

5. Ashcroft and Mermin (1976, pp. 330–7). For the mean-field theory in phase transition, see Toda et al. (1983, § 4.5), Chaikin and Lubensky (1995, Ch. 4).

6. Anderson (1963, p. 9).

7. Ashcroft and Mermin (1976, pp. 348–9), Anderson (1984, pp. 70–84).

8. Anderson (1984, p. 85). Fermions are particles with half-integral spins; electrons are fermions.

9. Anderson (1963, p. 96).

10. The editors of *The New Palgrave: A Dictionary of Economics* said recently: "By the mid-1960s the A–D [Arrow–Debreu] model really did appear, what many people still consider it to be, the keystone of modern neoclassical economics." The remark appears in the preface to the volume entitled *Allocation, Information, and Markets*, which was published in 1989.

11. Arrow (1983, p. 135).

12. Arrow and Hahn (1971, Chs. 1–5). The topic is standard in intermediate microeconomic textbooks.

13. Most models assume free disposal of unwanted commodities, toxic or not. Waste disposal can be incorporated into economic models by setting a market-adjustable price on pollution. The idea has been implemented in the 1990 Clean Air Act, in which the emission of sulfur dioxide by utility companies is treated as a kind of "property right" that can be traded in the open market. So far the program is successful in reducing sulfur emission and acid rain.

14. The technical *competition* differs from its everyday meaning, which suggests price wars, promotion campaigns, brand names, and innovative products. In theory, such oligopolist competition constitutes market failure and imperfect competitiveness. In perfectly competitive markets, a price war is impossible because no one can set prices. Advertisement is a waste of resources because all consumers already have perfect knowledge of all products and their minds are firmly made up. Innovation is discouraged, for without protection against imitators jumping in immediately, there is little justification for the effort of invention. That is why technology is regarded as an exogenous factor and the advancement of technology a series of shocks perturbing the economy. In short, competition is so perfect it instantly wipes out any conspicuous advantage and reduces all individuals to the equal status of passivity.

15. The Invisible Hand Theorem, as dubbed by Samuelson, is the first theorem of welfare economics. The second theorem states that a Pareto-efficient equilibrium with any actual resource allocation can be achieved by a redistribution of initial endowments before the market process begins. See, for example, Varian (1992, Ch. 13). Since the initial

endowments are exogenous factors in economic theories, the second theorem of welfare economics effectively pushes the consideration of income distribution out of economics. On the basis of the theorem some economists argue that whatever desired distribution is better achieved by a lump-sum transfer of payment that creates a certain distribution of initial endowment, for example, a negative income tax, the market, which is what economics is about, will take over to make the distribution Pareto-optimal.

16. Solow (1989, p. 77).
17. Hart and Holmström (1987, p. 72).
18. See, e.g., Varian (1992, § 21.1).
19. The size of the economy does not figure explicitly in the general equilibrium models. However, it is tacit in the assumptions of the infinite divisibility of commodities and the exclusion of increasing returns to scale.
20. Varien (1992, § 21.4).
21. Synoptically, the reason why characters of totally unrelated individuals cannot be compared is explained in two steps: The first involves the way a single individual is characterized; the second turns on the concept of unrelatedness. Ordinarily, we compare individuals by using predicates; if snow is white and polar bears are white, then they are similar in color. The common predicates hide the convention of predication that implicitly relates the individuals. When the convention is made explicit in some scientific theories and abolished in the sharp formulation of unrelated individuals, the means of comparison collapses. If individuals are to be compared, then additional means must be introduced, either by some explicit interaction or by some convention.

The root of the incomparability is the general conceptual structure in which we make definite descriptions of objects. Usually, we describe an object by a definite predicate, such as being white. However, the predicate is a convention; we could have chosen another adjective. The exposure of the convention and the redress for its arbitrariness are an important advance in science. This is the major conceptual change brought by relativistic physics. It is interesting that shortly before Einstein's paper of 1905, economists who argued about cardinal and ordinal utilities adumbrated similar ideas. Today utility is represented in the same general way as physical predicates are represented in many fundamental physical theories – by what physicists call symmetry.

The following argument applies to objects. To avoid any shade of subjectivism, I treat economic units strictly as things. *Preference order* will mean simply a lineal order that is the objective characteristic of a thing. "**x** is preferred to **y**" is replaced by "**x** ranks higher than **y**."

The preference order is the objective characteristic of an economic individual. To facilitate analysis, economists represent it by a *utility function*, which systematically assigns a value of utility to each possible commodity bundle such that the utility of bundle **x** is greater than or equal to the utility of bundle **y** if and only if either **x** ranks higher than **y** in the preference order or the two rank equally. A utility function is a complicated predicate for a complicated character, the preference order.

There are many alternative utility functions for a preference order, because all that the utility function needs to represent is an ordering. Suppose in a preference order **x** ranks higher than **y**, which ranks higher than **z**. Then the utility function that assigns the values (3, 2, 1) is a representation of the order (**x**, **y**, **z**), the function that assigns (12, 7, 3) is another, the function that assigns (50, 11, 2) is a third, and so on. Generally, for a given utility function, a *monotone transformation* that preserves the order yields another utility function that is equally good in representing the same preference order. Since a whole class of utility function is possible, the choice of a particular one is a matter of convenience.

The representation of a preference order by utility functions is analogous to the representation of a geographic configuration by coordinate systems. By invoking a particular utility function, we can use predicate statements such as "The utility of bundle **x** is 10 and the utility of **y** is 5" instead of relational statements such as "**x** ranks higher than **y**." Similarly, by invoking a particular cartographic coordinate system, we can say "*A* is at

20°E 30°N and B is at 20°E 25°N" instead of "A is north of B." The coordinates of A and B will change in another cartographic system, but their relative positions do not change, for they belong to the objective state of affairs.

In physicists' terminology, both the utility and the cartographic representations are instances of *symmetry*. The utility functions, the monotone transformations, and the preference order form the triad of a symmetry: Conventional representations, symmetry transformations, and the objective state of affairs that is invariant under the transformations. A utility function is a conventional representation for the objective preference order, and various representations are connected by monotone transformations. In short, the character type of economic individuals has a monotonic symmetry. (For more discussion of the structure and interpretation of symmetry, see Auyang [1995, §§ 6 and 15].)

Symmetry is not limited to geometric configurations; it is used for qualities that are not spatiotemporal, for instance, the internal structures of elementary particles. Generally, we have an objective state of an individual, for instance, the preference order of an economic individual or the Minkowskian world of special relativity. The objective state is definitely described in terms of a complicated predicate, a utility function or a coordinate system. There are many valid predicates for an objective state, and the rules for translating among them are stipulated by the symmetry transformations, the monotone transformations for economic individuals, or the Lorentz transformations of special relativity. The choice of a specific predicate is arbitrary, provided it conforms to the translation rules. The *constrained arbitrariness of definite predicates* underlies the incomparability of the characters of totally unrelated individuals.

The rationale for symmetry is that definite predicates always involve certain conventional elements and therefore say too much. Symmetry transformations erase the extraneous information carried in the predicates and extract the feature of the objective state, which physicists call "coordinate-free." Why then do we bother with the predicates? Because they enable us to free the elements of a complicated objective state from their relational structure, to describe them singly, thus facilitating measurement, understanding, and analysis. The number system underlying utility functions has rich structures, some of which are not required for the representation of the preference order. The monotone transformations selectively erase the unwanted features, so that only those features that are invariant under the transformations are retained as objective. Economists call the feature invariant under the monotone transformations *ordinal utility*. Similarly, the feature that is invariant under another group of transformations, the positive linear transformations, is called *cardinal utility*. The debate on which is more appropriate for microeconomics is analogous to that whether the Galilean or Lorentz transformations are more appropriate in representing mechanics. Both ordinal and cardinal utilities frustrate interhousehold comparison of utilities, for despite their substantive difference, they share the general form of representation.

So far we have considered a single individual. Now we turn to a collection of unrelated or independent individuals. The question is to articulate clearly the idea of *independence*. I will focus on physics because it reveals the steps of argument omitted in economics. (For more detailed discussion, see Auyang, 1995, § 26.)

Most physical theories study a single system, so that the symmetry is *global* in the sense that it applies to the system as a whole, or to the whole world if it is the topic of the theory. A global coordinate system allows us to describe the whole system in a uniform way. Interacting field theories, including general relativity and quantum field theories, are different. These theories specifically aim to study the interaction among various spatial-temporal parts of the world, so that each part must be regarded as an individual. The basic individuals in field theories are technically called *Events*. Each Event has its own character and is uniquely identified by a spatiotemporal position in the field. The character of an Event is complicated; for instance, an Event in general relativity is a Minkowski space, which is no simpler than a preference order. Since the theories aim to study the fundamental interactions of the physical world, they must ensure that the

Events are independent in the absence of the interactions. Thus they carefully purge the extraneous relations among the Events instilled by our customary way of thinking. As a global convention, the global symmetry is such an extraneous relation.

A coordinate system is a conventional predicate. In global symmetry, a predicate, once chosen, is enforced throughout the world. This is similar to our ordinary way of thinking. We can ascribe any predicate to snow, but once we have decided on "white," then it applies to snow everywhere. We can translate to another language and call snow *bai*; the new convention is again globally enforced. This is the idea behind global symmetries, where the coordinate systems and the transformations among them are effective all across the world. The idea is intolerable in the physical theories of fundamental interaction. The adoption of a specific convention and the translation to another involves information, and the instantaneous promulgation of the information throughout the world violates the laws of relativity, which demand that information be transmitted from Event to Event with finite speed.

A globally enforced predicate hides a relational structure. If two things are separately described as white, we say they have the same color, thus establishing the relation of similarity between them. To prevent the fundamental interactions from being contaminated by the relational structure hidden in global coordinate systems, field theories *localize the symmetry*. Consequently coordinate systems and the transformations among them are defined for each Event separately. For instance, in general relativity the symmetry of special relativity is localized to each spatiotemporal point, so that each Event is endowed with its private Minkowskian structure complete with coordinate systems and Lorentz transformations. The transformation of the coordinate system for one Event is performed independently of those for other Events, so that the convention of picking a particular system is left to each Event. This is usually expressed in the saying that the light cone is free to rotate from point to point in general relativity. More generally, the convention of definite predication is decentralized to each individual.

Now the situation in physics resembles that in economics. There are many Events, each with its own character describable in terms of its own coordinate system, just as there are many economic individuals, each with its own preference order describable in terms of its own utility function. Guess what: The characters of different Events cannot be compared. The reason for the incomparability is the same as in economics. The Events cannot be compared not because their characters are different; two things, one weighing ten pounds and the other fifteen pounds, are comparable because they are described in a common convention of predication. The Events are incomparable because there is no standard of comparison when the convention for choosing predicates is free to change from Event to Event. If snow in the North Pole is called red by a convention and snow in the South Pole green by another convention, there is no way to decide whether they are of the same color. Similarly, the utilities of different individuals are incomparable because the utility function is a conventional predicate and the convention varies from individual to individual.

In physics as in economics, the concept that individuals are incomparable because of their idiosyncratic conventions of predication is rigorously defined. Looking at the mathematics, one must concede the incomparability. However, mathematics also shows that the individuals can be made comparable by additional constructions. Economics and physics are empirical theories, which must decide what mathematical concepts are appropriate to represent their topics. Does the notion of incomparable individuals adequately capture the salient features of what the sciences purport to study? Are additional concepts for comparison justified?

Physicists argue that the incomparable Events make no physical sense; a heap of independent Events does not constitute a world. They demand that the objective state of the world is invariant under the transformations of coordinate systems for individual Events. The demand is satisfied only if the theoretical structure is expanded to include an interaction field, for example, gravity, in general relativity. The interaction mediates among the conventions of the Events, thus instituting a *physically significant* way to compare their characters. The introduction of the interaction field also modifies the characters of the Events by making them relatable, so that the Events become the source of interaction. The result is the description of an integrated world without hidden global conventions.

The situation in economics is different. In the general equilibrium model for perfectly-competitive market, the only relevant relation among individuals is taken care of via the price parameter in the independent individual approximation. No further relation is considered and no comparison of utility is required. Therefore there is no need to expand the theoretical structure to take account of human discourses that facilitate comparison, if one is not inclined to consider such questions as the distribution of well-being.

22. That interindividual utility comparison is not needed in some microeconomic models does not imply that it is generally impossible or superfluous. In real life, we are not the kind of solipsistic individuals depicted in economic models. We work with others, persuade, and are persuaded, constantly adjusting our behaviors and modifying our judgments as a result of intercourse. Through social interactions we come to have a feel for the preferences of others. The inadequacy of solipsism is argued by many philosophers. Ludwig Wittgenstein argued that a language private to an individual is impossible; to speak a language is to engage in a social practice (1953, §§ 241–81). Donald Davidson argued that in attributing desires and preferences to others, the attributor must have used his own values in a way that provides a basis of comparison (1986). Doubtlessly our communication is far from perfect and misunderstanding is frequent. We make lots of mistakes in our attribution, but the overall record of the attributions is not too dismal; we do succeed in understanding each other so a certain extent. The intelligibility of ordinary discourse, in which such attributions are prevalent, shows a large degree of similarity among us. Only against this background of similarity can we talk about our differences.

Utilitarianism requires a way of comparing the magnitudes of utilities so that the utilities of all individuals can be summed and the total maximized. John Rawls's theory of justice requires a way of comparing the values of utility to find out the worst-off individual whose welfare is to be protected in social changes. In welfare economics, John Hicks's principle of compensating the parties injured in advancing social welfare implicitly assumes some standard to evaluate injuries, without which compensation is empty talk. All these theories collapse if the utilities of different individuals cannot be compared. The avoidance of interindividual utility comparison is fundamental to Arrow's conclusion that it is impossible to derive a satisfactory social welfare function from individual utility functions without dictatorship.

Since the 1970s, there has been some effort to find means for interindividual utility comparison. Arrow introduced the notion of "extended sympathy" as a basis for social choice. Amartya Sen developed a social welfare functional that includes interindividual utility comparison and began a systematic study of the consequence of comparisons on social welfare theories. The means to compare utility are conceptual tools to address certain socioeconomic problems. The tools do not imply particular solutions, and even less the goodness or desirability of the solutions. The value judgment is a totally separate consideration. Unfortunately, the two continue to be confused in many economics books that exclude interindividual utility comparison on the pretext that it is "normative" or even "unscientific." See, for example, Robbins (1984, p. 141). Amartya Sen remarks, "For reasons that are not altogether clear, interpersonal utility comparisons were then diagnosed as being themselves 'normative' or 'ethical'" (1987, p. 30). For a bibliography and more recent results on interpersonal utility comparison, see Elster and Roemer (1991).

23. Samuelson (1947, p. xix).

24. Lewontin (1980).

25. See Maynard Smith (1987), Sober (1993, Ch. 4), and the papers in Brandon and Burian (1984) and Sober (1984).

26. Emlen (1987). In some cases, the substitution of a single allele has clear-cut phenotypic manifestations and noticeable causal effects, for instance, in the case of an allele that results in a disease. However, even in these rare cases there may be complicated side

effects. Many announcements of the discovery of a gene for this or that disease were quietly withdrawn for lack of solid evidence. Various genetic interactions, including antagonistic pleiotropy, and their effects on evolution, are reviewed in Hoffmann and Parsons (1991, Ch. 5). Ruthen (1993) reviews many cases in which the claims of geneticists were invalidated.

27. Lewontin (1974, p. 13).
28. Darwin adopted Herbert Spencer's phrase "survival of the fittest" and added it to the fifth edition of the *Origin of Species* as a synonym for "natural selection." The move was an attempt to defuse the false impression suggested by "natural selection" that nature acts as some kind of deity. *Fitness* is used only once in the first edition of the *Origin*, where it is synonymous to "adaptedness" and *fittest* means "best adapted." Dunbar said, "It is to adaptation, *not* to fitness, that Darwin's term 'fit' (or 'fittest') refers" (1982).
29. Lewontin (1974, p. 22; his italic).
30. See, for example, Lewontin (1970), (1978). Mayr's account of the logic of Darwin's theory of natural selection mentions neither adaptation nor the condition of living (1982, pp. 479–81). Maynard Smith cited multiplication, variation, and heredity as the essential elements of evolution (1987, p. 121).
31. E. Wilson (1975, p. 68). The fitness r_x of the allele x is defined in the rate equation $dn_x/dt = r_x n_x$, where n_x is the percentage of the allele x in the population, and x ranges over all possible alleles for the locus. According to Wilson, "The theory of evolution by natural selection is embodied in this definition of the different rates of increase of alleles on the same locus."
32. Reviewing the notion of fitness, R.I.M. Dunbar said: "It is this formulation that lies at the very root of the criticism of circularity, for it offers us no necessary reason why different variants should leave different numbers of offspring other than the fact that they do indeed do so" (1982). E. Wilson remarked: "Discussion of evolution by natural selection often seems initially circular: the fittest genotypes are those that leave more descendants, which, because of heredity, resemble the ancestors; and the genotypes that leave more descendants have greater Darwinian fitness. Expressed this way, natural selection is not a mechanism but simply a restatement of history." He went on to discuss the mathematics that shows "some of the basic laws of population genetics turn rather trivially on the same tautological statement" (1975, p. 67). Wilson did not explain how this "initial" image is improved.

Some philosophers charged that "the survival of the fittest" is tautological and devoid of empirical content (e.g., Popper, 1972, pp. 267–72, who later recanted). The charge has been answered many times, but its irritation is felt in the continuing reference to the tautological problem in the evolutionary literature; see, e.g., Brandon (1990), Sober (1993). Evolutionists often counter that all theories are somehow tautological or circular, as Newton's second law is a tautology. The quibble on the meaning of tautology has missed the point of the criticism. It is true that a mathematical structure encapsulated by a set of axioms can be regarded as a tautology; so can mathematical empirical theories with a set of tightly integrated concepts. However, each concept in these tautologies has many independent instantiations and consequences, and their interconnection makes a single instantiation significant. Metaphorically, their "circles" cover vast territories that manifest their power, generality, and significance. In contrast, the "circle" of "survival of the fittest" with an exogenously given Darwinian fitness spectrum has been reduced to a point with no content. It is trivial or empty.

Several philosophers have addressed the criticism by novel interpretations of fitness and overall adaptedness. They all posit fitness as an absolute property of individual organisms. Some have defined fitness as the propensity or dispositional property of an organism to produce a certain number of offspring. Others asserted that fitness is a nonphysical property supervenient on organisms. (See Sober [1993], for review and reference.) I fail to see how our understanding of evolution is enhanced by the posit of unexplained dispositional properties that have no manifestation other than the statistics of actual proliferation.

Mills and Beatty (1979) said that propensity is a dispositional property with an onto-logical status; that the paradigm case of a propensity is a subatomic particle's propensity to decay in a certain period. I disagree. Quantum mechanics never attribute any myste-rious ontological disposition to subatomic particles. It only talks about the *probability* of the particle's decay, not the propensity. Probability is theoretical and is always converted into statistical terms in empirical contexts, so that a certain proportion of an ensemble of particles decays in a time span, full stop.

33. Various optimization models are lucidly presented in Alexander (1982). Critical reviews of optimization theories are found in Oster and Wilson (1978) and Parker and Maynard Smith (1990).

34. The example of clutch size is found in Stearns (1992, Ch. 7).

35. Oster and Wilson (1978, p. 292). This critical chapter sometimes indicates that optimiza-tion is better viewed not as a natural process but as a mathematical method, which contradicts the interpretation of evolution as an optimizing process. However, the in-terpretation is prominent and is adopted uncontested elsewhere in the book. It is also common in the biological literature on optimization models.

36. Lewontin (1978), (1980); Gould and Lewontin (1978); Maynard Smith (1978). See Mayr (1988, P. III) for a recent account of the critique with more references.

37. Oster and Wilson (1978).

38. Jacob (1977).

39. For instance, the market is as ruthless in eliminating inefficiency as natural selection, but it has not eliminated the inefficient *QWERTY* typewriting keyboard that we are all using. The keyboard configuration was originally designed to slow the flying fingers of typists to prevent the problem of jamming mechanical keys. It survives, although the jamming problem was solved long ago and it is hindering productivity growth in this electronic age. To understand its present monopoly, one must examine the effect of its history, which is the crux of evolution. Such path-dependent considerations have no place in optimization models. Thus the models provide at best a biased view of evolution.

40. Maynard Smith (1982, p. 1), (1978).

41. Parker and Hammerstein (1985). See also Maynard Smith (1978) and Oster and Wilson (1978).

42. Grafen (1991, p. 6).

43. G. Williams (1985); Dawkins (1989).

44. See, for example, Dennett (1995), R. Wright (1994).

45. Dawkins (1989, p. 44).

46. Darwin (1871, p. 394). The statement was added in the second edition.

5. Interacting Individuals and Collective Phenomena

1. Schelling (1978, Ch. 4).

2. Anderson (1963, Ch. 3); Ashcroft and Mermin (1976, Ch. 22).

3. See, for example, Ashcroft and Mermin (1976, Ch. 23).

4. Williamson (1985, p. 18).

5. Hart and Holmström (1987); Kreps (1990, Ch. 16). Williamson (1985) gives a detailed analysis of behavioral assumptions and various kinds of contracts, especially those in-volving asset specificity.

6. Coase (1988, p. 3).

7. Coase (1937) first emphasized the importance of transaction costs. Transaction-cost eco-nomics and the importance of asset specificity are discussed in detail in Williamson (1985). The nature of the firm and the problems of integration are reviewed in Holmström and Tirole (1989); see also Tirole (1988, Ch. 1).

8. The often-cited explanation for large firms, the technological advantage of large-scale production, cannot be the whole story. The same technology can be implemented in different administrative organizations, as parts can be purchased instead of produced in

house. Moreover, technology does not explain vertically integrated firms that undertake disparate although related businesses. Many oil refineries integrate backward into crude oil production and forward into distribution of refined products. They can instead signed outside contracts to purchase the crude and sell the products, processes that use totally different technologies.

9. Tirole (1988, Ch. 11); Kreps (1990, Chs. 11, 12). Aumann (1987) gives a critical review of the history of game theory and its economic applications.

10. The behavior of oligopolists is a standard topic in the "new" industrial organization theory. Dynamic price competition and tacit collusion are discussed in Tirole (1988, Ch. 6). See also Kreps (1990, Ch. 14). These models differ from the "old" industrial organization study, which is mainly empirical.

11. For instance, in 1982 there were about 260,000 firms in manufacturing, the largest sector in the U.S. economy, but the largest 100 firms owned about half of the total assets. About 90 percent of the total sales of cigarettes, 86 percent of the sales of breakfast cereal foods, and 77 percent of the sales of beer and malt beverages were produced by the leading 4 firms in the respective industries (Scherer, 1990, pp. 58–62).

12. The computer tournament organized by R. Axelrod (1984) involved a game in which the prisoner's-dilemma game is repeated two hundred times. Each program was matched to all entries, including itself. The winner was a simple strategy called *tit-for-tat,* which entails cooperating in the first period and afterwards playing whatever strategy its opponent adopted in the previous period.

13. The following discussion on evolutionary game theory follows Maynard Smith (1982, Ch. 2). For more concise introductions, see Parker and Hammerstein (1985) and Parker and Maynard Smith (1990).

14. The concept of structured population is most clearly explained in D. Wilson (1980, Ch. 2).

15. The "trait group" was introduced by David Wilson to contrast with permanent groups, which are reproductively isolated groups that last indefinitely and send out propagules with a composition resembling their own. The significance of permanent groups is hotly contested.

16. The material discussed here is usually called group selection and kin selection in the literature. Michod (1982) and D. Wilson (1983) are good reviews. In kin-selection models, apparently altruistic behaviors are explained by the benefit reaped by the genes that the organism shares with its relatives. Many theories have shown that the results of these models are equivalent to that obtained in group-selection models that do not invoke genes (Michod, 1982). Perhaps the most popular concept in these topics is the *inclusive fitness* introduced by W. D. Hamilton; the inclusive fitness of a character is its fitness in isolation plus all its effects, each weighted by a factor accounting for its degree of relatedness to the affected organisms. It has mainly been associated with kin selection because of the context of its introduction. However, Hamilton (1975) argues that it is actually more general; kinship is only one way of accounting for the relatedness of the actor and the recipient. D. Wilson states, "Given the polemic nature of the controversy over group selection and the consequence of switching from one definition to the other, evolutionists can be expected to remain polarized for the foreseeable future" (in Keller and Lloyd, 1992, p. 148).

17. An extensive review of the empirical findings on cooperative breeding and their possible explanations is found in Emlen (1991).

18. The field data on the myxoma virus and their analysis can be found in May and Anderson (1983). Lewontin (1970) interprets the results in terms of the effect of population structure.

6. Composite Individuals and Emergent Characters

1. Plato, *Theaetetus,* 204. Mill (1843, Bk. III, Ch. 6, § 1; Ch. 10, § 3). The distinction between resultant and emergent properties was first made by G. H. Lewes in *Problems of Life and Mind* (1874). A brief history of the idea of emergence is given in Stoeckler (1991).

2. See Clery (1996) for a brief review.

3. A predicate p is supervenient on a set of predicates S if and only if p does not distinguish any entities that cannot be distinguished by S. Supervenience was first used to describe characteristics such as goodness and beauty. As a notion of dependency, it has become popular in the philosophies concerning levels of organization. For instance, some philosophers of biology argue that fitness is supervenient on organismic traits.

I find little role for supervenience in scientific theories. Microexplanations are causal explanations, and once causal explanations are available, they render supervenience superfluous if not false. Macroscopic properties generally do not supervene on microscopic properties. Broken symmetry explicitly shows the appearance of macroscopic features that make distinctions forbidden by the laws of microphysics (§ 23). For example, the temporal irreversibility of thermodynamic processes breaks the time-reversal symmetry of fundamental physical laws (§ 40). Novel distinctions emerge partly because we carve nature differently while conceptualizing on different spatial and temporal scales. Macropredicates sometimes go beyond the system to include external contingencies that make salient features stand out. Other times they exclude atypical system configurations that are real in the microdescription. Thus micro- and macropredicates do not always describe identical systems. The material difference between the systems in the micro- and macrodescriptions may be so slight that we roughly call the systems the same, but it often ushers in enough structural discrepancies to spoil the supervenience of macropredicates.

4. Chaikin and Lubensky (1995, § 3.6); Binney et al. (1992, Ch. 1); Anderson (1984, Ch. 2); Huang (1987, Ch. 16).

5. Anderson (1984, Ch. 2); Anderson and Stein (1984).

6. Anderson (1984, Ch. 2). More discussion of the structure of the ammonia molecule is found in Feynman et al. (1963, III-9).

7. Ashcroft and Mermin (1976, Ch. 33) discusses magnetism, including the mean field theory and the Ising model. A lucid and nontechnical account is found in Purcell (1963).

8. In mean field theory, direct spin–spin coupling is neglected. Instead, the neighboring spins of a particular spin constitute a mean field or a molecular field, which affects the spin in the same way as an external magnetic field. The mean field is determined self-consistently from the behaviors of all particular spins responding to it.

9. Binney et al. (1992, Ch. 1).

10. K. Wilson (1979); Huang (1987, Ch. 18); Chaikin and Lubensky (1995, § 5.5). Graphic results of computer simulations for the Ising model near the critical point can be found in K. Wilson (1979) and Binney et al. (1992, p. 19).

11. Binney et al. (1992, Ch. 1); Huang (1987, Ch. 16); Anderson (1984, Ch. 2); Chaikin and Lubensky (1995, § 5.4).

12. An intuitive example of fractals is a rugged coastline. If we estimate the length of a certain coastline from its photos taken from a satellite and an airplane, we will get a greater length from the photo taken from the airplane, after adjustment for the difference in the scales of the two maps. The photo taken at low altitude shows more capes and coves, which are smoothed out in the high-altitude survey. A land survey of the same coastline yields an even greater length because it reveals even more minute twists. Consequently the length of the coastline does not scale as straight lines scale. Nevertheless, since the coastline is self-similar – the blow-up picture of a short stretch of coastline appears similar to the coarse picture of a longer stretch – its scaling does obey certain mathematical rules with jagged scaling factors. Similar scaling behavior occurs in the fluctuation structure of critical phenomena. Many books on fractals came out in the last decade. Perhaps the most well-known is the one by Benoit Mandelbrot, who drew fractals to the attention of the scientific community.

13. Gould (1989) and Levinton (1992) give interesting accounts of the Cambrian Explosion.

14. At the end of the nineteenth century, this possibility was suggested by the proponents of orthogenesis, who argued that evolution was determined by forces internal to the organism (Bowler 1983, pp. 268–70). Biologists who specialize in growth and development have also argued that the interaction among genes defines a developmental track that

constrains evolution. The constraint is not absolute; some species, for instance, the marsu-pial tree frogs, exhibit modifications in the most crucial stages of development (Levinton 1992).

15. See, for example, Stanley (1981).

16. Gould and Eldredge (1993) review the development of the idea of punctuated equilib-rium. Kerr (1995) reports on the recent results on the fossil record.

17. Mayr (1988, p. 403); Ayala (1985, p. 77). More problems of macroevolution are discussed in Mayr (1988, Ch. 23).

18. Kauffman (1993, pp. 10, 29). Most of the material discussed in this section is taken from this reference.

19. Many examples of antagonistic pleiotropy are given in Hoffmann and Parsons (1991, Ch. 7).

20. Arrow (1951, p. 59). A social-choice function is dictatorial if it conforms to a single citizen's preferences regardless of those of others; imposed, if it taboos certain preferences.

21. See Stein (1989a). Anderson wrote a series of seven "reference frame" articles on the history of spin glasses, the first of which appeared in the January 1988 issue of *Physics Today*. When a liquid is cooled rapidly so that it has no time to crystallize into a solid, it enters a glass phase, where the disorder of the liquid is frozen into a metastable state. Spin glasses are magnetic systems. They are called glasses because the magnetic atoms are ran-domly frozen into a nonmagnetic host material, for instance, a small percentage of irons in a host of copper. Spin glasses exhibit both ferromagnetism and antiferromagnetism. Furthermore, whether the energy consideration favors a parallel or an antiparallel bond between any two spins is randomly decided by the way in which the spins are frozen when the material is prepared. Consider, for example, three spins in a triangular relation; each spin has only two orientations, up or down. One of the three bonds has lower energy if the spins it relates are antiparallel; the two other bonds have lower energies if the two spins they respectively relate are parallel. In this case it is impossible to orient the three spins in such a way as to minimize the energies of all three bonds. The system is frustrated.

22. Kauffman (1993, Chs. 2 and 3). A more popular account is found in Kauffman (1995).

23. Kauffman (1993, Ch. 5).

24. Kauffman (1993, p. xv).

25. Opinion in the 1990 Santa Fe workshop on artificial life, reported in the research news in *Science*, **247**: 1543 (1990).

26. Blanchard (1987).

27. These are standard material for macroeconomic textbooks. See, for example, Dornbusch and Fischer (1993), especially Chs. 3, 4, and 8.

28. Economists emphasize the distinction between *endogenous* variables determined within the economy and *exogenous* variables determined without and study the variation of the former with respect to the latter. The distinction is not emphasized in physics, where almost any variable can be put under control and made exogenous.

Thermodynamics distinguishes between extensive and intensive variables. *Extensive* variables such as energy and volume depend on the size of the system; *intensive* variables such as temperature and pressure do not. The equilibrium state of a thermodynamic system can be completely specified by its extensive variables, which are functionally related by the equilibrium condition. *Equilibrium* obtains with the maximization of a quantity called the entropy. The entropy, being a function of the extensive variables, is itself an extensive variable; it is doubled if all extensive variables of the system are doubled.

The intensive variables, defined as the partial derivatives of extensive variables, de-scribe how one extensive variable changes with another, keeping all the rest constant. The equilibration of the relevant intensive variables is obtains when two systems in contact enter into equilibrium. For example, temperature is the partial derivative of energy with respect to entropy, with volume and the number of particles kept constant. If two sys-tems at different temperatures are brought into thermal contact, energy is exchanged and consequently their temperatures change. The exchange stops and equilibrium is reached only when their temperatures become equal. The relations among the intensive and extensive variables are called the *equations of states*, which are most useful in characterizing specific systems. For instance, the equation of states for ideal gases known as the ideal gas law states that the product of pressure and volume is proportional to the product of the temperature and the number of particles.

Macroeconomic systems are more complicated than thermodynamic systems; thus the conceptual structure of macroeconomics is not as neat. However, there are certain similarities. In macroeconomics, aggregate variables such as output and employment are extensive; marginal quantities and multipliers are intensive. *Marginal quantities* are the partial derivatives of aggregate variables. For instance, the partial derivative of consumption with respect to output is the marginal propensity to consume; the partial derivative of output to employment is the marginal productivity of labor. Like thermodynamic intensive variables, the marginal quantities are often responsible for equilibration. However, the most important intensive variables, price level and interest rate, are introduced independently and not as derivatives. They are responsible for maintaining equilibrium across various markets, a complication absent in physics.

Among the aggregate variables, the output has a special status and can be compared to the entropy. The production function, which expresses the output in terms of employment and capital stock, is like the function expressing entropy in terms of energy and volume. Each holds the primary position in its theory. Mathematically, both output and entropy are homogeneous functions to the first order. If employment and capital stock are both multiplied by a factor, the output is multiplied by the same factor. Thus $Y(\lambda N, \lambda K) = \lambda Y(N, K)$, where Y is the output, N the employment, and K the capital stock. Similarly, $S(\lambda U, \lambda V, \lambda N) = \lambda S(U, V, N)$, where S is the entropy, U the energy, V the volume, and N the number of particles. This condition encapsulates the constant return to scales and is basic to most economic models.

Thermodynamics has a maximum principle, the second law asserting the maximization of entropy in equilibrium. The principle operates on the macrolevel, since entropy is a macrovariable. There is no macrolevel optimization principle in economics. However, whenever possible, economists invoke the optimizing behaviors of individual households and firms to find the microquantities to be aggregated.

29. Arrow (1987).
30. Greenwald and Stiglitz (1987, p. 120).
31. Lucas (1987, pp. 107ff.).
32. Lucas and Sargent (1979, p. 11). Philosophers of science have lots of fun arguing about the epistemology of unverifiable principles. Falsificationists criticize the positing of an empirically irrefutable principle as unscientific. Holists, arguing that any hypothesis can be protected from unfavorable empirical evidence by modifying other hypotheses in the theory, side with relativists in countering that the falsifiability criterion for scientific propositions is defunct. Moderate relativists would call new classical economics one of many research programs, one that takes the market-clearing principle as its hard core; if the core seems to be threatened, change something in the protective belt, for example, the decisions of the unemployed. Positivists dismiss the argument on the validity of a principle as pointless; the only justification of a hypothesis is that it yields results that agree with observations. The positivist position, popularized among economists by Friedman, seems to be taken up by Lucas and Sargent. Their opponents would charge that the positivist position contradicts their reductionist position. If predictive record is all that counts, then there is no reason to insist that a macrotheory should be based on individual behaviors or market clearing or any other principle. Their charge that macroeconomics is fundamentally flawed is unjustified; the worst they can say is that it has a bad predictive record. The record of their own theory does not appear to be any better.
33. Modigliani (1977).
34. Among the new Keynesian models are the implicit contract theory describing employer–worker relations, efficient wage theory accounting for the effect of price-dependent quality, insider–outsider models explaining why the unemployed have little effect on the

wages of the employed. There are many other new Keynesian models, some of which are collected in Mankiw and Romer (1991). Hahn and Solow (1995) give models for various types of market imperfection.

35. Solow (1980).
36. Tobin (1972, p. 9).
37. Arrow (1987).
38. Arrow (1987).
39. The Rational expectations hypothesis is often interpreted as a consistency axiom of economic models, stipulating that expectations made within an economy are consistent with the economic structure that results from the expectations.
40. Hayek (1948, Ch. 1). Note that Hayek's true and false individualisms are both different from the individualism that stresses autonomy and self-development. Hayek said that "the cult of distinct and different individuality," as expressed in the writings of Johann von Goethe and Wilhelm von Humboldt and Mill's *On Liberty*, "has nothing to do with true individualism but may indeed prove a grave obstacle to the smooth working of an individualistic system" (p. 26).

7. The Temporality of Dynamic Systems

1. Pericles' funeral oration is found in Bk. II, ¶¶ 35–46, of Thucydides' *The Peloponnesian War*.
2. Quentin's remark is found in Faulkner (1931, p. 81). The episode of Benjy's drive to the square occurs at the end of the novel. Kant's distinction between the subjective and objective time sequences is a major point in his argument for the concept of causality in the "Second Analogy" of the *Critique of Pure Reason*. Heidegger said: "Precisely that Dasein which reckons with time and lives with its watch in its hand – this Dasein that reckons with time constantly says 'I have no time.' Does it not thereby betray itself in what it does with time, in so far as it itself is, after all, time? Losing time and acquiring a clock for this purpose!" (1924, p. 15). The transformation of temporality to the ordinary conception of time is analyzed in the final chapter of *Being and Time*. His *The Concept of Time*, a lecture given earlier, is more accessible. Nietzsche said: "It is altogether impossible to *live* at all without forgetting. Or, to express my theme even more simply: *there is a degree of sleeplessness, of rumination, of the historical sense, which is harmful and ultimately fatal to the living thing, whether this living thing be a man or a people or a culture*" (1874, p. 62, his italic).
3. Sherover (1971, p. 283).
4. Aristotle said: "This is what time is: number of motion in respect of the before and after" (*Physics* 219b). Augustine said: "Time is nothing more than distention, but of what thing I know not, and the marvel is, if it is not of the mind itself.... It is in you, O my mind, that I measure my time" (*Confession* Bk. 11, Chs. 26, 27). McTaggart (1908) argued for the unreality of time and the incoherence of becoming. Henri Bergson said: "The flux of time is the reality itself, and the things which we study are the things which flow" (1907, p. 344). Hans Reichenbach said: "There is no other way to solve the problem of time than the way through physics.... If there is a solution to the philosophical problem of time, it is written down in the equations of mathematical physics" (1956, pp. 16f). Ilya Prigogine said: "These new forms of dynamics [quantum and relativistic mechanics] – by themselves quite revolutionary – have inherited the idea of Newtonian physics: a static universe, a universe of *being* without *becoming*" (1980, p. 4). There are many good anthologies of theories of time. See, for example, Sherover (1975), Poidevin and Macbeath (1993).

Time is a major theme in the philosophy of physics, partly because of reflections on relativity and the temporal irreversibility of thermodynamics. It does not appear to be a theme in the philosophies of biology and economics, and it is overshadowed by human significance and the meaning of historical development in the philosophy of history. In

the case of biology, reflection on population genetics is as ahistorical as the theory itself. Phylogenetic inference or the reconstruction of the tree of evolution is historical,
but the philosophical controversy over it centers around the notion of simplicity and the
adequacy of a sweeping hypothesis, with the notion of time barely visible.

Scientists occasionally reflect on the general nature of time. Physicists seem to be most
active in this area, probably because of the tradition dating back to the Newton–Leibniz
debate. See, for example, the conference proceeding edited by Halliwell et al. (1994).

5. Newton gave the most famous enunciation of dynamic substantival time: "Absolute, true,
and mathematical time, of itself, and from its own nature, flows equably without relation to anything external, and by another name is called duration" (*Principia*, Scholium
to "Definitions"). Note, however, that although Newton said that time flows, temporal passage plays no role in Newtonian mechanics or other mechanics. The metaphoric
impression of flow may arise from time's being the independent variable in dynamic
theories. Static substantival time is evoked by those who take a four-dimensional view
without change. For instance, various stages of Quine's "physical objects" occupy various
slices of substantival time (1976).

6. Some philosophers, for instance, Leibniz and Hegel, argued that time is ideal. Others,
including Bergson, Husserl, Heidegger, and Merleau-Ponty, regarded time as a structure
of the whole of human consciousness and life experience. Kant associated time mainly
with theoretical reason, and he took it to be a form of intuition and perception.

In the weakest sense, experiential time marks the difference between abstract and empirical entities. Abstract entities such as numbers are atemporal, but things and processes
that we know through experiences and experiments are unintelligible without the notion
of time. This weak sense of empirical time is fully compatible with the objectivity of time.

The causal theories of time were developed by positivist philosophers. See Sklar (1985,
Chs. 9 and 10) for a critical review. I think the causal theory of time has missed the weakest and most important sense of empirical time that distinguishes abstract and physical
entities. It is fine to posit a set of abstract elements and define various abstract relations
among them, calling them by the names of "event" and "causal connection." But what
have these abstract elements and relations to do with the objects and interactions in the
world? An event is more complicated than an abstract and unanalyzable *x*, and there is
more to a causal connection than an abstract binary reflexive and symmetric relation.
Is it possible to define empirically verifiable event and causal connection without using
temporal notions? When we try to flesh out the abstract terms to make them answerable
to the meaning of event and causal connection applicable to the world, we invariably find
that they hide temporal presuppositions. The causal theories of time fail to reduce time.
They at best succeed in reducing some temporal notions to other temporal concepts.

7. The inseparable connection between time and change is commonly although not universally acknowledged. The disagreement hinges on the confusion between general and
substantive concepts. The *general* concepts of time and change are inseparable because
they are integrated in the general concept of things; the individuation of successive states
of a thing by temporal indices embodies the possibility of change. However, the *substantive* descriptions of time and change are not inseparable; it is possible that the successive
states have similar characteristics, so that time ticks away while no substantive change
is observed. Arguments of time without change refer only to substantive changes. For
instance, Sydney Shoemaker's argument invokes cyclical processes, and cyclicity has presupposed the general idea of change (Shoemaker, 1969).

8. Hobbes, *Elements of Philosophy* (1655), collected in (1839, Vol. 1, p. 136).

9. Currie and Steedman (1990, p. 141).

10. Some people say that time is spatialized simply because it is represented geometrically
in physical theories. Insofar as time is regarded as continuous and having a lineal order,
it is susceptible to geometrical treatment. Geometry has become such an abstract and
encompassing branch of mathematics it can be applied to things not remotely spatial. For
more discussion on time in relativistic physics and its interpretation, see Auyang (1995,
esp. § 25).

11. *Beethoven, the Man and the Artist as Revealed in His Own Words*, ed. by F. Kerst and H. E. Krehbiel, Dover (1964), No. 54. Mozart's words are from a letter quoted in E. Holmes's *Life of Mozart*, Everyman's Library. The authenticity of the letter is doubtful.

12. Aristotle, *Physics*, Book V, Ch. 1.

13. Tenseless theories of time can be divided into "old" and "new" versions; see Oaklander and Smith (1994). Positions in the tensed camp are more diverse and are more often held by philosophers more concerned with analyzing ordinary language than making revisions. Q. Smith (1993) is a recent defense.

14. Kaplan (1978). The meaning of temporal indexicals is more complicated than that of other indexicals because of complex tense and other temporal modifiers. Besides the context of utterance, they also require the independent specification of other temporal indicators in the sentence.

15. Quine (1960, pp. 193f.).

16. 16. Kaplan (1978).

8. Complexity in the Temporal Dimension

1. Many books, articles, and conference proceedings on dynamical systems and chaos appeared in the last decade. Lichtenberg and Lieberman (1992) contains a comprehensive treatment. Ott (1993) puts more emphasis on dissipative systems and strange attractors. The first chapter of each gives an overview of the major ideas, including most of the material discussed here. See also the lecture series sponsored by the Sante Fe Institute. Wightman (1985) discusses the major problems and puts them in the context of the historical development of dynamical theories. Hao (1990) collects many classic articles on dynamics and chaos, with a long introduction to the key concepts and an extended bibliography. There are many popular introductions to chaos, for example, Crutchfield et al. (1986) and Lorenz (1993).

2. Lichtenberg and Lieberman (1992, § 7.2). Ott (1993, § 2.2).

3. For a review of the convection experiments, see Libchaber (1987).

4. May (1986) reviews the application of the logistic equation to ecological modeling and many field and laboratory data. Renshaw (1991) gives a detailed account of relevant mathematical modeling in population biology.

5. Organisms are often classified as r-strategists or K-strategists. The r-strategists, adapted to unstable environments, are opportunists that breed as fast as possible to take advantage of available resources and suffer decimation when the environment deteriorates. They are usually small and short-lived. The K-strategists that colonize stable habitats maintain an equilibrium population by opting for longevity and a relatively low fertility rate.

6. Renshaw (1991, p. 4).

7. Prigogine (1980). Anderson (1981) criticizes Prigogine's view.

8. The idea of chaos is vigorously marketed. Some authors try to find a "noble ancestry" for it, but distortion often results. The most commonly cited precursor is Hesiod's account of cosmogony: "Verily first of all did *chaos* came into being." Hesiod's *chaos* has nothing to do with unpredictability, sensitivity to initial conditions, or even the ordinary meaning of chaos. The root of the Greek noun *chaos* means "gape, gap, yawn." The idea that the earth and the sky were originally one undifferentiated mass was common to many of Hesiod's sources. The appearance of *chaos* as a gap separating the two is the first step toward a differentiated and orderly world (Kirk et al., 1983, pp. 34–5). The is the sense John Milton adopted in *Paradise Lost* in interpretation of Genesis 1:2: "In the beginning how the heavens and earths/Rose out of chaos." The interpretation of *chaos* as disordered matter came rather late, around the time of the Stoics. The search for pedigree is multi-cultural. An editor of an anthology on chaos quoted the Taoist Zhuang: "The Emperor of Center was called Hundun (Chaos)." *Hundun* does not mean chaos at all. The quotation occurs in a parable: Unlike others who have seven orifices for seeing, hearing, eating, and breathing, Hundun had none. His friends Hasty and Sudden tried to improve him

by boring the orifices. They drilled one orifice a day and on the seventh day Hundun died. *Hundun*, which means "primal wholeness," is used to describe the innocence of children. Closer to the technical meaning of chaos is a saying from the Chinese *Book of Change*, roughly translated as "A slip by a thousandth of a centimeter leads to an error of a thousand kilometers."

9. Ott (1993, pp. 26ff.).

10. Lichtenberg and Lieberman (1992, p. 303).

11. Lorenz (1993, Ch. 3 and Appendix 1); Ott (1993, § 2.4.1); Crutchfield et al. (1986, p. 52); Ferraz-Mello (1992). The chaotic behaviors of planetary motion occur mainly in such parameters as eccentricity and inclination. They do not imply catastrophic instabilities of the solar system, such as the earth's crossing the orbit of Venus and smashing into the sun.

12. See Ott (1993, § 3.8), which also gives the references for many experiments.

13. Lichtenberg and Lieberman (1992, § 7.2).

14. Lebowitz and Penrose (1973) presents a lucid introduction to ergodic theory. Various stochastic concepts are discussed concisely in Lichtenberg and Lieberman (1992, § 5.2c). A more mathematical treatment is found in Arnold and Avez (1968).

15. The Bernoulli system is defined in Arnold and Avez (1968, p. 8). Ornstein and Weiss (1991) is a comprehensive review of the Bernoulli flow and its isomorphism to various mechanical systems.

16. See, for example, Chatin (1975).

17. See the reviews by Lebowitz and Penrose (1973), Penrose (1979), and Sklar (1993).

18. Lichtenberg and Lieberman (1992, §§ 1.3, 1.4); Ott (1993, Ch. 7). Chernikov et al. (1988) gives a lucid qualitative discussion.

19. Wightman (1985, p. 19).

20. Russell (1913). His *Human Knowledge* (1948) contains chapters entitled "Causal Laws" and "Space–time and Causality."

21. If a cause is a necessary condition for an effect, the effect would not occur if the cause did not, but the effect may not follow if the cause occurs. This criterion is compatible with chancy happenings; the throw of a die is the necessary but insufficient condition for a 6. If a cause is the sufficient condition for an effect, the effect invariably follows if the cause occurs, but the same effect can follow alternative causes. This criterion is applicable to deterministic evolution; for a system defined by a dynamical rule, the initial state is the sufficient condition for the occurrence of subsequent states. For a critical review of conditional analysis, see Mackie (1974, Chs. 2 and 3), where the INUS condition is also discussed.

22. Mill (1843, Bk. III., Ch. 5, § 3).

23. Earman (1986) reviews various formulations of determinism and investigates determinism in various physical theories. Weatherford (1991) presents a defense of determinism and contains references to the determinism–free will debate. Suppes (1993) argues against determinism.

24. Porter (1986, p. 12); Hacking (1990, p. viii). Hacking (1990, Ch. 18) criticizes Ernst Cassirer's account of the history of determinism, which is similar to Porter's.

25. James (1884, p. 117).

26. Suppes (1993).

27. Laplace (1814, p. 4). This popular essay on probability became the preface to the second edition of his technical treatise *Analytic Theory of Probability*, the first edition of which appeared in 1812.

28. Hacking (1990, p. 1).

9. Stochastic Processes and Their Explanations

1. See, for example, Papoulis (1991, § 2.2), or any textbook on probability.

2. Bernoulli, *Art of Conjecturing*, Part IV (1713), quoted in Datson (1987, p. 302).

3. Aristophanes, *Frogs*. In action, the scale only compared the weights of the verses of Aeschylus and Euripides, but Dionysus' explanations suggested that the weights can be added.

4. See, for example, Papoulis (1991, Ch. 4).

5. See, for example, Papoulis (1991, Ch. 10); Kampen (1992, Ch. 3).

6. See, for example, Kubo et al. (1985, Ch. 1); Kampen (1992, IV.1, VIII. 3, 4, 8).

7. Labor productivity or output per worker-hour in the United States today is almost ten times that at the end of the Civil War, although the growth rate of productivity has declined sharply since the mid-1970s. Only in recent years have economists come to grips with the endogenous mechanisms underlying productivity growth, especially the increasing return to scales due to the spillover effect of technological innovation. (See, e.g., the winter 1994 issue of the *Journal of Economic Perspectives*.)

8. Some reasons are offered for the relative stability of the postwar economy: the increase in the size of the government; the growth in personal income tax; the buffering of fluctuations in disposable income; the creation of the Federal Deposit Insurance Cooperation, which helps to prevent financial crisis. Economists are split on whether the government's discretionary policies stabilize or destabilize the economy; whether its effort to "fine-tune" the economy actually makes matters worse. For a comparison of the natures and causes of prewar and postwar recessions, see Gordon's introduction to Gordon (1986).

9. Sargent and Sims (1977) gives the time-series approach to business cycles. Granger's notion of causality was developed in the 1960s; Granger (1980) reviews it together with various means of finding and testing it in time-series data. For a recent review and comparison with other versions of probabilistic causation, see Vercelli (1992). For time-series analysis, see Kennedy (1992, Chs. 16 and 17).

10. Goodwin (1990) surveys the application of chaotic dynamics to economic problems (Richard Goodwin was one of those who worked on nonlinear economic models in the early days). Boldrin (1988) is a more technical review. See also the articles in Anderson et al. (1988).

11. The real business cycle model presented here is largely based on Kyland and Prescott (1982). A textbook treatment is found in Blanchard and Fischer (1989, Ch. 7). McCallum (1992) presents a simplified version with a critique. McGrattan (1994) reviews the more recent works.

12. McCallum (1992, p. 174).

13. Wright (1955).

14. Gillespie (1991, Ch. 4).

15. Gillespie (1991, p. 142).

16. See Kimura (1983), which addresses many criticisms raised in Lewontin (1974, Ch. 5). Golding (1994) includes up-to-date data and theoretical models.

17. For instance, in the 1980s elementary particle physicists came up with the theory that protons decay with a half-life in excess of 10^{33} years. To detect this extremely rare decay process, huge experimental tanks of water were set up surrounded by detectors. Physicists hoped for a few counts of decay per year, even though there are more than 10^{30} protons in the tank. These counts must be differentiated from stray neutrinos and other noise, although the experiments were hidden in deep mines that shield against cosmic neutrinos. Years passed. With each year the theoretical half-life was extended, but the limit of experimental resolution is fast approached. Even if protons do decay with a sufficiently long half-life, our experiments would not be able to differentiate the signal from the noise.

18. See Gillespie's paper in Golding (1994).

19. Lewontin (1970, p. 267); Gillespie (1991, p. 63); Lewontin (1991, p. 146).

20. The three spatial and temporal scales were first discussed by Nikolai Bogolubov in the generalization of the Boltzmann equation. See Liboff (1990, § 2.4.2).

21. Various classes of equations on the kinetic level are discussed in Balescu (1975, Ch. 11); a more physical and comprehensive presentation is found in Kubo et al. (1985). A clear exposition of the derivation of the hydrodynamic equations is given in Huang (1987, Ch. 5).

22. The derivation of the Boltzmann equation and the *Stosszahlansatz* is found in many standard statistical mechanics textbooks. See, for example, Balescu (1975, § 11.4, 11.5); Huang (1987, Chs. 3 and 4).

23. It is often called the hypothesis of molecular chaos. "Chaos" here is not to be confused with the chaos discussed in Ch. 8.

24. The rigorous derivation of Boltzmann's equation is due to Oscar Lanford. Lanford used Liouville's equation with a special kind of initial conditions, notably the condition that the N-particle joint distribution is factorized into the product of single-particle distributions. Under these conditions he showed that the joint distributions converge almost everywhere to products of single-particle distributions obeying Boltzmann's equation for a time interval a fraction of the collision time. Penrose (1979, pp. 1978ff.) gives a concise discussion; Sklar (1993) is more extensive.

25. Lebowitz (1993, p. 17).

26. Huang (1987, Ch. 5).

27. Gigerenzer et al. (1989, p. xiii).

28. For more discussion on the differences among individual, statistical, and probabilistic statements, see Auyang (1995, Appendix A3).

29. The first statement appeared in Peirce's 1892 essay "The Doctrine of Necessity Examined" (1932, Vol. 6, p. 32). Referring to it in another essay a few months later, he introduced the name *tychism*. The second appeared in the essay "The Doctrine of Chance" published in 1878 (1932, Vol. 2, p. 409).

30. Probability causation was separately proposed by Hans Reichenbach, I. J. Good, Clive Granger, and Patrick Suppes. Sobel (1995) gives a critical review with consideration to applications in the social and behavioral sciences.

31. Suppes (1984, p. 10).

32. Suppes (1984, Ch. 3).

33. Lewontin (1974a, quote on p. 403).

34. The classic papers on scientific explanation are included in many anthologies on the philosophy of science. Hempel's original papers are collected in Hempel (1965, P. IV). Salmon (1984) covers probabilistic explanations carefully. Pitt (1988) includes most major variants and rivals.

35. Strictly speaking, statements that hold with probability 1 are not laws if a law is a true universally quantified statement, for probability 1 allows atypical cases that violate the proposition, provided that the statistical weight of the atypical cases is zero. Consequently some philosophers dismiss the laws of thermodynamics as "bogus."

36. See, e.g., Hempel (1965, Ch. 5); Salmon (1984, Ch. 2). Hempel called the explanation probabilistic, but "statistical explanation" is now more popular.

37. Attempts to explicate and develop the meanings of probability in various contexts are usually called theories of probability and distinct from the probability calculus, although they all use some ideas of the calculus. See Weatherford (1982) for a review of various philosophical interpretations of probability.

38. Some variation on the sense of the degree of believability of opinions or propositions is one of the oldest meanings of the word *probability*. As late as the time of Galileo, believability was based on the approval of authorities. By the time of Leibniz, the bases of belief became evidence and reason. Leibniz thought that probability is a logical relation among propositions, and he envisioned a new kind of logic based on probability in addition to the familiar deductive logic. Two centuries later, a similar idea was developed by Keynes, who argued that "probability describes the various degrees of rational belief about a proposition which different amounts of knowledge authorize us to entertain" (1921, p. 3). The probability of a hypothesis is not absolute but is relative to its supporting evidence. In a similar vein, the philosopher Rudolf Carnap tried to develop an inductive logic based on probability. It was unsuccessful. However, what has come to be known as the logical theory of probability does clarify at least one important sense in which the concept of probability functions in our everyday thinking and rational deliberation.

39. Suppes (1984, pp. 52, 93–9).

40. See, for example, Kyberg (1974) and Salmon (1984). Sklar remarked that the interpretation of probability as a dispositional property is most congenial to tychists (1993, pp. 101, 122–4).
41. Suppes (1984, p. 10).
42. Keynes (1937, pp. 113f.).
43. Weinberg (1987, p. 8). Quantum systems are so well behaved it is difficult to find quantum chaos, whereas chaos is quite common in classical systems.
44. Feynman (1965, p. 129).

10. Directionality, History, Expectation

1. Reichenbach (1956) identifies the direction of time with the direction of entropy increase. Horwich (1987) considers a range of problems on asymmetries in time in metaphysics, epistemology, and physics. Sklar (1993, Chs. 7, 8, and 10) concentrates on issues related to physics.
2. The idea of common cause was introduced by Reichenbach (1956) and elaborated by Salmon (1984). Horwich (1987) uses it to construct the idea of fork asymmetry, which he argues to be basic to all asymmetries in time.
3. See Sober (1988, Ch. 3, and the references cited there).
4. For a lucid explanation of the time reversal symmetry, see Sachs (1987, Chs. 1 and 2). Davies (1974, Ch. 1) also explains the relation of time reversal symmetry to relativity.
5. Here I follow the exposition of Lebowitz (1993) and his earlier papers cited therein. See also R. Penrose (1989, pp. 304–22). The explanation is mainly based on the Boltzmann entropy. Lebowitz explained the difference between the Boltzmann entropy and the Gibbs entropy. Davies (1974, Ch. 2) is more technical. Sklar (1993, 7.III) reviews various interpretations of time reversal.
6. In a *spin-echo experiment*, perfectly aligned nucleus spins in a crystal are set to rotate in a magnetic field at $t = 0$. Because of crystal inhomogeneity, some spins rotate faster than others, and soon they become randomly aligned. At $t = t_1$ the spins are flipped by a radio pulse so that those farthest ahead become the farthest behind. The spins continue to rotate. At $t = 2t_1$, they are again perfectly aligned just as they were at $t = 0$. The realignment after reversal obtains even when the spins interact with each other. See Brewer and Hahn (1984).
7. R. Penrose (1989, pp. 322–47).
8. Dobzhansky et al. (1977, p. 486).
9. Mayr (1982, p. 130), (1988, p. 149); Gould (1989, p. 277); Lewontin (1991, pp. 143–4). The role of historical narratives in evolutionary biology was observed by the philosopher T. A. Goudge in 1961. Some philosophical discussions of evolutionary narrative are found in the articles in Sumner et al. (1981) and Nitecki and Nitecki (1992).
10. Lewontin (1991, p. 143).
11. Unlike descriptions of the past states of natural systems, which are made in contemporary or timeless concepts applicable at all times, the human past cannot be described solely in modern or timeless terms. The historian must ascertain how past situations were conceptualized and perceived by those who lived in them, for people acted according to their perceptions and their actions influenced the course of events. Furthermore, laws, even statistical, are unavailable for human action and motivation, on which most historical explanations are based. The "because" in an answer to "Why did you do it?" does not indicate a law. The regularities implied in most historical statements are not laws but what we expect people to do in normal situations. Consequently, *verstehen* (empathic understanding) and hermeneutic interpretation of the life expressions of other people are crucial to historical research. Philosophers of science often regard verstehen versus deduction from timeless laws as the major difference between the human and the natural sciences.
12. Stone (1979, p. 91).

13. Hempel (1942). This paper allows only deterministic laws; the admission of statistical laws was a response to criticisms. Dray (1966) collects important papers on the appropriateness of covering-law explanations to history.

14. Mayr (1982, p. 72). After surveying the philosophical portraits of history and science, Lewontin said: "If one examines science as it is actually carried out, it becomes immediately clear that the assertion that it consists of universal claims as opposed to merely historical statements is rubbish. The actual body of scientific theory and practice is a heterogeneous collection of modalities, illustrated, for example, by the current state of biology" (1991).

15. The Hegelian and Marxist views of history as a dialectical process of thesis, antithesis, and synthesis are fully dynamic, but they are too abstract. Marx's doctrine was vulgarized and made more substantive in the Stalinist era. The resultant naive version of historical materialism posits historical laws that direct all nations to pass through fixed stages of economic development, each with its characteristic mode of production, level of technology, and class conflict. By the 1960s, the idea of historical law was abandoned by most Marxist historians. E. P. Thompson said: "We can see a *logic* in the responses of similar occupational groups undergoing similar experiences, but we cannot predict any *law*" (Preface to *The Making of the English Working Class*, 1963, his italics). Perry Anderson cited "mechanisms of single events or the laws of motion of whole structures" as "that which is not known" (Foreword to *Lineages of the Absolutist State*, 1974). Both are acknowledged by historians as outstanding works in history and are widely influential beyond the Marxist circle.

16. Bury's view is discussed in Oakeshott's article "Historical Continuity and Causality Analysis," in Dray (1966, pp. 200–202). This notion of contingency appears to be different from Gould's notion, the meaning of which is not clear (1989).

17. The peculiar construction of narrative statements using anticipatory predicates is discussed by A.C. Danto.

18. Gould (1980, p. 258).

19. Kitcher (1985, p. 8). Wilson introduced sociobiology, *the study of the biological base of all social behaviors*, in 1975 as a new synthesis that extrapolates adaptation models to a wide range of topics including those in psychology and sociology. The first chapter of his *Sociobiology*, which lays out the vision of the new synthesis, bears the title "The Morality of the Gene." The last chapter discusses human social behaviors and institutions. It argues that many social institutions such as male dominance are genetically based and can only be changed at an unknown cost. Reactions were polarized and violent. Some biologists argued that many of Wilson's doctrines, especially those concerning human behaviors, are based on faulty reasoning and little if any empirical evidence. Many papers on the sociobiological debate are collected in Caplan (1978) and Barlow and Silverberg (1980). Human sociobiology has today developed into *new social Darwinism*, which includes evolutionary psychology, evolutionary morality, and evolutionary epistemology. (See, for example, R. Wright, 1994 and Dennett, 1995. Hogan, 1995a, reviews the new social Darwinists).

20. Wilson (1975, p. 562, his italics). The accounts are expanded and less cautiously stated in his *On Human Nature*.

21. Biologists, realizing that not all shared characters reveal genetic similarity, distinguish between homology and analogy. A *homologous* character of a group of organisms is a shared character that they derive from a common ancestor. An *analogous* character is a shared character that is unrelated to common descent. For example, birds' wings, humans' arms, and whales' flippers look different and serve different functions, but they are homologous; the bones in the three cases are arranged in the same pattern because they are inherited from an ancestor with similarly structured forelimbs. In contrast, the wings of birds, bats, and insects are superficially alike and serve similar functions, but they are analogous; they evolved independently and have no common genetic origin. Only homologous characters that are similar because they are derived from a common ancestor imply genetic similarity; analogous characters that are similar because they serve similar functions do not.

    To ascertain whether two characters are homologous or analogous is a tedious and delicate task, and the results are often uncertain. The distinction is especially subtle and important for human behaviors, because we depend heavily on nurture and are strongly influenced by our social environment. Since the adaptation models are unsupported by genetic models, they cannot differentiate whether the adaptation is biological or social. By ascribing all examples of indoctrinability to the working of the conformer gene, Wilson simply assumed that they are homologous without considering the possibility that they may be analogous.

Indoctrinability is only one of many cases in which superficially similar examples are arbitrarily lumped together. For instance, Wilson cited numerous examples of vertebrates risking themselves in the defense against predators and argued that they are the result of kin selection that maximizes inclusive fitness (1975, pp. 121–2). Kitcher examined many of the cited cases in detail and found that they are disparate phenomena. He criticized Wilson's method: "There is a suggestion that certain forms of behavior have been selected because of a particular benefit they bring to those who exhibit them, but neither the behavior nor the benefit is precisely specified.... Cases are lumped together as if there were some single system of evolutionary accounting that could apply to them all" (1985, pp. 159–65).

22. Marshall's remark is found in the Preface of the first edition of his *Principles of Economics*. Diamond (1994, p. 3).
23. Samuelson (1947, p. 311); Machlup (1991, pp. 10, 24). The classified typology of the distinctions between "statics" and "dynamics" takes up four pages in Machlup (1991).
24. Samuelson adopted Ragnar Frisch's formulation "A system is dynamical if its behavior over time is determined by functional equations in which 'variables at different points of time' are involved in an 'essential' way" (1947, p. 314).
25. For control and variational theories and their application to economics, see Intriligator (1971, esp. Ch. 11). A clear discussion of the Ramsey model for consumption and accumulation of capital is found in Blanchard and Fischer (1989, §§ 2.1, 2.2).
26. The derivation of the Lagrange equation, which is the generic name of many physical equations of motion, from the variational principle can be found in many intermediate mechanics textbooks. See, for example, Mandl and Shaw (1984, Ch. 2) for a concise discussion of Lagrangian field theory and standard references.
27. See, for example, Varien (1992, Ch. 11) and Kreps (1990, Ch. 3) for risk in microeconomics. Stochastic control and risk in macroeconomics are found in Sargent (1987, §§ 1.6, and 1.7) and Blanchard and Fischer (1989, § 6.2).
28. Knight (1921).
29. Keynes (1936, Ch. 12); (1937, p. 119).
30. Hahn and Solow (1995, p. 143).

11. Epilogue

1. Abelson and Sussman (1985, p. xvi).
2. Lycan (1990) collects some classic papers on mind–brain identity and functionalism, together with more recent references. Dualism and behaviorism dominated at one time or another in the history of ideas, but they are mostly out of favor today, when the battle for mind is waged between brain and computer.
3. Aristotle, *Nicomachean Ethics*, Bk. IV, Ch. 3. Aristotle's crown of virtue is the greatness of soul, sometimes translated as pride. Pride should not be confused with *hybris*, which was a totally distinct notion used, for instance, in the *Odyssey* to describe the wanton insolence of Penelope's suitors.
4. The currently popular antiscience mood is discussed in, for example, Holton (1993) and Maddox (1995).

Bibliography

Abelson, H. and Sussman, G. J. (1985). *Structure and Interpretation of Computer Programs.* Cambridge: MIT Press.

Agazzi, E., ed. (1991). *The Problem of Reductionism in Science.* New York: Kluwer Academic.

Alexander, R. M. (1982). *Optima for Animals.* London: Edward Arnold.

Anderson, P. W. (1963). *Concepts in Solids.* Reading, MA: Benjamin.

Anderson, P. W. (1972). More is Different. *Science,* **177**, 393–6.

Anderson, P. W. (1981). Some General Thoughts about Broken Symmetry. In *Symmetries and Broken Symmetries in Condensed Matter Physics*, ed. N. Boccara, pp. 11–20. Paris: IDSET.

Anderson, P. W. (1984). *Basic Notions of Condensed Matter Physics.* Menlo Park, CA: Benjamin.

Anderson, P. W., Arrow, K., and Pines, D. eds. (1988). *The Economy as an Evolving Complex System.* Redwood City, CA: Addison-Wesley.

Anderson, P. W. and Stein, D. L. (1984). Broken Symmetry, Emergent Properties, Dissipative Structures, Life and Its Origin: Are They Related? Reprinted in Anderson (1984), pp. 262–84.

Arnold, V. I. and Avez, A. (1968). *Ergodic Problems of Classical Mechanics.* Redwood City, CA: Addison-Wesley.

Arrow, K. J. (1951). *Social Choice and Individual Values*, 2nd ed. (1963). New York: Wiley & Son.

Arrow, K. J. (1983). *Collected Papers of Kenneth J. Arrow*, Vol. 2. *General Equilibrium.* Cambridge, MA: Harvard University Press.

Arrow, K. J. (1987). Economic Theory and the Hypothesis of Rationality. In Eatwell et al. (1987), Vol. II, pp. 69–74.

Arrow, K. J. and Hahn, F. (1971). *General Competitive Analysis.* San Francisco: Holden-Day.

Arthur, W. B. (1990). Positive Feedbacks in the Economy. *Sci. Am.,* **262(2)**, 92–99.

Ashcroft, N. W. and Mermin, N. D. (1976). *Solid State Physics.* Philadelphia: Saunders College.

Aumann, R. J. (1987). Game Theory. In Eatwell et al. (1987), Vol. 2, pp. 460–79.

Auyang, S. Y. (1995). *How Is Quantum Field Theory Possible?* New York: Oxford University Press.

Axelrod, R. (1984). *The Evolution of Cooperation.* New York: Basic Books.

Ayala, F. J. (1985). Reduction in Biology. In Depew and Weber (1985), pp. 65–80.

Balescu, R. (1975). *Equilibrium and Nonequilibrium Statistical Mechanics.* New York: Wiley.

Barlow, G. W. and Silverberg, J., eds. (1980). *Sociobiology: Beyond Nature/Nurture?* Boulder, CO: Westview Press.

Bechtel, W. and Richardson, R. C. (1993). *Discovering Complexity.* Princeton, NJ: Princeton University Press.

Bergson, H. (1907). *Creative Evolution.* New York: Henry Holt (1937).

Binney, J. J., Dowrick, N. J., Fisher, A. J., and Newman, M. E. J. (1992). *The Theory of Critical Phenomena.* New York: Oxford University Press.

Blanchard, O. J. (1987). Neoclassical Synthesis. In Eatwell et al. (1987), Vol. 3, pp. 634–6.

Blanchard, O. J. and Fischer, S. (1989). *Lectures on Macroeconomics*. Cambridge, MA: MIT Press.

Blanchard, O. J. and Watson, M. W. (1986). Are Business Cycles All Alike? In *The American Business Cycle*, ed. R. J. Gordon, pp. 123–82. Chicago: University of Chicago Press.

Blaug, M. (1985). *Economic Theory in Retrospect*, 4th ed. New York: Cambridge University Press.

Blaug, M. (1992). *The Methodology of Economics*, 2nd ed. New York: Cambridge University Press.

Bohman, J. (1991). *New Philosophy of Social Science*. Cambridge, MA: MIT Press.

Boldrin, M. (1988). Persistent Oscillations and Chaos in Dynamic Economic Models: Notes for a Survey. In Anderson, Arrow and Pines (1988), pp. 49–76.

Bowler, P. J. (1989). *Evolution, the History of an Idea*, 2nd ed. Berkeley: University of California Press.

Boyd, R., Gasper, P., and Trout, J. D., eds. (1991). *The Philosophy of Science*. Cambridge, MA: MIT Press.

Brandon, R. N. (1990). *Adaptation and Environment*. Princeton, NJ: Princeton University Press.

Brandon, R. N. and Burian, R. M., eds. (1984). *Genes, Organisms, Populations*. Cambridge, MA: MIT Press.

Brewer, R. and Hahn, E. (1984). Atomic Memory. *Sci. Am.*, **251(6)**, 50–7.

Brewster, D. (1855). *Memoirs of the Life, Writings, and Discoveries of Sir Isaac Newton*. New York: Johnson Reprint (1965).

Brush, S. G. (1976). *The Kind of Motion We Call Heat*. Amsterdam: North-Holland.

Brush, S. G. (1983). *Statistical Physics and the Atomic Theory of Matter*. Princeton, NJ: Princeton University Press.

Buchdahl, G. (1967). *Metaphysics and the Philosophy of Science*. Cambridge, MA: MIT Press.

Caplan A., ed. (1978). *The Sociobiology Debate*. New York: Harper & Row.

Carnap, R. (1928). *The Logical Structure of the World*, 2nd ed. (1961). Berkeley: University of California Press (1967).

Carnap, R. (1966). *Philosophical Foundations of Physics: An Introduction to the Philosophy of Science*. New York: Basic Books.

Chaikin, R. M. and Lubensky, T. C. (1995). *Principles of Condensed-Matter Physics*. New York: Cambridge University Press.

Chandler, A. D., Jr. (1977). *The Visible Hand*. Cambridge, MA: Harvard University Press.

Charles, D. and Lennon, K., eds. (1992). *Reduction, Explanation, and Realism*. New York: Oxford University Press.

Chatin, G. J. (1975). Randomness and Mathematical Proof. *Sci. Am.*, **232(5)**, 47–52.

Chernikov, A. A., Sagdeev, R. Z., and Zaslavsky, G. M. (1988). Chaos: How Regular Can It Be? *Phys. Today*, **41(11)**, 27–35.

Clery, D. (1996). Closing in on Superconductivity. *Science*, **271**, 288–9.

Coase, R. (1937). The Nature of the Firm. Reprinted in *Nature of the Firm*, ed. O. E. Williamson and S. G. Winter. New York: Oxford University Press.

Coase, R. (1988). *The Firm, the Market, and the Law*. Chicago: University of Chicago Press.

Cook, K. S. and Levi, M. (1990). *The Limits of Rationality*. Chicago: University of Chicago Press.

Cornwell, J. ed. (1995). *Nature's Imagination*. New York: Oxford University Press.

Cowan, G. D., Pines, D., and Meltzer, D., eds. (1994). *Complexity: Metaphor, Models, and Reality*. Redwood City, CA: Addison-Wesley.

Crutchfield, J. P., Farmer, J. D., Parkard, N. H., and Shaw, R. S. (1986). Chaos. *Sci. Am.*, **255(6)**, 46–57.

Currie, M. and Steedman, I. (1990). *Wrestling with Time*. Manchester: Manchester University Press.

Cyert, R. M. and March, J. G. (1963). *A Behavioral Theory of the Firm*. Englewood Cliffs, NJ: Prentice Hall.

Darwin, C. (1859). *The Origin of Species*. Chicago: Encyclopedia Britannica (1952).

Darwin, C. (1871). *The Descent of Man*. Chicago: Encyclopedia Britannica (1952).

Datson, L. J. (1987). Rational Individuals versus Laws of Society: From Probability to Statistics. In Krüger et al. (1987), Vol. 1, pp. 296–304.

Davidson, D. (1986). Judging Interpersonal Interests. In *Foundations of Social Choice Theory*, eds. Elster and Hylland, pp. 195–212. New York: Cambridge University Press.

Davies, P. C. W. (1974). *The Physics of Time Asymmetry*. Berkeley: University of California Press.

Dawkins, R. (1989). *The Selfish Gene*, 2nd ed. New York: Oxford University Press.

Deane, P. (1978). *The Evolution of Economic Ideas*. New York: Cambridge University Press.

Dennett, D. C. (1995). *Darwin's Dangerous Idea*. New York: Simon & Schuster.

Depew, D. J. and Weber, B. H., eds. (1985). *Evolution at a Crossroads*. Cambridge, MA: MIT Press.

Depew, D. J. and Weber, B. H. (1995). *Darwinism Evolving*. Cambridge, MA: MIT Press.

Descartes, R. (1684). *Rules for the Direction of the Mind*. In *The Philosophical Writings of Descartes*, Vol. 1, pp. 7–77. New York: Cambridge University Press.

Diamond, P. A. (1994). *On Time*. New York: Cambridge University Press.

Dobzhansky, T. (1970). *Genetics of Evolutionary Process*. New York: Columbia University Press.

Dobzhansky, T., Ayala, F. J., Stebbins, C. L., and Valentine, J. W. (1977). *Evolution*. San Francisco: Freeman.

Dornbusch, R. and Fischer, S. (1993). *Macroeconomics*, 5th ed. New York: McGraw-Hill.

Dray, W. H., ed. (1966). *Philosophical Analysis and History*. New York: Harper & Row.

Dunbar, R. I. M. (1982). Adaptation, Fitness, and the Evolutionary Tautology. In King's College Sociobiology Group (1982), pp. 9–28.

Dunbar, R. I. M. (1988). The Evolutionary Implications of Social Behavior. In *The Role of Behavior in Evolution*, ed. H. C. Plotkin, pp. 165–88. Cambridge, MA: MIT Press.

Dupré. J., ed. (1987). *The Latest and the Best: Essays on Evolution and Optimality*. Cambridge, MA: MIT Press.

Dupré, J. (1993). *The Disorder of Things*. Cambridge, MA: Harvard University Press.

Eatwell, J., Milgate, J. M., and Newman, P., eds. (1987). *The New Palgrave: A Dictionary of Economics*. 4 vols. London: Macmillan.

Earman, J. (1986). *A Primer of Determinism*. Dordrecht: Reidel.

Einstein, A. (1916). *Relativity, the Special and General Theory*. New York: Crown (1961).

Einstein, A. (1954). *Ideas and Opinions*. New York: Crown.

Elster, J. and Roemer, J. E., eds. (1991). *Interpersonal Comparisons of Well-Being*. New York: Cambridge University Press.

Emlen, J. M. (1987). Evolutionary Ecology and the Optimality Assumption. In Dupré (1987), pp. 163–78.

Emlen, S. T. (1991). Evolution of Cooperative Breeding in Birds and Mammals. In Krebs and Davies (1991), pp. 310–37.

Endler, J. A. (1986). *Natural Selection in the Wild*. Princeton, NJ: Princeton University Press.

Ereshefsky, M., ed. (1992). *The Units of Evolution*. Cambridge, MA: MIT Press.

Falconer, D. S. (1989). *Introduction to Quantitative Genetics*, 2nd ed. London: Longmans.

Farnsworth, E. A. (1990). *Contracts*, 2nd ed. Boston: Little, Brown.

Faulkner, W. (1931). *The Sound and the Fury*. New York: Penguin.

Ferraz-Mello, S., ed. (1992). *Chaos, Resonance and Collective Dynamical Phenomena in the Solar System*. Dordrecht: Kluwer Academic.

Feynman, R. P. (1965). *The Character of Physical Law*. Cambridge: MIT Press.

Feynman, R. P. (1985). *QED*. Princeton, NJ: Princeton University Press.

Feynman, R. P., Leighton, R. B., and Sands, M. (1963). *The Feynman Lectures on Physics*. 3 vols. New York: Addison-Wesley.

Fisher, R. A. (1922). On the Dominance Ratio. *Proc. Royal Soc. Edinburgh*, **42**, 321–41.

Frey, P. W., ed. (1977). *Chess Skill in Man and Machine*. New York: Springer-Verlag.

Friedman, M. 1953. The Methodology of Positive Economics. In *Essays in Positive Economics*, pp. 3–43. Chicago: University of Chicago Press.

Futuyma, D. 1992. History and Evolutionary Processes. In Nitecki and Nitecki (1992), pp. 103–30.

Galbraith, J. K. (1972). *The Great Crash 1929*, 3rd ed. Boston: Houghton Mifflin.

Galilei, G. (1629). *Dialogue Concerning the Two Chief World Systems*. Berkeley: University of California Press (1967).

Galison, P. and Stump, D. J., eds. (1996). *The Disunity of Science*. Stanford, CA: Stanford University Press.

Gary, M. R. and Johnson, D. S. (1979). *Computers and Intractability*. New York: W. H. Freeman.

Ghiselin, M. T. (1974). A Radical Solution to the Species Problem. *Systematic Zoology*, **23**, 536–44.

Gigerenzer, G., Swittinik, Z., Porter, T., Datson, L., Beatty, J., and Krüger, L. (1989). *The Empire of Chance*. New York: Cambridge University Press.

Gillespie, J. H. (1991). *The Causes of Molecular Evolution*. New York: Oxford University Press.

Golding, B., ed. (1994). *Non-Neutral Evolution*. New York: Chapman & Hall.

Goodwin, R. M. (1990). *Chaotic Economic Dynamics*. New York: Oxford University Press.

Gordon, R. J., ed. (1986). *The American Business Cycle*. Chicago: University of Chicago Press.

Gould, S. J. (1980). Sociobiology and the Theory of Natural Selection. In Barlow and Silverberg (1980), pp. 257–69.

Gould, S. J. (1983). The Hardening of the Modern Synthesis. In *Dimensions of Darwinism*, ed. M. Grene, pp. 71–93. New York: Cambridge University Press.

Gould, S. J. (1989). *Wonderful Life*. New York: Norton.

Gould, S. J. and Eldredge, N. (1993). Punctuated Equilibrium Comes of Age. *Nature*, **366**, 223–7.

Gould, S. J. and Lewontin, R. C. (1978). The Spandrels of San Marco and the Panglossian Paradigm: A Critique of the Adaptationist Programme. *Proc. R. Soc. London*, **205**, 581–98.

Grafen, A. (1991). Modelling in Behavioral Ecology. In Krebs and Davies (1991), pp. 9–31.

Granger, C. W. J. (1980). Testing for Causality. *J. Economic Dynamics and Control*, **2**, 28–48.

Gravelle, H. and Rees, R. (1981). *Microeconomics*. London: Longman.

Greenwald, B. and Stiglitz, J. E. (1987). Keynesian, New Keynesian, and New Classical Economics. *Oxford Economic Papers*, **39**, 119–32.

Greenwood, P. J., Harvey, P. H., and Slatkin, M., eds. (1985). *Evolution, Essays in Honor of John Maynard Smith*. New York: Cambridge University Press.

Grossman, S. J. and Hart, O. D. (1986). The Costs and Benefits of Ownership: A Theory of Vertical and Lateral Integration. *J. Political Econ.*, **9**, 691–719.

Hacking, I. (1990). *The Taming of Chance*. New York: Cambridge University Press.

Hahn, F. (1982). Stability. In *Handbook of Mathematical Economics*, Vol. II, ed. K. Arrow and M. D. Intriligator, pp. 745–94. Amsterdam: North-Holland.

Hahn, F. (1989). Information Dynamics and Equilibrium. In *The Economics of Missing Markets, Information, and Games*, ed. F. Hahn, pp. 106–26. New York: Oxford University Press.

Hahn, F. and Solow, R. (1995). *A Critical Essay on Modern Macroeconomic Theory*. Cambridge, MA: MIT Press.

Halliwell, J. J., Pérez-Mercader, J., and Zurek, W. H., eds. (1994). *Physical Origins of Time Asymmetry*. New York: Cambridge University Press.

Halmos, P. R. (1944). The Foundation of Probability. *Am. Math. Monthly*, **51**, 493–510.

Hamilton, W. D. (1975). Innate Social Aptitude of Man: An Approach from Evolutionary Genetics. In *Biosocial Anthropology*, ed. R. Fox, pp. 133–55. London: Malaby Press.

Hanson, N. R. (1958). *Patterns of Discovery*. New York: Cambridge University Press.

Hao, Bai-Lin, ed. (1990). *Chaos II*. Singapore: World Scientific.

Hart, O. and Holmström, B. (1987). Theory of Contracts. In *Advances in Economic Theory*, ed. T. F. Bewley, pp. 71–156. New York: Cambridge University Press.

Hausman, D. M. (1992). *The Inexact and Separate Science of Economics*. New York: Cambridge University Press.

Hausman, D. M., ed. (1994). *The Philosophy of Economics*, 2nd ed. New York: Cambridge University Press.

Hayek, F. A. (1948). *Individualism and Economic Order*. Chicago: University of Chicago Press.

Heap, S. H. (1989). *Rationality in Economics*. Oxford: Basil Blackwell.

Heidegger, M. (1924). *The Concept of Time*. Oxford: Blackwell (1992).

Heller, M. (1990). *The Ontology of Physical Objects: Four-Dimensional Hunks of Matter*. New York: Cambridge University Press.

Hempel, C. G. (1942). The Function of General Laws in History. *J. Philos.*, **39**, 35–48.

Hempel, C. G. (1965). *Aspects of Scientific Explanation*. New York: Free Press.

Hicks, J. (1976). Revolution in Economics. In *Method and Appraisal in Economics*, ed. S. J. Latsis. New York: Cambridge University Press.

Hicks, J. (1976a). Some Questions of Time in Economics. In *Evolution, Welfare, and Time in Economics*, ed. A. M. Tang. Lexington, MA: Lexington Books.

Hicks, J. (1979). *Causality in Economics*. New York: Basic Books.

Hicks, J. (1985). *Methods of Dynamic Economics*. New York: Oxford University Press.

Hobbes, T. (1839). *English Works by Thomas Hobbes*, ed. W. Molesworth. London: John Bohn.

Hoddeson, L., Braun, E., Teichmann, J., and Weart, E., eds. (1992). *Out of the Crystal Maze*. New York: Oxford University Press.

Hoffmann, A. A. and Parsons, P. A. (1991). *Evolutionary Genetics and Environmental Stress*. New York: Oxford University Press.

Holmström, B. and Tirole, J. (1989). The Theory of the Firm. In *Handbook of Industrial Organization*, Vol. 1, ed. R. Schmalensee and R. D. Willig, pp. 61–134. Amsterdam: North Holland.

Holton, G. (1993). *Science and Anti-Science*. Cambridge, MA: Harvard University Press.

Horgan, J. (1995). From Complexity to Perplexity. *Sci. Am.*, **272(6)**, 104–9.

Horgan, J. (1995a). The New Social Darwinists. *Sci. Am.*, **273(4)**, 174–81.

Horwich, P. (1987). *Asymmetries in Time*. Cambridge, MA: MIT Press.

Huang, K. (1987). *Statistical Mechanics*. 2nd ed. New York: Wiley.

Hubbard, R. and Wald, E. (1993). *Exploding the Gene Myth*. Boston: Beacon Press.

Hull, D. L. (1977). The Ontological Status of Species as Evolutionary Units. In *Foundational Problems in the Special Sciences*, ed. R. Butts and J. Hintikka, pp. 91–102. New York: Reidel.

Hull, D. L. (1989). *The Metaphysics of Evolution*. Albany: State University of New York Press.

Hutchison, T. (1992). *Changing Aims in Economics*. Cambridge: Blackwell.

Intriligator, M. D. (1971). *Mathematical Optimization and Economic Theory*. Englewood Cliffs, NJ: Prentice-Hall.

Jacob, F. (1977). Evolution and Tinkering. *Science*, **196**, 1161–6.

Jacquemin, A. (1987). *The New Industrial Organization*. Cambridge, MA: MIT Press.

James, W. (1884). The Dilemma of Determinism. Collected in *The Will to Believe*, pp. 114–40. Cambridge, MA: Harvard University Press (1979).

Jancel, R. (1963). *Foundations of Classical and Quantum Statistical Mechanics*. Elmsford, NY: Pergamon Press.

Kampen, N. G. van (1992). *Stochastic Processes in Physics and Chemistry*, 2nd ed. Amsterdam: North-Holland.

Kant, I. (1781). *Critique of Pure Reason*, trans. N. Kemp-Smith. New York: St. Martin's Press.

Kant, I. (1784). Idea for a Universal History with a Cosmopolitan Intent. In *Perpetual Peace and Other Essays*, pp. 29–40. Indianapolis: Hackett.

Kaplan, D. (1978). Dthat. In *Syntax and Semantics*, Vol. 9, ed. P. Cole, pp. 221–52. New York: Academic Press.

Kauffman, S. A. (1993). *The Origin of Order*. New York: Oxford University Press.

Kauffman, A. (1995). *At Home in the Universe*. New York: Oxford University Press.

Keller, E. F. and Lloyd, E. A., eds. (1992). *Keywords in Evolutionary Biology*. Cambridge, MA: Harvard University Press.

Kellert, S. M. (1993). *In the Wake of Chaos*. Chicago: University of Chicago Press.

Kennedy, P. (1992). *A Guide to Econometrics*, 3rd ed. Cambridge, MA: MIT Press.

Kerr, R. A. (1995). Did Darwin Get it All Right? *Science*, **267**, 1421–2.

Keynes, J. M. (1921). *A Treatise on Probability*. London: MacMillan.

Keynes, J. M. (1936). *The General Theory of Employment, Interest, and Money*. Cambridge, MA: Harvard University Press (1964).

Keynes, J. M. (1937). After the General Theory. Reprinted in *Collected Writings*, Vol. 14, pp. 109–23.

Kimura, M. (1983). *The Neutral Theory of Molecular Evolution*. New York: Cambridge University Press.

King's College Sociobiology Group., ed. (1982). *Current Problems in Sociobiology.* New York: Cambridge University Press.

Kirk, G. S., Raven, J. E., and Schofield, M. (1983). *The Presocratic Philosophers.* 2nd ed. New York: Cambridge University Press.

Kitcher, P. (1985). *Vaulting Ambition.* Cambridge, MA: MIT Press.

Knight, F. (1921). *Risk, Uncertainty, and Profit.* Boston: Houghton Miffin.

Krebs, J. R. and Davies, N. B., eds. (1991). *Behavioral Ecology.* Oxford: Blackwell Scientific.

Kreps, D. M. (1990). *A Course in Microeconomic Theory.* Princeton, NJ: Princeton University Press.

Kripke, S. (1972). Naming and Necessity. In *The Semantics of Natural Language,* ed. G. Harman and D. Davidson. New York: Reidel.

Krüger, L. L., Daston, J., and Heidelberger, M., eds. (1987). *The Probabilistic Revolution.* Cambridge, MA: MIT Press.

Kubo, R., Toda, M., and Hashitsume, N. (1985). *Statistical Physics II: Nonequilibrium Statistical Mechanics.* Berlin: Springer-Verlag.

Kuznets, S. (1963). Parts and Wholes in Economics. In *Parts and Wholes,* ed. D. Lerner, pp. 41–72. New York: The Free Press of Glencoe.

Kyberg, H. (1974). Propensities and Probabilities. *Br. J. Philos. Sci.,* **25**, 358–75.

Kyland, F. E. and Prescott, E. C. (1982). Time to Build and Aggregate Fluctuations. *Econometrica,* **50**, 1345–70.

Kyland, F. E. and Prescott, E. C. (1990). Business Cycles: Real Facts and a Monetary Myth. *Fed. Res. Bank of Minneapolis Quarterly Rev.* Spring, pp. 3–18.

Landau, L. D. and Lifshitz, E. M. 1980. *Statistical Physics,* 3rd ed. Oxford: Pergamon Press.

Laplace, P. S. (1814). *A Philosophical Essay on Probabilities.* trans. F. S. Truscott and F. L. Emory. New York: Wiley (1917).

Lebowitz, J. L. (1993). Boltzmann's Entropy and Time's Arrow. *Phys. Today,* **46(9)**, 32–8.

Lebowitz, J. L. and Penrose, O. (1973). Modern Ergodic Theory. *Phys. Today,* **26(2)**, 23–6.

Leijonhufvud, A. (1992). Keynesian Economics: Past Confusions, Future Prospects. In Vercelli and Dimitri (1992), pp. 16–37.

Lerner, D., ed. (1963). *Parts and Wholes.* New York: Macmillan.

Levinton, J. S. (1992). The Big Bang of Animal Evolution. *Sci. Am.,* **267(5)**, 84–91.

Lewontin, R. C. (1970). The Units of Selection. *Annu. Rev. Ecology and Systematics,* **1**, 1–14.

Lewontin, R. C. (1974). *The Genetic Basis of Evolutionary Change.* New York: Columbia University Press.

Lewontin, R. C. (1974a). The Analysis of Variance and the Analysis of Causes. *Am. J. Human Genetics,* **26**, 400–11.

Lewontin, R. C. (1978). Adaptation. *Sci. Am.,* **239(3)**, 213–30.

Lewontin, R. C. (1980). Adaptation. Reprinted in Sober (1984), pp. 235–51.

Lewontin, R. C. (1991). Facts and Factitious in the Natural Sciences. *Critical Inquiry,* **18**, 140–53.

Lewontin, R. C. (1992). *Biology as Ideology.* New York: Harper Collins.

Libchaber, A. (1987). From Chaos to Turbulence in Bénard Convection. *Proc. Roy. Soc. London,* **A413**, 633.

Liboff, R. L. (1990). *Kinetic Theory, Classical, Quantum, and Relativistic Descriptions.* Englewood Cliffs, NJ: Prentice-Hall.

Lichtenberg, A. J. and Lieberman, M. A. (1992). *Regular and Chaotic Dynamics,* 2nd ed. New York: Springer-Verlag.

Lorenz, E. (1993). *The Essence of Chaos.* Seattle: University of Washington Press.

Lucas, R. E. (1987). *Modern Business Cycles.* New York: Basil Blackwell.

Lucas, R. E. and Sargent, T. J. (1979). After Keynesian Macroeconomics. *Federal Reservè Bank of Minneapolis Quarterly Review,* **3(2)**, 1–16.

Lukes, S. (1973). *Individualism.* New York: Harper & Row.

Lycan, W. G., ed. (1990). *Mind and Cognition.* Cambridge: Basil Blackwell.

Machlup, F. (1991). *Economic Semantics,* 2nd ed. New Brunswick, NJ: Transaction Publisher.

Macpherson, C. B. (1962). *The Political Theory of Possessive Individualism, Hobbes to Locke*. New York: Oxford University Press.

Maddox, J. (1995). The Prevalent Distrust of Science. *Nature*, **378**, 435–7.

Mandl, F. and Shaw, G. (1984). *Quantum Field Theory*. New York: Wiley.

Mankiw N. G. and Romer, D., eds. (1991). *New Keynesian Economics*. Cambridge, MA: MIT Press.

Martin, M. and McIntyre, L. C., eds. (1994). *Readings in the Philosophy of Social Science*. Cambridge, MA: MIT Press.

Mates, B. (1986). *The Philosophy of Leibniz*. New York: Oxford University Press.

May, R. M. (1986). When Two and Two Do Not Make Four: Nonlinear Phenomena in Ecology. *Proc. Roy. Soc. London*, **B228**, 241–266.

May, R. M. and Anderson, R. M. (1983). Parasite-Host Coevolution. In *Coevolution*, ed. D. J. Futuyma and M. Slatkin, pp. 186–206. Sunderland, MA: Sinauer.

Maynard Smith, J. (1969). The Status of Neo-Darwinism. In *Towards a Theoretical Biology*, Vol. 2. ed. C. H. Waddington. Chicago: Aldine.

Maynard Smith, J. (1978). Optimization Theory in Evolution. *Annu. Rev. Ecology Systematics*, **9**, 31–56.

Maynard Smith, J. (1982). *Evolution and the Theory of Games*. New York: Cambridge University Press.

Maynard Smith, J. (1983). Current Controversies in Evolutionary Biology. In *Dimensions of Darwinism*, ed. M. Grene, pp. 273–86. New York: Cambridge University Press.

Maynard Smith, J. (1987). How to Model Evolution. In Dupre (1987), pp. 119–32.

Mayr, E. (1959). Where Are We? *Cold Spring Harbor Symposia Quant. Biol.*, **24**, 1–14.

Mayr, E. (1982). *The Growth of Biological Thought*. Cambridge, MA: Harvard University Press.

Mayr, E. (1988). *Toward a New Philosophy of Biology*. Cambridge, MA: Harvard University Press.

McCallum, B. T. (1992). Real Business Cycle Theories. In Vercelli and Dimitri (1992), pp. 167–82.

McGrattan, E. R. (1994). A Progress Report on Business Cycle Models. *Federal Reserve Bank of Minneapolis Quarterly Review*, Fall, 2–12.

McTaggart, J. M. E. (1908). The Unreality of Time. *Mind*, **18**, 457–84.

Michod, R. E. (1982). The Theory of Kin Selection. *Annu. Rev. Ecology and Systematics*, **13**, 23–55.

Michod, R. E. (1986). On Fitness and Adaptedness and Their Role in Evolutionary Explanation. *J. Hist. Biol.*, **19**, 289–302.

Michod, R. E. and Sanderson, M. J. (1985). Behavioral Structure and the Evolution of Cooperation. In Greenwood et al. (1985), pp. 95–104.

Mill, J. S. (1836). On the Definition of Political Economy and the Method of Investigation Proper to It. In Hausman (1994), pp. 52–68.

Mill, J. S. (1843). *A System of Logic*. In *Philosophy of Scientific Method*, ed. E. Nagel. New York: Hafner Press (1974).

Mills, S. and Betty, J. (1979). The Propensity Interpretation of Fitness. *Philos. of Sci.*, **46**, 263–86.

Mirowski, P. (1989). *More Heat than Light: Economics as Social Physics, Physics as Nature's Economics*. New York: Cambridge University Press.

Modigliani, F. (1977). The Monetarist Controversy or, Should We Forsake Stabilization Policies? *Am. Econ. Rev.*, **67(2)**, 1–19.

Nagel, E. (1960). The Meaning of Reductionism in the Natural Sciences. In *Philosophy of Science*, ed. A. Danto and S. Morgenbesser. New York: Meridian Book.

Nagel, E. (1979). *The Structure of Science*. 2nd ed. Indianapolis: Hackett.

Nelson, R. R. and Winter, S. G. (1982). *An Evolutionary Theory of Economic Change*. Cambridge, MA: Harvard University Press.

Newton, I. (1687). *Mathematical Principles of Natural Philosophy*. Chicago: Encyclopedia Britannica (1952).

Newton, I. (1704). *Opticks*. Chicago: Encyclopedia Britannica (1952).

Nietzsche, F. (1874). *On the Uses and Disadvantages of History for Life*. In *Untimely Meditations*. New York: Cambridge University Press (1983).

Nitecki, M. H. and Nitecki, D. V., eds. (1992). *History and Evolution*. Albany: State University of New York Press.

Oaklander, L. N. and Smith, Q., eds. (1994). *The New Theory of Time*. New Haven, CT: Yale University Press.

Oderberg, D. S. (1993). *The Metaphysics of Identity over Time*. New York: St. Martin's Press.

Ollman, B. (1976). *Alienation*, 2nd ed. New York: Cambridge University Press.

Oppenheim, P. and Putnam, H. (1958). Unity of Science as a Working Hypothesis. In Boyd et al. (1991), pp. 405–28.

Ornstein, D. S. and Weiss, B. (1991). Statistical Properties of Chaotic Systems. *Bull. Am. Math. Soc.*, **24**, 11–116.

Oster, G. F. and Wilson, E. O. (1978). *Caste and Ecology in the Social Insects*. Princeton, NJ: Princeton University Press.

Ott, E. (1993). *Chaos in Dynamical Systems*. New York: Cambridge University Press.

Papoulis, A. (1991). *Probability, Random Variables, and Stochastic Processes*, 3rd ed. New York: McGraw-Hill.

Parker, G. A. and Hammerstein, P. (1985). Game Theory and Animal Behavior. In Greenwood et al. (1985), pp. 73–94.

Parker, G. A. and Maynard Smith, J. (1990). Optimality Theory in Evolutionary Biology. *Nature*, **348**, 27–33.

Peirce, C. S. (1932). *Collected Papers of Charles Sanders Peirce*, eds. C. Hartshorne and P. Weiss. Cambridge, MA: Harvard University Press.

Penrose, O. (1979). Foundations of Statistical Mechanics. *Rep. Prog. Phys.*, **42**, 1939–2006.

Penrose, R. (1989). *The Emperor's New Mind*. New York: Penguin Books.

Perelson, A. S. and Kauffman. S. A., eds. (1991). *Molecular Evolution on Rugged Landscapes*. Redwood City, CA: Addison-Wesley.

Phelps, E. S. (1990). *Seven Schools of Macroeconomic Thought*. New York: Oxford University Press.

Pitt, J. C., ed. (1988). *Theories of Explanation*. New York: Oxford University Press.

Plato, J. von. (1994). *Creating Modern Probability*. New York: Cambridge University Press.

Plotkin, H. C., ed. (1988). *The Role of Behavior in Evolution*. Cambridge, MA: MIT Press.

Poidevin, R. and Macbeath, M., eds. (1993). *The Philosophy of Time*. New York: Oxford University Press.

Polanyi, K. (1944). *The Great Transformation*. Boston: Beacon Press (1957).

Popper, K. R. (1972). *Objective Knowledge*. New York: Oxford University Press.

Porter, T. M. (1986). *The Rise of Statistical Thinking, 1820–1900*. Princeton, NJ: Princeton University Press.

Prigogine, I. (1980). *From Being to Becoming*. New York: Freeman.

Prout, T. (1969). The Estimation of Fitness from Population Data. *Genetics*, **63**, 949–67.

Purcell, E. (1963). Parts and Wholes in Physics. In Lerner (1963), pp. 11–21.

Quine, W. V. O. (1951). Two Dogmas of Empiricism. In *From a Logical Point of View*, pp. 47–64. Cambridge, MA: Harvard University Press.

Quine, W. V. O. (1960). *Word and Object*. Cambridge, MA: MIT Press.

Quine, W. V. O. (1976). Worlds Away. In *Theories and Things*, pp. 124–8. Cambridge, MA: Harvard University Press.

Quinton, A. (1973). *The Nature of Things*. London: Routledge & Kegan Paul.

Radford, R. A. (1945). The Economic Organization of a P.O.W. Camp. *Economica*, **12**, 189–201.

Raup, D. M. (1966). Geometric Analysis of Shell Coiling. *J. Paleontol.*, **40**, 1178–90.

Rawls, J. (1971). *A Theory of Justice*. Cambridge, MA: Harvard University Press.

Reichenbach, H. (1956). *The Direction of Time*. Berkeley: University of California Press.

Renshaw, E. (1991). *Modelling Biological Populations in Space and Time*. New York: Cambridge University Press.

Roberts, J. (1987). Perfectly and Imperfectly Competitive Markets. In Eatwell et al. (1987), Vol. 3, pp. 837–41.

Robbins, L. (1984). *The Nature and Significance of Economic Science*, 3rd ed. New York: New York University Press.

Rosenberg, A. (1992). *Economics, Mathematical Politics or Science of Diminishing Returns?* Chicago: University of Chicago Press.

Rosenberg, A. (1994). *Instrumental Biology or the Disunity of Science*. Chicago: Chicago University Press.

Ruelle, D. (1991). *Chance and Chaos*. Princeton, NJ: Princeton University Press.

Russell, B. (1913). On the Notion of Cause. *Proc. Aristotelian Soc.*, **13**, 1–26.

Russell, B. (1948). *Human Knowledge*. New York: Touchstone Books.

Sachs, R. G. (1987). *The Physics of Time Reversal*. Chicago: University of Chicago Press.

Salmon, W. C. (1984). *Scientific Explanation and the Causal Structure of the World*. Princeton, NJ: Princeton University Press.

Samuelson, P. A. (1947). *Foundations of Economic Analysis*, Enlarged ed. (1983). Cambridge, MA: Harvard University Press.

Sargent, T. J. (1987). *Macroeconomic Theory*, 2nd ed. Boston: Academic Press.

Sargent, T. J. and Sims, C. A. (1977). Business Cycle Modelling without Pretending to Have Much a Priori Economic Theory. In *New Methods in Business Cycle Research*, ed. C. A. Sims, pp. 45–159. Minneapolis: Federal Reserve Bank of Minneapolis.

Schelling, T. C. (1978). *Micromotives and Macrobehavior*. New York: Norton.

Scherer, F. M. (1990). *Industrial Market Structure and Economic Performance*, 3rd. ed. Chicago: Rand McNally.

Schumpeter, J. A. (1954). *History of Economic Analysis*. New York: Oxford University Press.

Sen, A. (1980). Plural Utility. *Proc. Aristotelian Soc.*, **1980–81**, 193–215.

Sen, A. (1987). *Ethics and Economics*. Oxford: Blackwell.

Shackle, G. L. S. (1972). *Epistemics and Economics: A Critique of Economic Doctrines*. New York: Cambridge University Press.

Sherover, C. M. (1971). *Heidegger, Kant, and Time*. Bloomington: Indiana University Press.

Sherover, C. M., ed. (1975). *The Human Experience of Time*. New York: New York University Press.

Shimony, A. (1987). The Methodology of Synthesis: Parts and Wholes in Low-Energy Physics. In *Kelvin's Baltimore Lectures and Modern Theoretical Physics*, ed. R. Kargon and P. Achinstein. Cambridge: MIT Press.

Shoemaker, S. (1969). Time Without Change. *J. Philos.*, **66**, 363–81.

Simon, H. A. (1969). *The Sciences of the Artificial*. Cambridge, MA: MIT Press.

Simon, H. A. (1983). *Reason in Human Affairs*. Stanford, CA: Stanford University Press.

Simon, H. (1995). Near Decomposability and Complexity: How a Mind Resides in a Brain. In *The Mind, the Brain, and Complex Adaptive Systems*, ed. H. J. Morowitz and J. T. Singer, pp. 25–44. Redwood City, CA: Addison-Wesley.

Simons, P. (1987). *Parts: A Study in Ontology*. New York: Oxford University Press.

Sklar, L. (1985). *Philosophy and Spacetime Physics*. Berkeley: University of California Press.

Sklar, L. (1993). *Physics and Chance*. New York: Cambridge University Press.

Smith, A. (1776). *An Inquiry into the Nature and Causes of the Wealth of Nations*. New York: Modern Library (1937).

Smith, Q. (1993). *Language and Time*. New York: Oxford University Press.

Sobel, M. E. (1995). Causal Inference in the Social and Behavioral Sciences. In *Handbook of Statistical Modeling for the Social and Behavioral Sciences*. ed. G. Arminger, C. C. Clogg, and M. E. Sobel, pp. 1–38. New York: Plenum.

Sober, E., ed. (1984). *Conceptual Issues in Evolutionary Biology*, 2nd ed. (1994). Cambridge, MA: MIT Press.

Sober, E. (1988). *Reconstructing the Past*. Cambridge, MA: MIT Press.

Sober, E. (1993). *Philosophy of Biology*. Boulder, CO: Westview Press.

Sober, E. and Lewontin, R. C. (1982). Artifact, Cause, and Genic Selection. *Philos. Sci.*, **49**, 157–80.

Sober, E. and Wilson, D. A. (1994). A Critical Review of Philosophical Work on the Unit of Selection Problem. *Philos. Sci.*, **61**, 534–55.

Solow, R. M. (1980). On Theories of Unemployment. *Am. Econ. Rev.*, **70**, 1–11.

Solow, R. M. (1989). How Economic Ideas Turn to Mush. In *The Spread of Economic Ideas*, ed. D. C. Colander and A. W. Coats, pp. 75–84. New York: Cambridge University Press.

Stanley, S. M. (1981). *The New Evolutionary Timetable*. New York: Basic Books.

Stearns, S. C. (1992). *The Evolution of Life Histories*. New York: Oxford University Press.

Stein, D. L., ed. (1989). *Lectures in the Sciences of Complexity*. Redwood City, CA: Addison-Wesley.

Stein, D. L. (1989a). Spin Glasses. *Sci. Am.*, **261(1)**, 52–9.

Stiglitz, J. E. (1992). Methodological Issues and the New Keynesian Economics. In Vercelli and Dimitri (1992), pp. 38–86.

Stoeckler, M. (1991). History of Emergence and Reductionism. In Agazzi (1991), pp. 71–90.

Stone, L. (1979). The Revival of Narrative: Reflections on a New Old History. In *The Past and the Present Revisited*, pp. 74–96. New York: Routledge & Kegan Paul (1987).

Strawson, P. F. (1959). *Individuals*. London: Methuen.

Sumner, L. W., Slater, J. G., and Wilson, F., eds. (1981). *Pragmatism and Purpose*. Toronto: University of Toronto Press.

Suppes, P. (1984). *Probabilistic Metaphysics*. New York: Basil Blackwell.

Suppes, P. (1993). The Transcendental Character of Determinism. *Midwest Studies in Philosophy*, **XVIII**, 242–57.

Szathmáry, E. and Maynard Smith, J. (1995). *The Major Evolutionary Transitions*. San Francisco: Freeman.

Tirole, J. (1988). *The Theory of Industrial Organization*. Cambridge, MA: MIT Press.

Tobin, J. (1972). Inflation and Unemployment. *Am. Econ. Rev.*, **62(1)**, 1–18.

Toda, M., Kubo, R., and Saito, N. (1983). *Statistical Physics I*. New York: Springer-Verlag.

Trout, J. D. (1991). Reductionism and the Unity of Science. In Boyd et al. (1991), pp. 386–92.

Varian, H. R. (1992). *Microeconomic Analysis*, 3rd ed. New York: Norton.

Vercelli, A. (1992). Causality and Economic Analysis: A Survey. In Vercelli and Dimitri (1992), pp. 392–421.

Vercelli, A. and Dimitri, N., eds. (1992). *Macroeconomics, a Survey of Research Strategies*. New York: Oxford University Press.

Wake, M. H. (1993). Morphology and Evolutionary Theory. *Oxford Survey in Evolutionary Biology*, **8**, 289–346. New York: Oxford University Press.

Weatherford, R. (1982). *Philosophical Foundations of Probability Theory*. London: Routledge & Kegan Paul.

Weatherford, R. (1991). *The Implications of Determinism*. New York: Routledge.

Weinberg, S. (1977). The Search for Unity, Notes for a History of Quantum Field Theory. *Daedalus*, **106(4)**, 17–35.

Weinberg, S. (1987). Newtonianism and Today's Physics. In *300 Years of Gravitation*, ed. S. W. Hawking and W. Israel, pp. 5–16. New York: Cambridge University Press.

Weinberg, S. (1995). Reductionism Redux. *New York Rev. Books*, **XLII(15)**, 39–42.

Weyl, H. (1949). *Philosophy of Mathematics and Natural Science*. Princeton, NJ: Princeton University Press.

Whitrow, G. J. (1972). Reflections on the History of the Concept of Time. In *The Study of Time*, Vol. 1, ed. J. T. Fraser, F. C. Haben, and G. H. Müller, pp. 1–11. Berlin: Springer-Verlag.

Wightman, A. S. (1985). Regular and Chaotic Motions in Dynamical Systems: Introduction to the Problems. In *Regular and Chaotic Motions in Dynamical Systems*, eds. G. Velo and A. S. Wightman. New York: Plenum.

Wilcox, D. J. (1987). *The Measure of Times Past*. Chicago: University of Chicago Press.

Williams, G. C. (1966). *Adaptation and Natural Selection*. Princeton, NJ: Princeton University Press.

Williams, G. C. (1985). A Defense of Reductionism in Evolutionary Biology. In *Oxford Survey in Evolutionary Biology*, **2**, 1–27. New York: Oxford University Press.

Williamson, O. E. (1985). *The Economic Institutions of Capitalism*. New York: The Free Press.

Wilson, D. S. (1980). *The Natural Selection of Populations and Communities*. Menlo Park, CA: Benjamin.

Wilson, D. S. (1983). The Group Selection Controversy: History and Current Status. *Annu. Rev. Ecology Systematics*, **14**, 159–88.

Wilson, E. O. (1975). *Sociobiology*. Cambridge, MA: Harvard University Press.

Wilson, K. G. (1979). Problems in Physics with Many Scales of Length. *Sci. Am.*, **241(2)**, 158–79.

Wittgenstein, L. (1953). *Philosophical Investigations*. New York: Macmillan.

Wright, R. (1994). *The Moral Animal*. New York: Pantheon.

Wright, S. (1931). Evolution in Mendelian Populations. *Genetics*, **16**, 97–159.

Wright, S. (1955). Classification of the Factors of Evolution. *Cold Spring Harbor Symposia of Quantitative Biology*, **20**, 16–24.

Name Index

Subject Index